# CELLULOSE – BIOMASS CONVERSION

Edited by **Theo van de Ven** and **John Kadla**

## Cellulose – Biomass Conversion

http://dx.doi.org/10.5772/3471
Edited by Theo van de Ven and John Kadla

## Contributors

Toshitaka Funazukuri, Stefan Zepnik, Tilo Hildebrand, Stephan Kabasci, Hans-Joachim Ra-dusch, Thomas Wodke, Pedram Fatehi, Shereen A. Soliman, Yahia A. El-Zawahry, Abdou A. El-Mougith, Andrew S. Wieczorek, Damien Biot-Pelletier, Vincent J.J. Martin, Yutaka Tamaru, Ana M. López-Contreras, Ri-Sheng Yao, Feng-He Li, Notburga Gierlinger, Tobias Keplinger, Michael Harrington, Manfred Schwanninger, Dekui Shen, Rui Xiao, Sai Gu, Huiyan Zhang

**Published by InTech**

Janeza Trdine 9, 51000 Rijeka, Croatia



**Publishing Process Manager** Marija Radja
**Typesetting** InTech Prepress, Novi Sad
**Cover** InTech Design Team

Additional hard copies can be obtained from orders@intechopen.com

Cellulose – Biomass Conversion, Edited by Theo van de Ven and John Kadla
p. cm.
ISBN 978-953-51-1172-6

For EU product safety concerns:
IN TECH d.o.o., Prolaz Marije Krucifikse Kozulić 3, 51000 Rijeka, Croatia,
info@intechopen.com or visit our website at intechopen.com.

# Meet the editors

Professor Theo van de Ven holds the Sir William C. Macdonald Chair in the Department of Chemistry at McGill University. He is an expert in colloid and surface chemistry, both in fundamental aspects and applied to papermaking and cellulosic materials. He has published over 300 scientific papers, a book ("Colloidal Hydrodynamics", Acad. Press 1989) and several book chapters. He obtained the equivalent of a B.Sc and M.Sc. from the University of Utrecht, Holland, and his PhD from McGill University (1976).
Prof. van de Ven holds a Senior NSERC/FP Innovations Industrial Research Chair in "Colloid and Papermaking Chemistry". He is the Scientific Director of a Strategic NSERC Research Network in "Innovative Green Wood Fiber Products". Moreover he is the Director of a FQRNT Centre for Self-Assembled Chemical Structures (CSACS) and the Director of the Pulp and Paper Research Centre at McGill. He is also the Chair of FIBRE, a consortium of 8 research networks in forest innovation.

Professor John F. Kadla is the Canada research Chair in Advanced Biomaterials at the University of British Columbia, and the Scientific Director of a NSERC Strategic Research Network in "Biomaterials and Chemicals". He is an expert in lignocellulose chemistry and materials science. He has published over 200 scientific papers, numerous book chapters and presented more than 50 invited lectures. He obtained his B.Sc. in Biochemistry/Chemistry from the University of British Columbia, and his Ph.D. from North Carolina state University (1997).

# Contents

# Preface

Cellulose is only one of the components of biomass, although being the most abundant. To make useful chemicals or materials from cellulose requires as the first step the separation of cellulose from biomass. Various issues of cellulose extraction and its conversion are discussed in the chapters of this book on cellulose, the third and last one of a series of books on cellulose. This conversion of cellulose is an integral part of the biorefinery concept, an effort to derive optimum value from all biomass components, and as such compulsory reading for students and researchers in this area.

**Theo van de Ven**
Department of Chemistry at McGill University,
Canada

**John Kadla**
University of British Columbia,
Canada

Chapter 1

# Hydrothermal Conversion of Cellulose to Glucose and Oligomers in Dilute Aqueous Formic Acid Solution

Toshitaka Funazukuri

Additional information is available at the end of the chapter

http://dx.doi.org/10.5772/52007

## 1. Introduction

Many efforts have been made to convert various cellulosic materials into glucose for bio-ethanol production via enzymatic and acid hydrolysis [1-4]. Compared to common polysaccharides such as starch [5-7], guar gum [8], and pectin [9], cellulose is difficult to depolymerize because of the presence of inter- and intra-molecular hydrogen bonding. This bonding results in relatively longer reaction times for enzymatic hydrolysis, and lower yields of glucose from acid hydrolysis due to further decomposition of the products. In addition to these conventional methods, Adschiri et al. [10] and coworkers [11] employed supercritical water without any additives to hydrolyze cellulose in extremely short contact times. In this process, use of fine particles of the cellulosic material may be beneficial because accurate regulation of the contact times is needed. Unfortunately, because lignocelluloses include fibers, textiles, wood, and grass plants, which have widely varying compositions and structures, pulverizing and crushing are neither easy nor economical. From a practical point of view, with such cellulosic materials, relatively longer contact times, (i.e., on the order of several minutes) and mild temperatures (e.g., hydrothermal conditions) are attractive so that the conversion can be carried out in conventional reactors. Cellulose is in fact slowly degraded in water alone under hydrothermal conditions. Furthermore, the reaction temperature can be decreased to suppress decomposition of the desired products, and acid can be added to accelerate the reaction.

Many kinds of biomass, including wood, corn stover, and cotton fiber, have been pre-treated in dilute or concentrated acid solutions to moderately accelerate the hydrolysis of the samples [1-4, 12-26]. The most commonly used acid is sulfuric acid [12-22]. Organic acids such as formic acid and acetic acid are produced during the thermal treatment of lignocellulosic samples, making these potential pre-treatment candidates as well. However, hydrolysis of lignocellulosic samples in an organic acid solution has been limited [23-26].

In this chapter, the effectiveness of dilute acid solution is demonstrated in hydrothermal saccharification of cellulose based on the experimental results described in our previous study [27], and the formation step of each product was studied by examining the relationship between product yields.

## 2. Experimental

A schematic diagram of the semi-batch reactor set-up used in this study is shown in Figure 1. The set-up is similar to that employed for the hydrolysis of starch or poly(galacturonic acid) under hydrothermal conditions [6, 7, 9]. The reactor, which was made of stainless-steel tubing 6.7 mm I.D., 8 cm long), was connected to a preheating column (1/8-inch stainless-steel tubing of 2.17 mm I.D., 2 m long). Stainless-steel tubings of 0.5 mm I.D. served as the reactor outlet (11 cm long) and cold water supply for quenching the eluted product solution. These tubes were joined with a T-union, and then the line was further connected to tubing (0.5 mm I.D., 44 cm long) equipped with a cooling jacket to quench the solution, followed by a back pressure regulator (Model 880-81, JASCO, Tokyo, Japan) capable of adjusting pressure fluctuations within ±0.1 MPa using an electromagnetic high-frequency open-shut valve. The preheating column and the reactor were immersed in a molten salt bath whose temperature was maintained within ± 2 K. The cellulosic samples tested were cotton cellulose (dewaxed, standard sanitary cotton, Pharmacopoeia of Japan, Tokyo), filter paper (ashless, No. 7, Advantec, Tokyo, Japan) and microcrystalline cellulose powder (Avicel, Merck Japan).

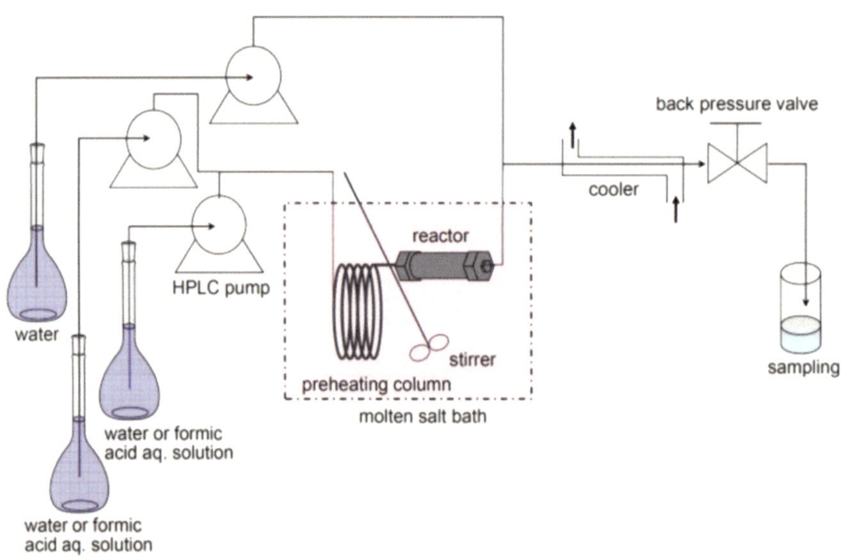


**Figure 1.** Experimental set-up.

A 0.5 g room-temperature cellulose sample wrapped softly with quartz wool (0.05 g) was placed in the reactor. A frit disk (2-μm pore size) was placed at the exit of the reactor to fix the quartz wool and cellulose sample, and the reactor and all of the lines were filled with ultrapure, degassed water supplied mainly at a constant flow velocity of 15 ml/min by two HPLC pumps. The reactor was then immersed in a molten salt bath maintained at a prescribed temperature. Reaction time was determined from the moment the reactor was immersed in the molten salt bath. It was found by measuring the inside reactor temperature with a thermocouple that the reactor temperature reached the prescribed value within two minutes. Room temperature ultrapure water provided at a flow rate of 5 ml/min and a pressure of 10 MPa by an HPLC pump at a constant flow mode was used to quench the eluted reaction solution. The product solution eluted from the back pressure regulator was collected at intervals from 1 to 10 min. The residence time of the fluid between the reactor inlet and the exit of the back pressure regulator was about 15 seconds [9]. Monogalacturonic acid as a tracer was pulse-injected by an HPLC injector (Rehodyne 7520, U.S.A.) with a 5 μL sample loop, installed in the line upstream the preheating column only when the residence time measurements were carried out, to determine the residence time of fluid in the reactor.

Glucose monomer, fructose, oligomers with degree of polymerization (DP) up to 9, and 1,6-anhydroglucose (levoglucosan) were quantitatively measured using high-performance anion-exchange chromatography (HPAEC, LC30, Dionex, Tokyo, Japan) with an electrochemical detector using a CarboPac PA column (Dionex Tokyo, Tokyo Japan). Oligomer yields (DP <6) were calibrated using the ratio of glucose to cellopentaose (Sigma, Tokyo, Japan) as a standard, and the yields of oligomers with DPs higher than 6 with cellobiose (Sigma, Tokyo, Japan) and cellotriose (Sigma, Tokyo, Japan). Secondary decomposition products were also measured by HPLC. The recovered solution was analyzed for total organic carbon (TOC) content using a total carbon analyzer (Model 5000A, Shimadzu, Kyoto, Japan). Oligomers having various DP values were identified using a MALDI-TOF mass spectrometer (AXIMA-CFR, Shimadzu, Kyoto, Japan).

Product yield and the amount of TOC in the solutions were defined as:

$$\text{Product yield}\left(\%\right)=100\times\frac{\text{Carbon of product component}\left(\text{g}\right)}{\text{Carbon of initial cellulose sample}\left(\text{g}\right)} \tag{1}$$

$$\text{TOC}\left(\%\right)=100\times\frac{\text{Carbon of soluble component}\left(\text{g}\right)}{\text{Carbon of initial cellulose sample}\left(\text{g}\right)} \tag{2}$$

$$\text{Conversion } x\left(-\right)=\frac{\text{Yield}\left(\%\right)\text{ of glucose or total sugar}}{100} \tag{3}$$

Total sugar yield (%) was defined as the sum of yields of glucose, fructose, and cellooligosaccharides with DP = 2 to 9. Note that cellulose samples, cotton cellulose, filter paper, and microcrystalline cellulose powder, were assumed to be pure cellulose.

# 3. Results and discussion

## 3.1. Comparison of cellulose types with pure water

Figure 2 compares yields of TOC and total sugar, which includes glucose, oligomers with DPs up to 9, and fructose, for cotton cellulose (CC), filter paper (FP) and cellulose powder (CP) at 543 K and 10 MPa in pure water. The formation rates in TOC and total sugar yields increase in order for CC to FP to CP. The three celluloses were almost completely solubilized and the total sugar yields reached about 60 % for CP and FP, and 48 % for CC. Note that almost no residual solids were left for all celluloses in the reactor after reaction completion. Although the TOC yields over time for FP and CC were not very different, the increase rates of total sugar yields for FP were much faster than those for CC. The difference between TOC and total sugar yields could be ascribed to be oligomers with DPs higher than 10. After 24 h, white fine particles precipitated were observed in the product solution although the solution was transparent and no precipitation when it was recovered soon after reaction completion.

Figure 3 shows yields of glucose and oligomers with DPs = 2 to 9 for three celulloses in pure water at pressure of 10 MPa, and 543 K for 30 min and 523 K for 60 min. The yields of glucose and oligomers substantially decreased with increasing DP for three celluloses, but the yields depended on the cellulose types. Except for glucose the yields from CP were the highest, and those from CC were the lowest. Each glucose yield from filter paper at 523 K and 543 K was the highest among the three types of cellulose, as compared with yields of oligomers having DP higher than 2. The reason is not known, and further studies are required.

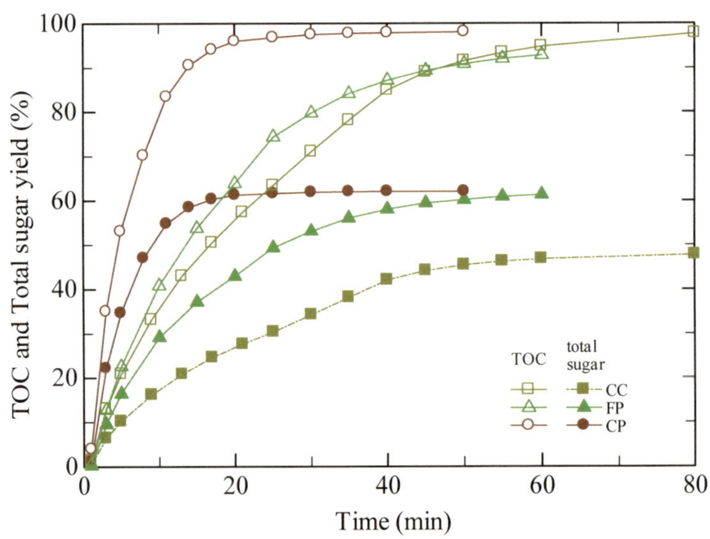


**Figure 2.** Comparison of cellulose types for TOC and total sugar yield at 543 K in water.

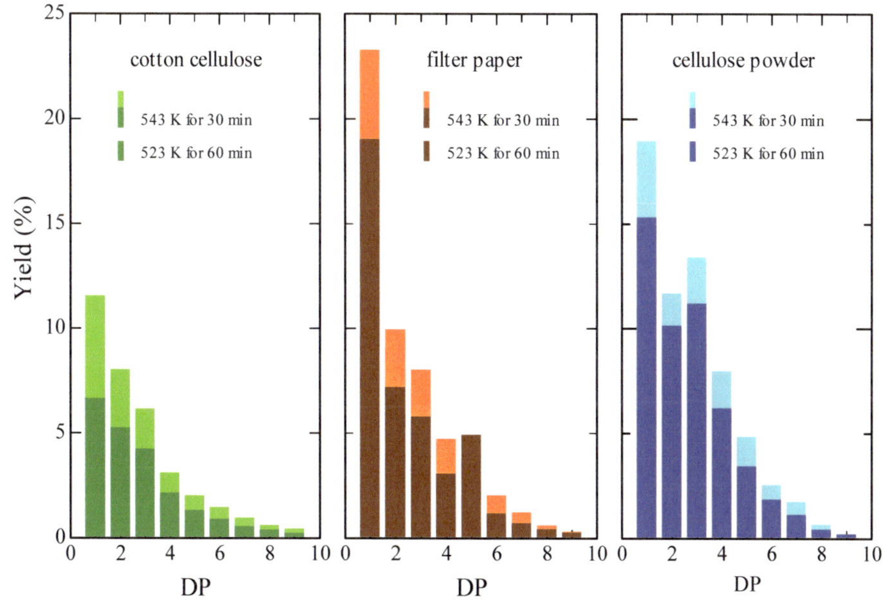


**Figure 3.** Comparison of cellulose types: cotton cellulose (CC), filter paper (FP), and cellulose powder (CP) for yields of monomer and oligomers with DP=2 to 9 at 543 K for 30 min and at 523 K for 60 min in water.

| Acid | Formic | Acetic | None |
|---|---|---|---|
| Conc (wt%) | 0.10 | 0.13 | - |
| Temperature (K) | 523 | 523 | 523 |
| Time (min) | 60 | 60 | 60 |
| Yield (100×gC/gC of initial sample) | | | |
| Glucose | 36.62 | 18.98 | 6.65 |
| DP = 2 | 16.07 | 10.96 | 5.24 |
| 3 | 14.26 | 10.95 | 4.22 |
| 4 | 8.26 | 5.94 | 2.13 |
| 5 | 4.21 | 3.57 | 1.32 |
| 6 | 2.34 | 1.99 | 0.90 |
| 7 | 1.40 | 1.49 | 0.55 |
| 8 | 0.48 | 0.61 | 0.38 |
| 9 | 0.17 | 0.33 | 0.22 |
| Fructose | 0.21 | 0.21 | 0.36 |
| Total sugar | 84.0 | 55.0 | 22.0 |
| 1,6-Anhydroglucose | 2.79 | 1.45 | 0.24 |
| 5-HMF | 1.31 | 1.56 | 0.07 |

**Table 1.** Product yields from cotton cellulose with aqueous formic acid and acetic acid solutions, together with water alone at 523 K and 60 min [27].

### 3.2. The presence of formic acid

Product yields for the hydrolysis of cotton cellulose at 523 K for 60 min in 0.1 wt% aqueous formic acid, 0.13 wt% acetic acid solution, and water alone are listed in Table 1. The addition of the acids was effective for significantly increasing the yields of glucose and oligomers with lower degrees of polymerization. The yield of total sugar, defined as glucose, fructose and oligomers with a DP up to 9, reached 84.0 % with formic acid and 55.0 % with acetic acid, and significantly improved compared to the 22.0 % obtained in the absence of acid. Yields of fructose, which is believed to be generated via isomerization of glucose under hydrothermal conditions [28], are low. Yields of 1,6-anhydroglucose (levoglucosan), obtained from the dehydroration of glucose, were slightly higher with formic acid than with acetic acid and water. 5-Hydroxymethylfurfural (5-HMF), a further undesirable decomposition product of glucose produced via dehydration [29] that acts as an inhibitor of the subsequent fermentation process [30,31], was found in 1 to 2 % yield with added acid, but only in a trace amount with water due to the lower conversion level. The results indicate that addition of formic acid is preferable to acetic acid because it provides higher yields of sugars and lower yields of decomposition products.

Figures 4 to 6 show yields of total sugar, glucose and cellobiose, respectively, over reaction time for CC in 0.1 wt% aqueous formic acid solution at 503 to 543 K and 10 MPa. The yields of the three components increased with increasing time and temperature. The maximum yield of 88 % for total sugar was attained after 20 min at the highest temperature (543 K). Above 523 K the total sugar yields seemed to reach almost 90 %, glucose and cellobiose yields did 40 % and 15 %, respectively, and both yields showed higher rates at higher reaction temperatures.

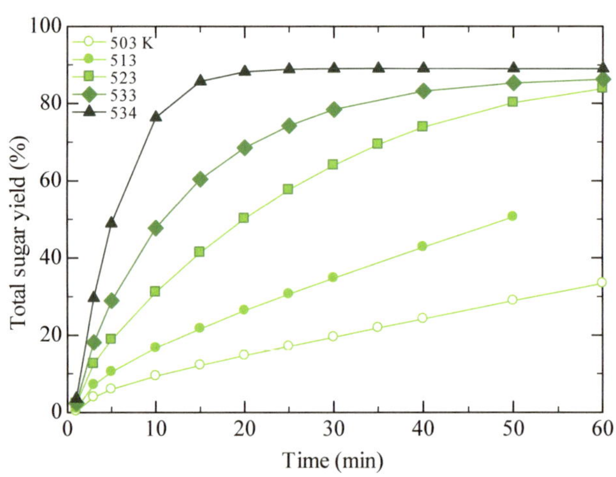


**Figure 4.** Total sugar yield over reaction time at temperatures from 503 to 543 K for cotton cellulose in 0.1 wt% aqueous formic acid solution [27].

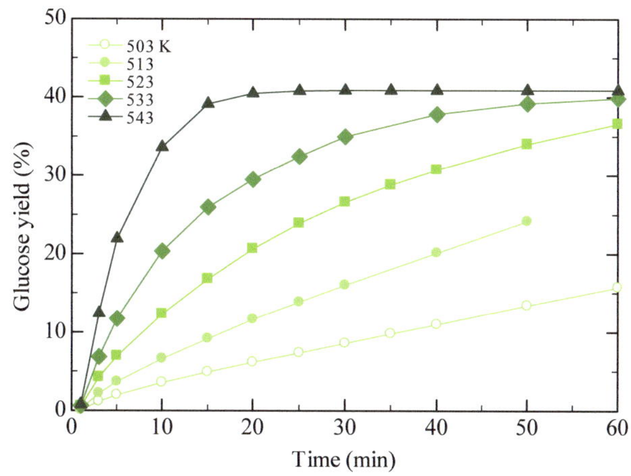


**Figure 5.** Glucose yield over reaction time at temperatures from 503 to 543 K for cotton cellulose in 0.1 wt% aqueous formic acid solution [27].

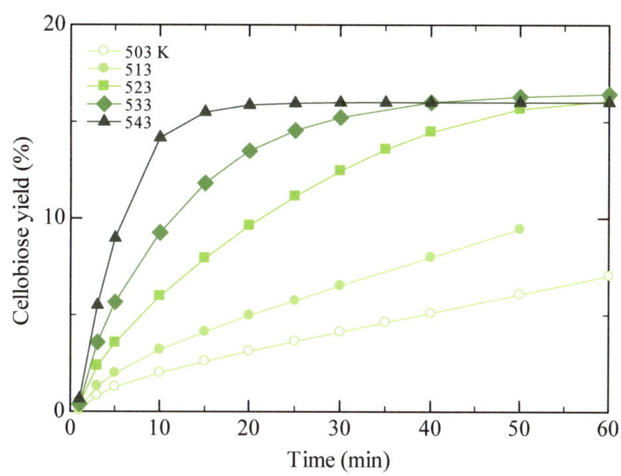


**Figure 6.** Cellobiose yield over reaction time at temperatures from 503 to 543 K for cotton cellulose in 0.1 wt% aqueous formic acid solution.

Figure 7 shows (1-$x$) over reaction time in the semi-logarithmic plot, where $x$ is the conversion based on total sugar yield or glucose yield, and the data are the same as in Figs 4 and 5. As depicted, over the main conversion, (1-$x$) ranges higher than 0.2, the data based on total sugar yield and glucose (not shown in figure) were well represented by each straight line at each temperature, and those can be expressed by the first order reaction kinetics in eq(4).

$$\frac{dx}{dt} = k(1-x) \tag{4}$$

where x is the overall first order reaction rate constant.

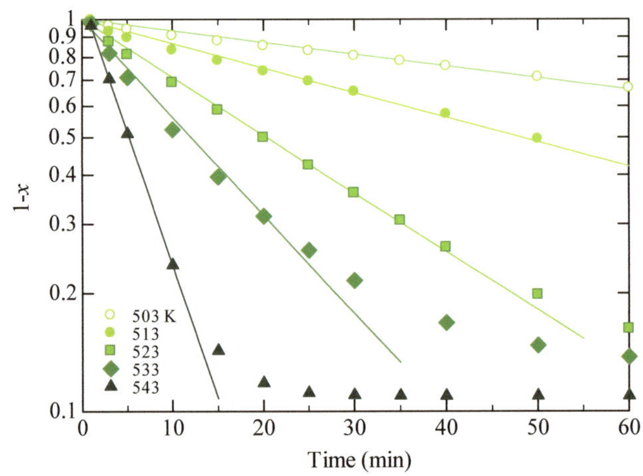


**Figure 7.** (1-$x$) over reaction time for data shown in Figure 4. $x$ is conversion based on total sugar yield [27].

Figure 8 shows Arrhenius plots for first order rate constants for conversions based on yields of total sugar and glucose. The pre-exponential factor and the activation energy are $4.157 \times 10^{14}$ 1/min and 161.6 kJ/mol for total sugar, and $9.656 \times 10^{13}$ 1/min and 159.6 kJ/mol for glucose, respectively. The activation energies are almost the same, and the pre-exponential factors for total sugar were about three times higher than those for glucose.

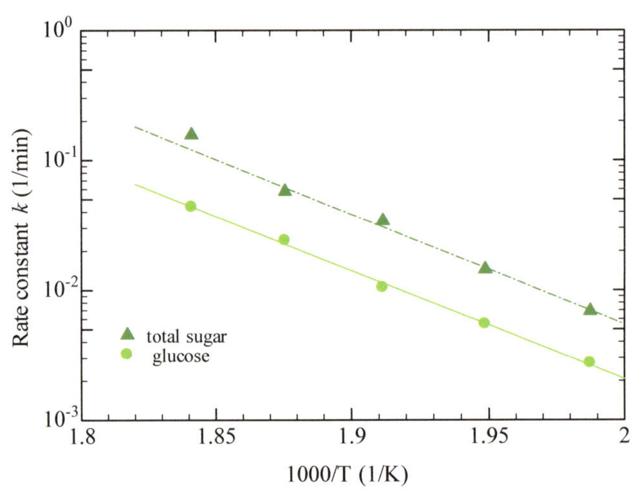


**Figure 8.** Arrhenius plots of rate constants $k$ for conversions based on total sugar yield and glucose, respectively, in 0.1 wt% aqueous formic acid solution [27].

Figure 9 shows the differences in ($k$ - $k_{water}$) vs square root of formic acid concentration for total sugar and glucose. Since cellulose was hydrolytically decomposed under hydrothermal conditions without any additives, the contribution of formic acid on the rate may be expressed by the difference. Cellulose degradation reaction can be considered to be two parallel reaction pathways: degradation in pure water with rate constant $k_{water}$ and hydrolysis in an aqueous dilute acid solution with $k$. Up to formic acid concentration of 1 wt%, the maximum concentration studied, the difference rates ($k$ - $k_{water}$) for conversions based on both total sugar and glucose yields were proportional to the square root of the concentration. This relationship may result from the fact that the concentration of [$H^+$] is proportional to the square root of the acid concentration in a dilute solution.

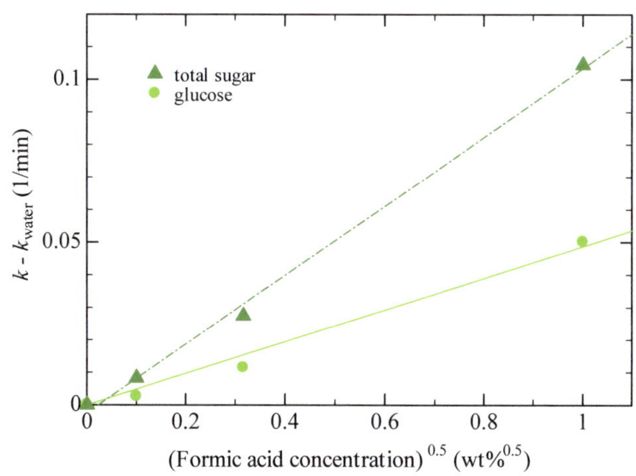


**Figure 9.** Rate constant difference ($k$ – $k_{water}$) vs. square root of formic acid concentration at 523 K for conversions based on glucose and total sugar, respectively [27].

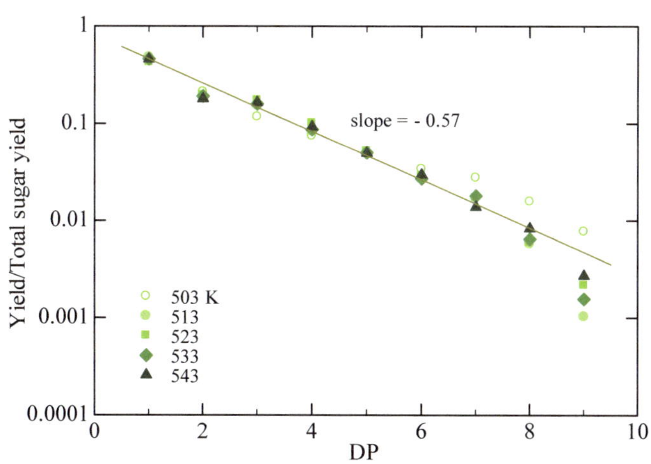


**Figure 10.** Ratio of yield to total sugar yield over degree of polymerization of cellooligosaccharides at various temperatures from 503 to 543 K for cotton cellulose in 0.1 wt% aqueous formic acid solution [27].

The yield ratio of product component to total sugar vs. degree of polymerization of cellooligosaccharide at temperatures from 503 to 543 K for various reaction times in a 0.1

wt% aqueous formic acid solution can be seen in Figure 10. Yield ratios as a function of formic acid concentration (0 to 1 wt%) for reactions run at 523 K are also shown in Figure 11. It is interesting that in a semi-logarithmic plot, most of the data points at various reaction times are almost overlapped at each temperature and formic acid concentration, and those are represented by straight lines with a slope of – 0.57. Only the ratios for oligomers with DP >7 at the lowest temperature of 503 K and at the highest acid concentration of 1 wt% at 523 K deviate from the line. The former could be due to lower conversion, whereas the latter may result from higher reaction rate at higher acid concentration. The fact that the ratios are expressed by a single straight line at various temperatures and concentrations, except under these two conditions, indicates that the hydrolytic depolymerization reaction could be controlled by the same reaction path or the same reaction stage in each case.

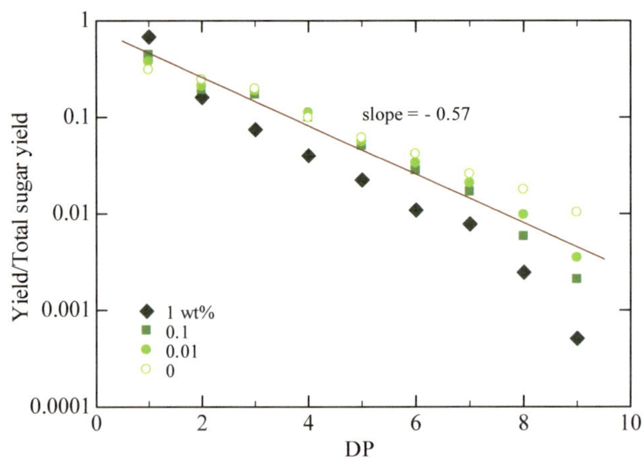


**Figure 11.** Ratio of yield to total sugar yield over degree of polymerization of monomer and oligomers at various acid concentrations at 523 K for cotton cellulose in aqueous formic acid solution [27].

The time change of the yield ratio of product components with a DP = 1 to 9 is plotted in Figure 12. After the first 10 min, the yield ratios for components with different DPs are nearly independent of reaction time, indicating that the formation rate of each component is constant. Furthermore, the values decrease with DP, presumably because of the differing solubilities of the components.

In Figures 13 and 14 the time changes of 5-hydroxymethylfurfural (5-HMF) yields are shown at temperatures from 503 to 543 K and formic acid concentrations from 0 to 1 wt%, respectively. Because 5-HMF can be produced from dehydration of the monosaccharide, it acts as an indicator of the further decomposition of the glucose produced in the reaction [6,7]. It is also undesirable because of its inhibitor effects on the proceeding fermentation process [30,31]. The yield of 5-HMF increased with increasing reaction time at most temperatures, but plateaued at 1 %, probably because of its further decomposition at the highest temperature. Acid concentration was also observed to have an effect on 5-HMF yield, which appeared to level off at 1.5 % at the highest acid concentration of 1 wt%.

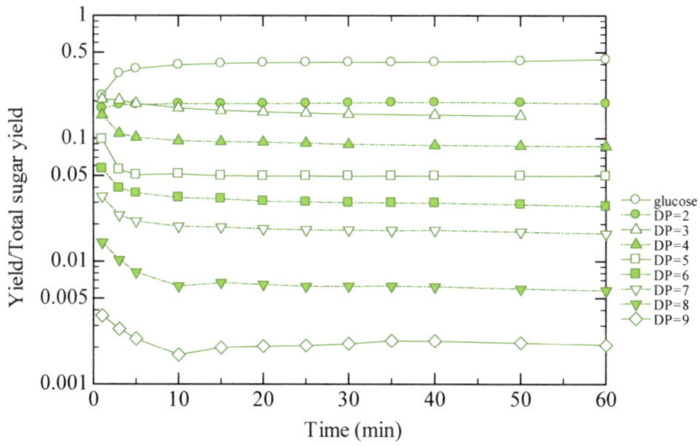


**Figure 12.** Ratio of each product yield to total sugar yield over reaction time for DP = 1 to 9 at 523 K in 0.1 wt% aqueous formic acid solution for cotton cellulose [27].

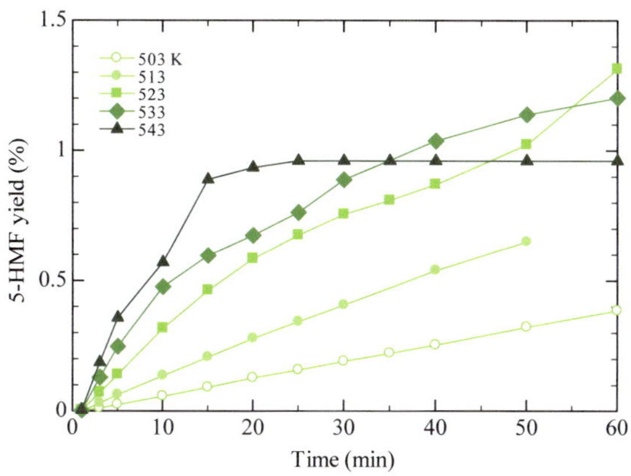


**Figure 13.** 5-HMF yield over reaction time at temperatures from 503 to 543 K in 0.1 wt% aqueous formic acid solution for cotton cellulose [27].

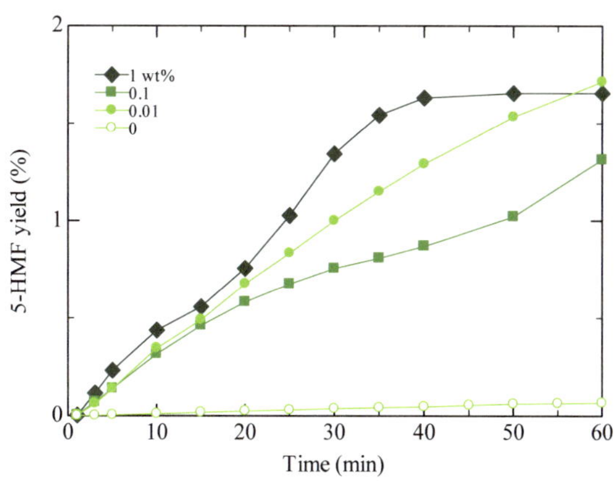


**Figure 14.** 5-Hydroxymethylfurfural yield over reaction time at 523 K and various formic acid concentrations up to 1 wt% for cotton cellulose [27].

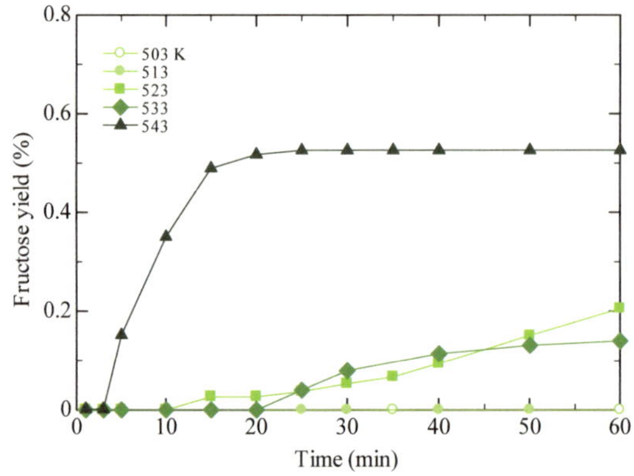


**Figure 15.** Fructose yield over reaction time at temperatures from 503 to 543 K in 0.1 wt% aqueous formic acid solution for cotton cellulose.

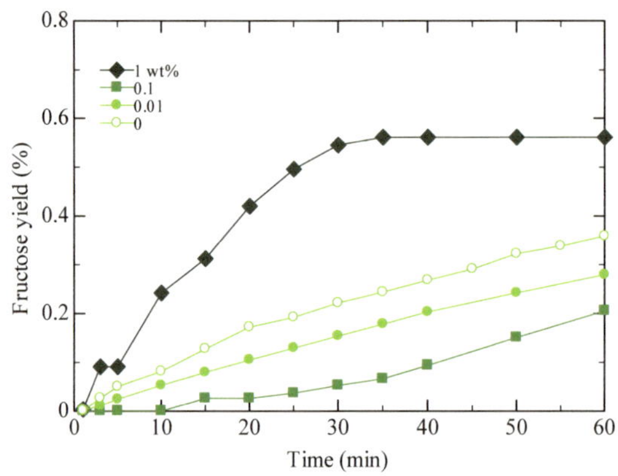


**Figure 16.** Fructose yield over reaction time at 523 K and various formic acid concentrations up to 1 wt% for cotton cellulose.

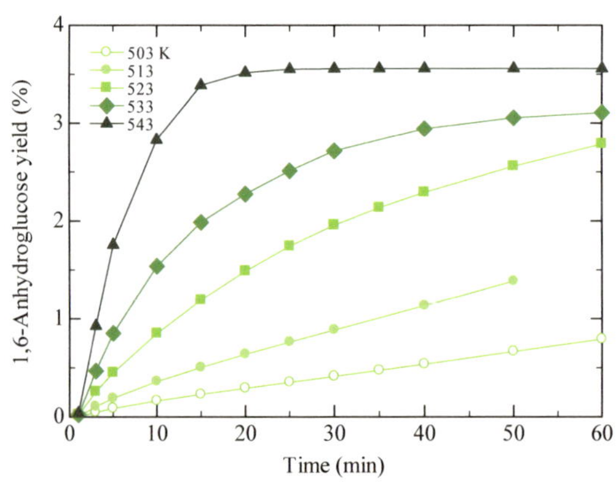


**Figure 17.** 1,6-Anhydroglucose yield over reaction time at temperatures from 503 to 543 K in 0.1 wt% aqueous formic acid solution for cotton cellulose.

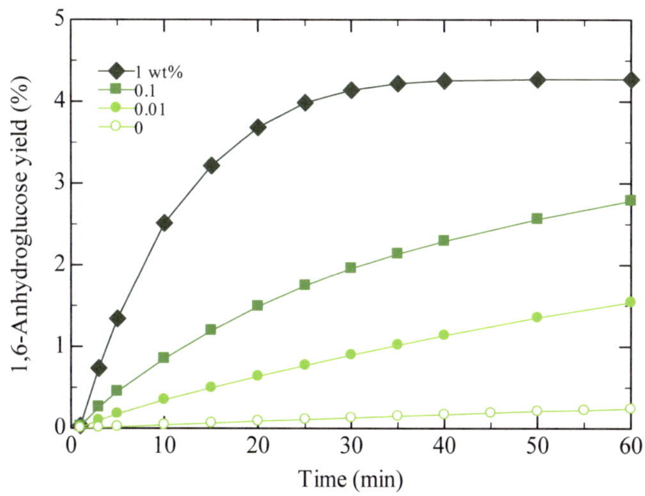


**Figure 18.** 1,6-Anhydroglucose yield over reaction time at 523 K and various formic acid concentrations up to 1 wt% for cotton cellulose.

Figures 15 and 16 show the effect of temperature at 0.1 wt% formic acid concentration and that of formic acid concentration at 523 K, respectively, on fructose yields over reaction time. Fructose yields were lower than 1 % at all conditions, and the yields did not increase with temperature and formic acid concentration. This may result from the formation via isomerisation, not direct hydrolysis of cellulose and/or its oligomers.

Figures 17 and 18 show the effects of temperature and formic acid concentration, respectively, on yield of 1,6-anhydroglucose, which is formed from dehydration of glucose. Differently from the time change in fructose yields, the yields at the highest temperature of 543 K and the highest formic acid concentration of 1 wt% simply increased with time, and then levelled off at each reaction condition. The yields also increased with temperature and formic acid concentration, as has been seen for glucose.

Figure 19 shows product yield vs. glucose yield at 503 to 543 K for various reaction times and 0.1 wt% formic acid concentration for oligomers with DP = 2 to 9. Although the yields of oligomers having different DPs were different, the yields were increasingly proportional to glucose yields, apparently, irrespective of reaction temperature and time. This may imply that the formation of each oligomer could be controlled by the same reaction step, not independently.

Figure 20 also shows cross yield plots of 5-hydroxymethylfurfural and 1,6-anhydroglucose. The yield of 5-HMF almost linearly increased with increasing 1,6-anhydroglucose yield except at higher yields of 1,6-anhydroglucose, and the slope was affected inversely by temperature. Both compounds were considered to be produced via dehydration reaction of glucose, and the yields of both compounds were dependent.

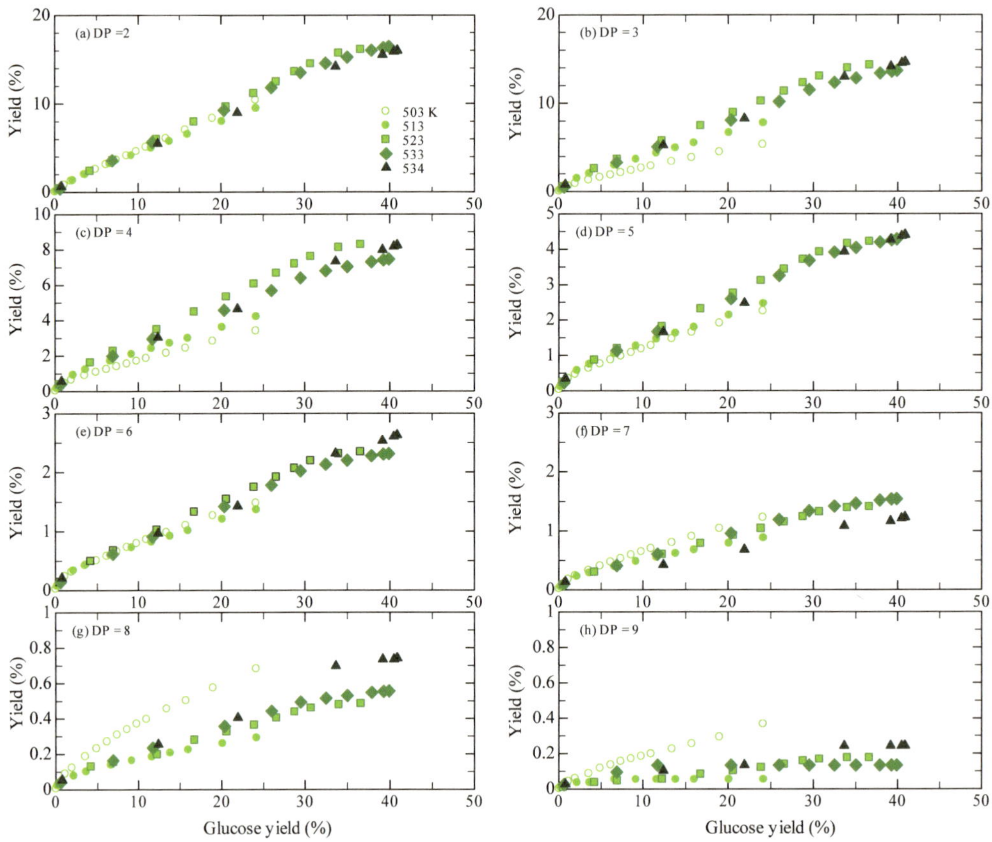


**Figure 19.** Oligomer yield vs. glucose yield at temperatures from 503 to 534 K for various reaction times at 0.1 wt% aqueous formic acid solution for cotton cellulose.

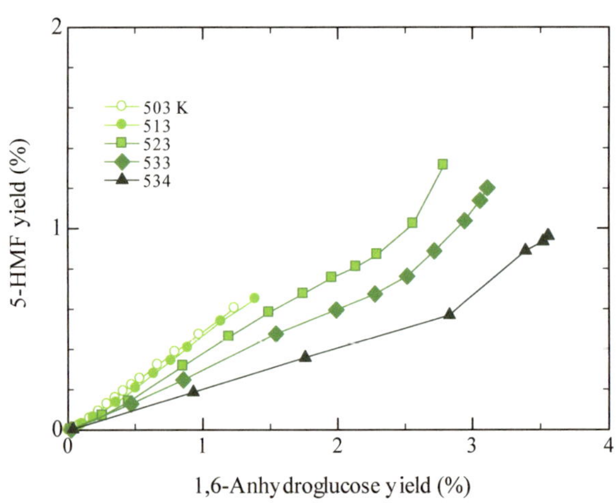


**Figure 20.** 5-HMF yield vs. 1,6-anhydroglucose yield at temperatures from 503 to 534 K for various reaction times at 0.1 wt% aqueous formic acid solution for cotton cellulose.

## 4. Conclusions

In a semi-batch reactor depolymerization of the three types of cellulosic samples was hydrothermally carried out at temperatures of 523 and 543 K and 10 MPa in pure water, and that of cotton cellulose at temperatures from 503 to 543 K and 10 MPa in a dilute aqueous formic acid solution at concentrations up to 1 wt%. The product yields and the rates were influenced by the cellulose types. The yields of major products such as glucose, fructose and oligomers having DPs up to 9, 5-HMF and 1,6-anhydroglucose were measured. The presence of a small amount of formic acid was significantly effective for increasing the yields and the reaction rates. The amount of unconverted material based on yields of glucose or total sugar (glucose, fructose and oligomers having DP up to 9) was represented by first-order reaction kinetics with nearly the same activation energies.

## Author details

Toshitaka Funazukuri
*Department of Applied Chemistry, Chuo University, Bunkyo-ku, Tokyo, Japan*

## 5. References

[1] Bobleter O, Prog. Polym. Sci. 1994; 19: 797-841.
[2] Bobleter O, Polysacchrides, edited by Dumitriu S., Marcel Dekker, New York 2005; Chapt. 40, 893-936.
[3] Olsson L., Jørgensen H., Krogh K. B. R., Roca C., ibid, Chapt. 42, 957-993.
[4] Wyman C. E., Decker S. R., Himmel M. E., Brady J. W., Skopec C. E., Viikari L., ibid, Chap 43, 995-1033.
[5] Nagamori M., Funazukuri T., J. Chem. Technol. Biotechnol., 2004; 79: 229-233.
[6] Miyazawa T., Ohtsu S., Nakagawa Y., Funazukuri T., J. Mater. Sci., 2006; 41: 1489-1494.
[7] Miyazawa T., Ohtsu S., Funazukuri T., J. Mater, Sci., 2008; 43: 2447-2451.
[8] Miyazawa T., Funazukuri T., Carbohydr. Res., 2006; 341: 870-877.
[9] Miyazawa T., Funazukuri T., Ind. Eng. Chem. Res., 2004; 43: 2310-2314.
[10] Adschiri T., Hirose S., Malaluan R., Arai K., J. Chem. Eng. Jpn, 1993; 26: 676-680.
[11] Sasaki M., Kabyemela B., Malaluan R., Hirose S., Takeda N., Adschiri T., Arai K., J. Supercrit. Fluids, 1998; 13: 261-268.
[12] Saeman J. F., Bubl J. L., Harris E. E., Ind. Eng. Chem., 1945; 37: 43-52.
[13] Harris E. E., Beglinger E., Ind. Eng. Chem., 1946; 38: 890-895.
[14] Gilbert N., Hobbs I. A., Levine J. D., Ind. Eng. Chem., 1952; 44: 1712-1720.
[15] Sharples A., Trans. Faraday Soc., 1957; 53: 1003-1013.
[16] Thompson D. R.; Grethlein H. E., Ind. Eng. Chem. Prod. Res. Dev., 1979; 18: 166-169.
[17] Church J. A., Wooldridge D., Ind. Eng. Chem. Prod. Res. Dev., 1981; 20: 371-378.
[18] Abatzoglov N., Bouchard J., Chronet E., Overrend R. P., Can. J. Chem. Eng., 1986; 64: 781-786.

[19] Bouchard J., Abatzoglou N., Chornet E., Overend R. P., Wood Sci. Technol., 1989; 23: 343-355.
[20] Mok W. S. L., Antal M. J. Jr., Varhegyi G., Ind. Eng. Chem. Res., 1992; 31: 94-100.
[21] Wyman C. E., Dale B. E., Elander R. T., Holtzapple M., Ladisch M. R., Lee Y. Y., Bioresour. Technol., 2005; 96: 1959-1966.
[22] Wyman C. E., Dale B. E., Elander R. T., Holtzapple M., Ladisch M. R., Lee Y. Y., Bioresour. Technol., 2005; 96: 2026-2032.
[23] Funazukuri T., Hirota M., Nagatake T., Goto M., Bioseparation Engineering, edited by Endo I., Nagamune T., Katoh S., Yonemoto T., Elsevier, Netherlands, 2000; 181-185.
[24] Xu F., Liu C. F., Geng Z. C., Sun J. X., Sun R. C., Hei B. H., Lin L., Wu S. B., Je J., Polym. Deg. Stab., 2006; 91: 1880-1886.
[25] Sun Y., Lin L., Pang C., Deng H., Peng H., Li J., He B., Liu S., Energy & Fuels, 2007; 21: 2386-2389.
[26] Xu J., Thomsen M. H., Thomsen A. B., J. Microbiol. Biotechnol., 2009; 19: 845-850.
[27] Asaoka Y., Funazukuri T., Res. Chem. Intermed, 2011; 37: 233-242.
[28] Speck J. C. Jr., Adv. Carbohydr. Chem., 1958; 13: 63-103.
[29] Kuster B. F. M., Starch/Stärke, 1990; 42: 314-321.
[30] Iwata Y., Hakko Kogaku Zasshi, 1949; 27: 304-306.
[31] Pfeifer P. A., Bonn G., Bobleter O., Biotechnol. Lett. 1984: 6: 541-546.

Chapter 2

# Cellulose Acetate for Thermoplastic Foam Extrusion

Stefan Zepnik, Tilo Hildebrand, Stephan Kabasci,
Hans-Joachim Ra-dusch and Thomas Wodke

Additional information is available at the end of the chapter

http://dx.doi.org/10.5772/56215

## 1. Introduction

Foam extrusion is a well established and widely used process to produce foamed products for packaging, construction, insulation or technical applications. Generally, extruded foams are produced by generating small gas bubbles in the polymer melt. These gas bubbles can be created either by means of chemical blowing agents (CBA) or physical blowing agents (PBA). The blowing agent is the primary factor which controls the foam density [1]. Along with the foam morphology this determines the end-use properties of the foam. In the case of chemical foaming the CBA, often as masterbatch or powder, is fed into the extruder together with the polymer. Due to heat dissipation during melt processing, the CBA decomposes or reacts either endothermically or exothermically and produces foaming gases such as water, carbon monoxide (CO), carbon dioxide ($CO_2$) or nitrogen ($N_2$). The resulting foam density is usually limited to 400-800 kg $m^{-3}$ as CBAs are too expensive to make foams below 400 kg $m^{-3}$ [1]. As a consequence, physical blowing agents (PBA) are used for low density foams. Basically, low boiling liquids such as water or gases like $CO_2$ or butane are used. The feeding process, the mixing of the polymer with the PBA as well as the control of the overall foaming process is more difficult for physical foaming. Specific machineries and knowledge about the complex physical foam extrusion process is often obligatory [2]. Understanding the relationship between the processing parameters and the polymer properties is fundamental to produce high quality foams [3]. For excellent foam extrusion behavior the polymer must fulfill specific thermal and rheological properties. Lots of research has been undertaken to investigate the melt rheology of conventional oil-based polymers with respect to their foamability [4-6]. Multi-axial stretching of the cell walls occurs during cell growth [5, 7]. As a consequence, adequate melt strength and melt extensibility are required. For effective dissolution of the blowing agent and for efficient mixing of the blowing agent with the polymer melt a specific pressure inside of the extruder is required. Therefore, a certain minimum melt viscosity is also necessary. If the melt viscosity is too high partial solidification of the melt during foaming can lead to insufficient foam expansion, heterogeneous foam morphologies and poor foam properties [6, 8].

It is well known that melt rheology is closely linked to the chemical structure, the molecular weight, and the polydispersity (PD) or the molecular weight distribution (MWD) of the polymer [5]. A broad MWD or high PD, the incorporation of side chains or branched structures as well as blending linear with branched or cross-linked polymers are favorable for good foam extrusion behavior. Blends of linear and branched polypropylene (PP), cross-linked polyethylene (PE) or blends of low-density polyethylene (LDPE) and high-density polyethylene (HDPE) are good examples for improved melt strength, melt extensibility, and consequently good foaming behavior [4, 9-13]. Chain extension of recycled poly(ethylene terephthalate) (PET) significantly improves the rheological properties and foaming behavior [14]. Blends of polystyrene (PS) with different molecular weights lead to bimodal MWD. These blends exhibit improved strain hardening and melt strength [15]. As melt rheology is strongly temperature-dependent, the thermal and thermoplastic properties of the polymer are important parameters to control nucleation and cell growth and to stabilize the overall foam network. The melt processing range and the corresponding window of foamability largely depend on the type of polymer used [3]. Basically, semi-crystalline polymers have smaller processing ranges than amorphous polymers due to specific windows of melting and crystallization. Biopolymers are typically sensitive to thermo-mechanical stress during processing. They often exhibit a small processing range and significant loss of their physical properties due to thermal degradation at high temperatures, strong shearing or long residence time in the molten state. A broad processing range without thermal degradation is needed to adjust the required melt properties of biopolymers for good foam extrusion. Therefore, the improvement of rheological properties as well as the broadening of the thermoplastic processing window is essential to increase the application of biopolymers for foamed products. In foam extrusion, the blowing agents used have an enormous effect on the thermal and rheological properties [4, 5, 16-18]. A dissolved blowing agent in the polymer melt can act as plasticizer. The addition of a blowing agent generally reduces the melt viscosity and the glass transition temperature of the polymer. However, the plasticizing effect largely depends on the type of blowing agent (physical or chemical), the concentration used, and the mutual behavior (solubility, diffusivity) of the blowing agent with the polymer melt. Hence, knowledge about the thermal and the rheological properties of polymer/blowing agent mixtures is as important as the properties of the polymer itself.

Today, polyethylene (PE), polypropylene (PP), and polystyrene (PS) are the standard polymers used for foam extrusion. These conventional polymers are derived from petrochemicals and are not bio-based. Nowadays, the reduction of waste including greenhouse gas emissions as well as minimizing material use and energy efficiency are key factors for establishing sustainable products. Due to non-renewable resources, conventional oil-based polymers used for foam extrusion do not fulfill all of these requirements. Therefore, much effort is being made to replace petrochemical-based polymeric foams by foams which are based on renewable resources such as starch or poly(lactic acid) (PLA) [19-26]. These biopolymers have several drawbacks especially if used for technical foamed products such as heat insulation. The low heat distortion resistance of PLA and the high moisture sensitivity of starch are the main disadvantages against using them as technical foams. Cellulose-based polymers are another promising group of biopolymers, which can be produced from a variety of raw materials, for example cotton, wood, and recycled paper. Organic cellulose esters such as cellulose acetate (CA), cellulose acetate butyrate (CAB), and cellulose acetate propionate (CAP) are bio-based polymers. They are produced through reaction of cellulose with organic acids, anhydrides or acid chlorides. Cellulose acetate is one of the oldest bio-based polymers in the world. It was first investigated by *Schützenberger* [27] in 1865. CA is produced in a two-step esterification process. At first, cellulose is activated and fully acetylated to cellulose triacetate (CTA). In a second step, a specific amount of water is added to CTA and the induced partial hydrolysis leads to partially substituted cellulose acetate [28, 29]. Cellulose acetates with a degree of substitution (DS) between 2.0 to 2.7 are usually used for thermoplastics [29]. The specific gravity of CA is about 1.30 g $cm^{-3}$ and the refractive index is 1.48 [28]. In Figure 1, the general structure of cellulose acetate is shown.

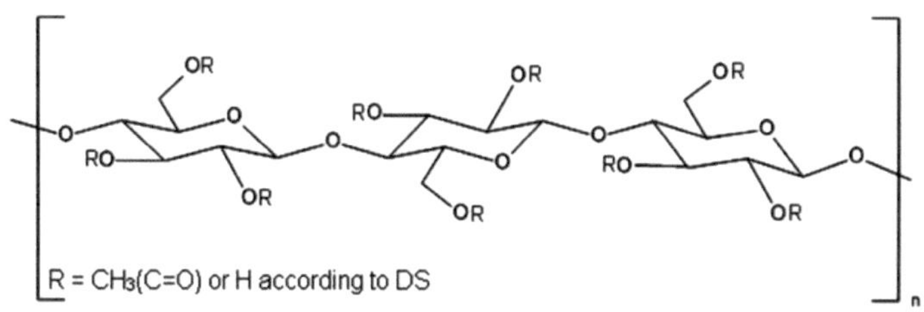


**Figure 1.** General conformation of cellulose acetate.

CTA is well known for its polymorphism as it can crystallizes in two-types of crystalline structures CTA I and CTA II [30]. Non-plasticized raw CA with a DS ranging from 1.75 to 2.7 shows also partially crystalline structures [30]. The crystallinity of these partially substituted CA is significantly lower due to less perfect and smaller crystallites [30]. External plasticization of partially substituted CA results usually in amorphous thermoplastic compounds [31] with high transparency and excellent optical properties. The thermal and mechanical properties are comparable to those of PS, styrene acrylonitrile copolymers (SAN) or poly(methyl methacrylate) (PMMA). Due to its properties, thermoplastic CA is a promising biopolymer for replacing PS in certain foam applications. Comparable to PS, the amorphous nature of plasticized CA is advantageous for foam extrusion applications due to the wider processing range accessible for foaming. However, pure CA exhibits strong interactions between its free hydroxyl groups (OH-groups) and has therefore a high glass transition temperature which is close to its thermal decomposition. As a result, unmodified CA cannot be processed using conventional thermoplastic processing technologies such as extrusion, injection molding or foaming. For melt processing, CA must be modified by means of blending or by external or internal plasticization. Blending of CA is difficult due to the high polarity and the strong intermolecular hydrogen bonds between the free OH-groups [29]. Even CA/CAB blends or CA/CAP blends are immiscible despite their similar chemical structures [29]. Furthermore, foam extrusion of blends is more complex and more difficult as compared to homopolymers due to different melt flow behavior of the blend polymers and the selective solubility behavior of the blowing agent in the blend polymers. As a result, external plasticization by means of low molecular weight plasticizers is state of the art for melt processing of CA [29]. Research activities in the field of thermoplastic CA were limited over the last decades due to the focus on petrochemical-based polymers or other bio-based polymers such as PLA [29]. Nowadays, thermoplastic CA is no more than a niche product despite its good properties. Thus, data is scarce relating to the foamability and foam extrusion behavior of thermoplastic CA [32-34]. The existing patents [35-37] primarily describe foaming processes either for open-cell foams or for foamed filter rods. No systematic research has been published concerning the usability of CA for thermoplastic foam technologies such as foam extrusion. The relationship between the properties of externally plasticized CA, its foam extrusion behavior, and the final foam properties has not been studied in detail yet. The aim of this chapter is to discuss the suitability of externally plasticized CA for foam extrusion. The influence of external plasticization on thermoplasticity, thermal and rheological properties of CA is presented with respect to foam extrusion requirements. The flow behavior of an externally plasticized CA melt loaded with a blowing agent was also studied. Finally, preliminary foam extrusion tests were carried out and the general foam extrusion behavior of externally plasticized CA is presented. For these investigations, raw CA was obtained from Acetati SpA (Italy) as white powder with a DS of 2.5 and a molecular weight $M_n$ around 50 000 g mol$^{-1}$. Table 1 shows the non-toxic, bio-based, and biodegradable plasticizers used for external plasticization. Three different plasticizer concentrations – 15, 20, and 25 weight percentages (wt.-%) – were selected.

| Plasticizer (Abbr.) | Formula | Molar mass [g mol$^{-1}$] | Molar volume [cm$^3$ mol$^{-1}$] | Boiling point (760 mmHg) [°C] |
|---|---|---|---|---|
| Allyl alcohol ethoxylate (AAE) | $C_5H_{10}O_2$ | 102.13 | 98.20 | no data |
| Glycerol diacetate (GDA) | $C_7H_{12}O_5$ | 176.17 | 149.30 | 245-280 |
| Glycerol triacetate (GTA) | $C_9H_{14}O_6$ | 218.21 | 187.95 | 250-270 |
| Ethylene glycol diacetate (EGDA) | $C_6H_{10}O_4$ | 146.14 | 130.48 | 180-190 |
| Triethyl citrate (TEC) | $C_{12}H_{20}O_7$ | 276.28 | 242.27 | 290-300 |
| Acetyl triethyl citrate (ATEC) | $C_{14}H_{22}O_8$ | 318.32 | 273.71 | 320-330 |

**Table 1.** Plasticizers used for external plasticization of CA and their corresponding properties.

## 2. Thermoplasticity and thermal properties

### 2.1. Influence of plasticizer on melt processing window

Melt processing performance of externally plasticized CA was investigated using an internal laboratory mixer with a chamber volume of around 370 cm$^3$ (Mixer W/N/B/S 350 Plasti-Corder® Lab-Station, Brabender). The processing temperature was set at 180 °C and the rotational speed of the blades was kept constant at 60 $min^{-1}$. To minimize thermo-mechanical induced degradation of this biopolymer, mixing time was fixed at 3 min. The CA obtained from Acetati SpA (Italy) was premixed with the selected plasticizer at room temperature in a powder mixer. The dry blends were fed into the preheated chamber of the internal mixer. Mass temperature and torque were measured during melt processing. The specific mechanical energy input (*SME*) in J $g^{-1}$ was calculated from the torque-time graphs according to Eq. (1) [38]

$$SME = \frac{\omega}{m} \cdot \int_0^{t_{max}} C(t)dt \quad (1)$$

where $\omega$ is the rotor speed, $m$ is the sample mass, $C(t)$ is the torque at time $t$ and $t_{max}$ is the mixing time. For melt processing of highly substituted CA, an appropriate amount of plasticizer is necessary due to the extremely narrow window between melting and thermal degradation of pure CA [31]. Figure 2 shows typical kneader graphs as a function of mixing time in dependence of plasticizer type using equal plasticizer contents.

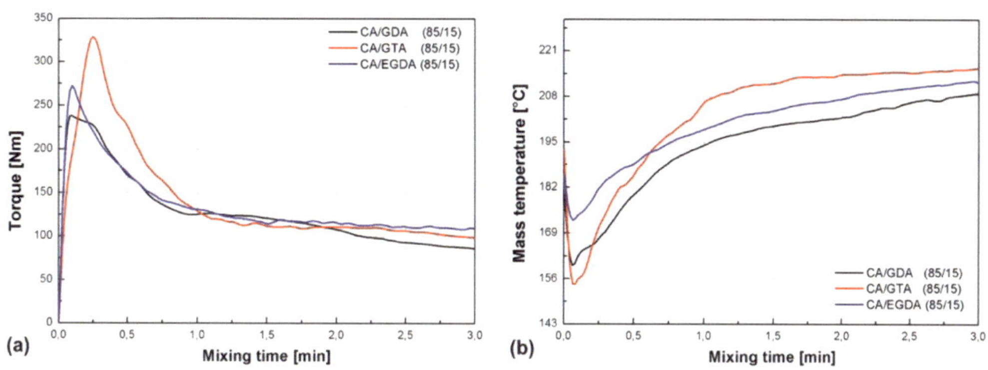


**Figure 2.** (a) Torque and (b) mass temperature of externally plasticized CA compounds as a function of mixing time and plasticizer type at constant plasticizer content (15 wt.-%).

When using a highly miscible plasticizer, improved melt processing of CA is achieved. This is expressed by lower torque and mass temperature during mixing. Highly miscible plasticizers can strongly decrease the intra- and intermolecular forces. Thus, higher chain mobility and lubricity is achieved. Consequently, lower torque for melting and mixing is required. This is very important for stable melt processing of CA as lower torque input means less shearing. Heat dissipation and overall thermo-mechanical stress are lower, and thus degradation of CA during mixing is minimized. Similar results can be achieved by increasing the plasticizer content of a less miscible plasticizer due to increasing plasticizer fraction in the CA matrix, which causes stronger lubricity [32, 39]. As seen in Figure 3, the calculated *SME* and the maximum mass temperature during melt processing confirm the assumption above. An increase in plasticizer content leads to less shear (lower *SME*) and less heat dissipation (lower mass temperature) during melt processing. As a conclusion, the more the plasticizer is compatible with CA, the better the plasticizing performance and the broader the window for melt processing of CA becomes. However, it is not possible to add any amount of plasticizer due to saturation effects and solubility limits in the biopolymer matrix.

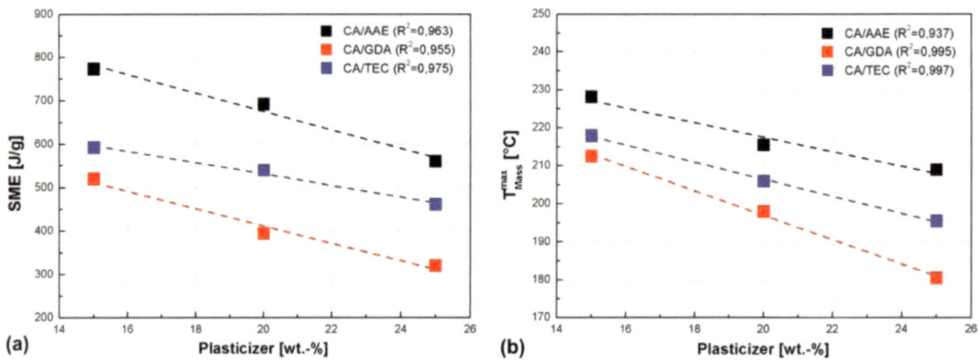


**Figure 3.** (a) *SME* input and (b) maximum mass temperature of selected externally plasticized CA compounds as a function of plasticizer type and content.

The obtained results for melt processing of externally plasticized CA are very important with respect to foam extrusion processes. The broader the window of melt processing of CA is, the easier it will be to adjust the rheology to the foam extrusion process and the wider the processing range for the foam extrusion process. The use of highly miscible plasticizers is favored as less amount of plasticizer is needed to achieve the desired properties. Less plasticizer evaporation during foam extrusion and less plasticizer migration during foam use may also occur when using a highly compatible plasticizer instead of a less miscible one.

## 2.2. Influence of plasticizer on thermal properties

As mentioned in the introduction, a broad melt processing range without thermal degradation of the biopolymer is necessary to adjust the required melt rheology for foam extrusion. Therefore, thermal stability of externally plasticized CA was studied using thermogravimetric analysis (TGA). A first run with a heating rate of 10 K $min^{-1}$ from 25 °C to 400 °C in nitrogen atmosphere was conducted to obtain general information on the thermal degradation behavior of externally plasticized CA. A second run with a heating rate of 10 K $min^{-1}$ from 25 °C to 400 °C in nitrogen atmosphere with an isotherm of 6 min at the boiling temperature range of the plasticizer was performed to get more information about plasticizer loss. The onset degradation temperature ($T_d^{onset}$) was determined according to DIN EN ISO 11358. Table 2 summarizes the obtained results for $T_d^{onset}$ and mass loss of the first degradation step for the non-plasticized raw CA and the externally plasticized CA compounds.

Figure 4 shows typical thermal degradation curves of pure CA and externally plasticized CA as a function of plasticizer content and plasticizer type. Pure, highly substituted CA degrades in one-step with a $T_d^{onset}$ of around 350 °C. CA plasticized with allyl alcohol ethoxylate (AAE) shows one-step degradation similar to pure CA. This plasticizer might influence the thermal stability of CA to a certain extent. A slight decrease in the onset degradation temperature of CA is observed, as seen in Figure 4 (b) and Table 2. In most cases two-step degradation of externally plasticized CA is observed. The first degradation step is closely linked to the boiling range and evaporation of the plasticizer whereas the second degradation step relates to CA (Table 2). The mass loss in the first degradation step correlates well with the incorporated plasticizer content. With an increase in plasticizer content an increase in mass loss of the first degradation step is observed. As a consequence, most of the plasticizers used do not influence the second degradation step and thermal stability of CA remains almost constant independent of plasticizer content. The plasticizer evaporation and its influence on thermal stability of CA are closely related to its miscibility with CA and its physical properties such as the boiling temperature range. With regard to foam extrusion, early loss of plasticizer during melt processing can lead to drastic changes in the viscosity and the melt strength. As a result, foam extrusion ability of the externally plasticized CA can be significantly reduced. All selected plasticizers, except EGDA, evaporate at temperatures (first degradation step) above the melt processing temperature of CA. The good compatibility and solubility of the selected plasticizers further minimize early evaporation during melt processing of CA. Thus, melt processing properties such as melt rheology can be maintained for the foam extrusion process.

| Compound | Plasticizer content [wt.-%] | $T_d^{onset}$ [°C] (DIN EN ISO 11358) | | Mass loss of 1. step [%] |
|---|---|---|---|---|
| | | 1. step | 2. step | |
| CA | 0 | - | 348 | - |
| CA/AAE | 15 | - | 345 | no loss of plasticizer |
| | 20 | - | 343 | |
| | 25 | - | 340 | |
| CA/GDA | 15 | 267 | 349 | 14.6 |
| | 20 | 258 | 347 | 20.2 |
| | 25 | 242 | 350 | 24.0 |
| CA/GTA | 15 | 266 | 349 | 16.1 |
| | 20 | 267 | 350 | 20.9 |
| | 25 | 248 | 349 | 24.3 |
| CA/EGDA | 15 | 186 | 346 | 16.2 |
| | 20 | 182 | 348 | 21.4 |
| | 25 | 178 | 349 | 25.7 |
| CA/TEC | 15 | 293 | 344 | 17.4 |
| | 20 | 284 | 349 | 20.7 |
| | 25 | 273 | 345 | 23.6 |
| CA/ATEC | 15 | 290 | 352 | 17.7 |
| | 20 | 279 | 351 | 21.0 |
| | 25 | 270 | 349 | 23.1 |

**Table 2.** Onset degradation temperature ($T_d^{onset}$) and mass loss of selected externally plasticized CA compounds as a function of plasticizer content (mass loss of 1. step was determined after the isotherm of 6 min at the boiling range).

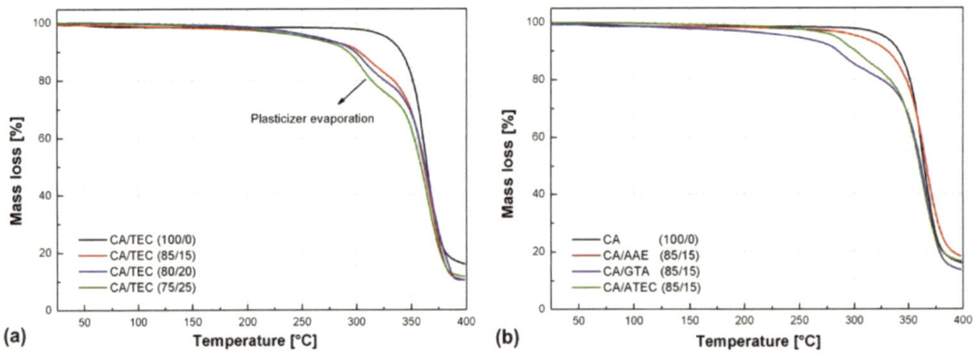


**Figure 4.** TGA curves (first run without an isotherm of 6 min at the boiling range) of pure non-plasticized CA and plasticized CA: (a) as a function of plasticizer content (TEC) and (b) as a function of plasticizer type at constant plasticizer content (15 wt.-%).

The thermoplasticity and the melt processing range of amorphous polymers are directly linked to their glass transition temperature ($T_g$). Glass transition temperatures of externally plasticized CA were measured by means of differential scanning calorimetry (DSC). Two heating cycles (up to 240 °C) and one cooling cycle (down to -50 °C) were conducted with a heating and cooling rate of 10 K $min^{-1}$ and an isotherm of 6 min at the end of each cycle. It is well known that plasticizers can have a tremendous effect on $T_g$ and consequently on thermoplasticity and melt processing performance of the polymer. Different theories and models exist, which describe the plasticizing principles [40]. In general, the plasticizer diffuses into the polymer matrix and weakens the intermolecular forces between the polymer chains due to shielding effects of functional groups along the chains. The chain mobility increases even at lower temperatures. Thus, glass transition temperature decreases whereas free volume, thermoplasticity, and flow behavior of the polymer increase. The plasticizer efficiency is often expressed by the extent to which a plasticizer reduces a polymer's $T_g$ [40]. The concentration dependence of the $T_g$ can be estimated using numerous models such as *Fox* or *Kelley-Bueche* [34, 40]. These models try to correlate the $T_g$ of the plasticized polymer with the $T_g$ of the pure polymer and the $T_g$ of the plasticizer (solvent) respectively. Due to its accuracy in extrapolating the $T_g$, the *Kelley-Bueche* model was used to calculate the $T_g$ of externally plasticized CA according to Eq. (2) [41]

$$T_g = \left[ \frac{\left(w_1 \cdot T_{g1}\right) + \left(K \cdot w_2 \cdot T_{g2}\right)}{w_1 + \left(K \cdot w_2\right)} \right] \tag{2}$$

where $T_g$, $T_{g1}$ and $T_{g2}$ are the glass transition temperatures of the plasticized polymer, the pure polymer and the plasticizer respectively, $w_1$ and $w_2$ are the weight fractions of the polymer and the plasticizer, K is a constant which can be derived from the densities $\rho_1$, $\rho_2$ and the thermal expansion coefficients $\alpha_1$, $\alpha_2$ of the polymer and the plasticizer. By applying the *Simha-Boyer* rule [41] ( $\Delta\alpha \cdot T_g \approx constant$), *K* can be simplified according to Eq. (3)

$$K = \frac{\rho_1 \cdot \alpha_2}{\rho_2 \cdot \alpha_1} = \frac{\rho_1 \cdot T_{g1}}{\rho_2 \cdot T_{g2}} \tag{3}$$

As can be seen in Figure 5 (a), calculated values are in good agreement with the experimental data from DSC. An increase in plasticizer content leads to a steady decrease in the $T_g$ of highly substituted CA due to increase in chain mobility and free volume. There seems to be a linear relationship between plasticizer content and reduction in glass transition temperature within the plasticizer concentration used (15 to 25 wt.-%). Similar results for the glass transition temperature of externally plasticized CA were found in the literature [34, 40, 42].

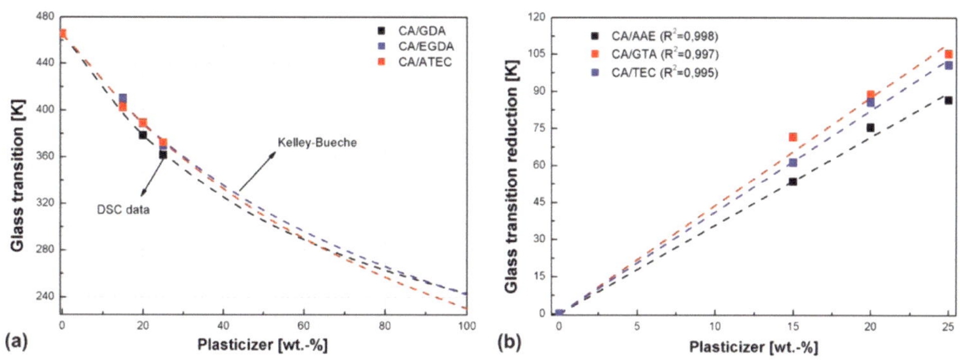


**Figure 5.** (a) $T_g$ of externally plasticized CA compounds as a function of plasticizer content and (b) efficiency of selected plasticizers in reducing the $T_g$ of CA.

The obtained thermal properties for externally plasticized CA confirm the improvements in melt processing from subchapter 2.1. Due to the significant decrease in glass transition while the thermal stability remains almost constant, a considerable improvement in thermoplasticity of externally plasticized CA is achieved. As a result, the melt processing range is enlarged and the thermal degradation of CA during processing is minimized. With respect to foam extrusion, these results are very favorable. The decrease of $T_g$ accompanied with an unchanged thermal stability of CA not only substantially widens the melt processing range but also the window for the foam extrusion, which will be shown in subchapter 4.

The plasticizer performance depends largely on its compatibility and miscibility with the polymer. Contrary to less compatible plasticizers (e.g. AAE), highly miscible plasticizers such as glycerol triacetate (GTA) or triethyl citrate (TEC) lead to stronger glass transition reduction, as seen in Figure 5 (b). This might be explained by stronger mutual interaction between the functional groups of the plasticizer and CA. A general characterization of the compatibility between a plasticizer and a polymer can be made from the *Hansen* three-dimensional solubility parameter [43], which is defined according to Eq. (4)

$$\delta_t = \sqrt{\delta_h^2 + \delta_d^2 + \delta_p^2} \tag{4}$$

where $\delta_t$ is the total *Hansen* solubility parameter, $\delta_h$ is the hydrogen bonding component, $\delta_d$ is the dispersion component and $\delta_p$ is the polar component. Table 3 shows the *Hansen* solubility parameter of pure CA and selected plasticizers. Highly substituted CA has a high solubility parameter, and thus it is a polar polymer comparable to polyamide (PA).

| Material | $\delta$ [(MPa)$^{0.5}$] | | | |
|---|---|---|---|---|
| | $\delta_t$ | $\delta_h$ | $\delta_d$ | $\delta_p$ |
| CA | 25.06 | 11.00 | 18.60 | 12.70 |
| GDA | 23.45 | 14.20 | 16.40 | 8.90 |
| GTA | 19,37 | 9.10 | 16.50 | 4.50 |
| EGDA | 19.51 | 9.80 | 16.20 | 4.70 |
| TEC | 20.98 | 12.00 | 16.50 | 4.90 |
| ATEC | 19.02 | 8.60 | 16.60 | 3.50 |

**Table 3.** *Hansen* solubility parameters of CA and selected plasticizers [43].

Following the rule "like dissolves like" it can be assumed that plasticizers with high polarity are favorable for effective plasticization of the polar CA. However, the simplified rule "like dissolves like" is only partially useful due to limitations for polar polymers such as CA or polyamide (PA). Interactions of polar polymers and plasticizers depend more on the presence and arrangement of functional groups along the chains being capable to form donor-acceptor interactions [43]. For good theoretical compatibility with CA, a high polar part of the solubility parameter (energy from dipolar intermolecular force) of the plasticizer is preferred. Due to the free OH-groups of partially substituted CA, hydrogen bonding also plays a significant role in solubility and miscibility. Therefore, a similar hydrogen part of the solubility parameter of the plasticizer is required for improved compatibility with CA. The selected citrate-based plasticizers as well as the acetate-based plasticizers show therefore excellent performance. However, the solubility of the plasticizer and its performance are also determined by a large number of other factors such as its diffusivity, its chemical structure (linear vs. branched) and functional groups. Additionally, molecular properties such as molecular weight or molar volume of the plasticizer are also very important for the plasticizer efficiency. Not only the plasticizer influences the compatibility but also the polymer, its chemical structure, chain flexibility or molecular architecture (amorphous vs. semi-crystalline) have an enormous effect on plasticizer solubility [43].

## 3. Rheological properties

### 3.1. Influence of plasticizer on melt flow behavior

As described in the introduction, the polymer's rheology is a key factor for stable foam extrusion processes. Therefore, the rheological properties of externally plasticized CA must be studied when discussing the foamability of this biopolymer. Melt flow behavior and melt viscosity were investigated as a function of plasticizer type and plasticizer content. The melt flow rate (MFR) was measured at 230 °C and 5 kg load. Melt viscosity was measured by means of high pressure capillary viscometer in a shear rate range typical for extrusion processes. The measurements were carried out at 230 °C with a rod die (L/D of 30/1) and a shear rate ranging from 10 to 3 000 $s^{-1}$. For fitting the experimental data, the well known *Cross Model* was used according to Eq. (5) [44]

$$\eta = \frac{\eta_0}{1 + \left(\lambda \cdot \dot{\gamma}\right)^{1-n}} \tag{5}$$

where $\eta_0$ is the zero shear viscosity, $\lambda$ and $n$ are fitting parameters and $\dot{\gamma}$ is the shear rate. Figure 6 (a) shows the melt flow rate (MFR) of externally plasticized CA as a function of plasticizer content and plasticizer type. The obtained relationship between MFR and plasticizer content is similar to the reduction in $T_g$ and decrease in *SME* and maximum mass temperature during melt processing. This means, an increase in plasticizer content leads to a steady increase in MFR due to increase in lubricity and chain mobility.

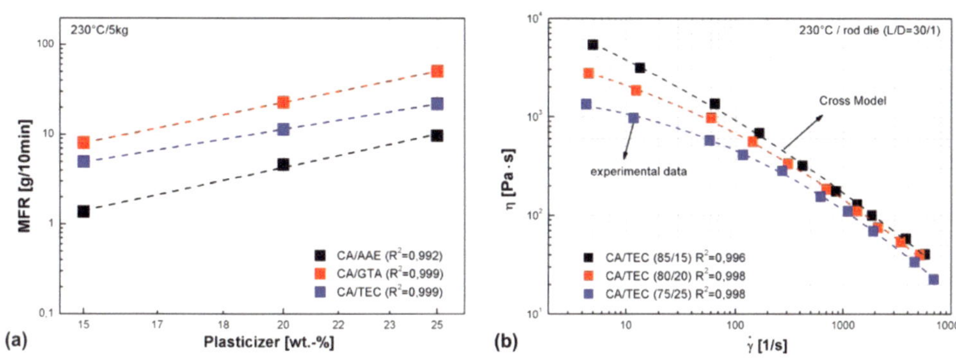


**Figure 6.** (a) MFR of externally plasticized CA compounds as a function of plasticizer type and plasticizer content and (b) melt viscosities of CA plasticized with TEC as a function of shear rate and plasticizer content.

Externally plasticized CA shows typical shear thinning behavior at high shear rates, as seen in Figure 6 (b). By increasing the plasticizer content, a steady decrease in melt viscosity at constant temperature is observed. From Figure 6 (b), one may be also concluded that the influence of plasticizer content diminishes at high shear rates above 1 000 $s^{-1}$ due to increased shear thinning. The melt viscosity results agree well with previous results obtained from melt processing (subchapter 2.1) and thermal characterization (subchapter 2.2). With respect to foam extrusion, melt viscosities and melt flow behavior of these externally plasticized CA compounds are in range of typical foam polymers such as PS, branched PE, and branched PP [10, 11, 15, 45, 46].

## 3.2. Influence of plasticizer on melt strength and melt extensibility

Besides melt viscosity, information about melt strength and melt elasticity is crucial for foam extrusion as both factors are important when discussing foam expansion, foam stability and foam collapse [47]. Therefore, rheotens tests were conducted using a rheograph with rheotens device from Göttfert. Melt strands were extruded through a single screw extruder using a rod die with an L/D of 30/2 and a constant throughput of 0.5 kg $h^{-1}$ at 220 °C. The extruded melt strands were grabbed and draw-down by a pair of rotating wheels fixed to the rheotens analyzer. The draw-down velocity of the extruded melt strand was then steadily increased with a constant acceleration rate of 24 mm $s^{-2}$. Detailed information on the rheotens test can be found in [48]. Figure 7 shows rheotens curves obtained for externally plasticized CA melts. The plasticizer content and the plasticizer type significantly influence the strength and drawability (extensibility) of the CA melt. An increase in plasticizer content leads to a decrease in draw-down force *F* (melt strength) while a tremendous increase in melt extensibility is observed. Highly compatible plasticizers behave like good solvents. As a result, intermolecular forces within the polymer network decrease whereas chain mobility increases. The polymer chains can then be easily disentangle and aligned in draw-down direction. As a consequence, the draw-down force *F* as well as the stress $\sigma$ decrease whereas the melt extensibility drastically increases. The typical draw resonance, which is seen in Figure 7 (a), is also reduced due to better flow and chain orientation in draw-down direction. The initial slope (melt stiffness) manually obtained from the stress curves also steadily decreases with increasing plasticizer content, as seen in Figure 7 (b).

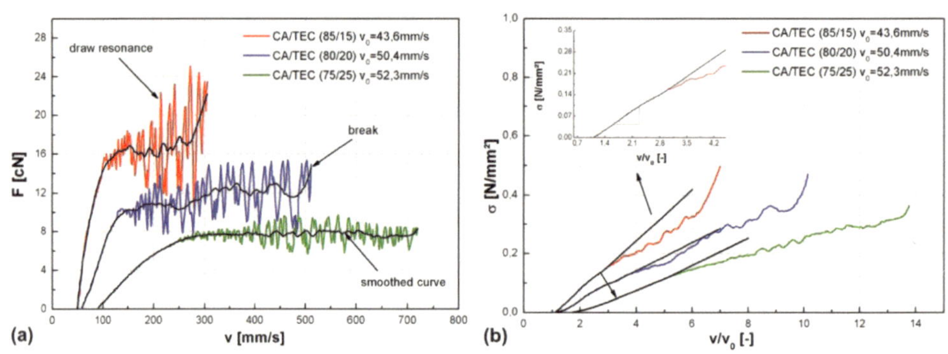


**Figure 7.** (a) Draw-down force *F* as a function of draw-down velocity *v* of CA plasticized with TEC in dependence of plasticizer content and (b) draw-down stress $\sigma$ as a function of draw ratio $v/v_0$ in dependence of plasticizer content.

Figure 8 shows the correlation between the melt flow rate (MFR) and the melt strength as well as the melt extensibility for CA plasticized with glycerol triacetate (GTA) and triethyl citrate (TEC).

With increasing melt flow rate (MFR) of externally plasticized CA, a continuous decrease of melt strength was found while melt extensibility steadily increases. An explanation for the good melt elasticity and melt strength could be the polydispersity of CA due to its natural character and the two-step esterification process [49, 50]. Polydispersity or bimodality, being attributed to a mixture of short and long chains and/or branched and linear chains, generally favors high melt strength and strain hardening [4]. The polydispersity index ($M_w/M_n$) of the pure CA used in this study was measured by means of gel permeation chromatography (GPC) with PS standards in THF solution. Three samples of the pure CA were measured. Table 4 shows the obtained values for $M_n$ and $M_w$ as well as $M_w/M_n$ for the non-plasticized raw CA and for selected externally plasticized CA compounds. The observed polydispersity index ranges from 3 to 5. This is comparable to branched PP [12, 51] or blends of different types of PS [15, 52].

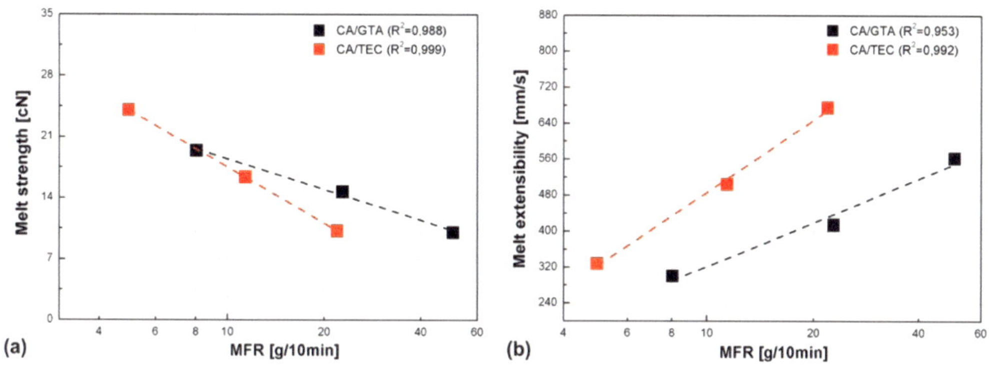


**Figure 8.** Correlation of the melt flow rate (MFR) with (a) the melt strength and (b) the melt extensibility of externally plasticized CA compounds as a function of plasticizer.

| Compound | Plasticizer content [wt.-%] | $M_n$ [g mol$^{-1}$] | $M_w$ [g mol$^{-1}$] | $M_w/M_n$ |
|---|---|---|---|---|
| CA (Sample 1) | 0 | 45 180 | 204 410 | 4.5 |
| CA (Sample 2) | | 56 045 | 266 460 | 4.8 |
| CA (Sample 3) | | 40 135 | 161 950 | 4.0 |
| CA/AAE | 15 | 53 948 | 225 300 | 4.2 |
| | 20 | 54 851 | 241 370 | 4.4 |
| | 25 | 51 769 | 251 590 | 4.8 |
| CA/GDA | 15 | 51 922 | 223 080 | 4.3 |
| | 20 | 43 790 | 197 480 | 4.5 |
| | 25 | 45 970 | 221 280 | 4.8 |
| CA/TEC | 15 | 44 695 | 173 730 | 3.9 |
| | 20 | 56 948 | 230 680 | 4.1 |
| | 25 | 44 860 | 201 770 | 4.5 |

**Table 4.** Number average molecular weight $M_n$, weight average molecular weight $M_w$ and polydispersity index of pure CA and selected externally plasticized CA compounds.

Due to the polydispersity and the strong intermolecular interaction between the free OH-groups of partially substituted CA, certain intrinsic melt strength of this biopolymer is given. The observed high melt extensibility for highly plasticized CA is excellent with respect to foam extrusion. It means that multi-axial stretching of the externally plasticized CA melt during foam expansion (cell growth) is possible without breakage of the thin cell walls. Nevertheless, the melt stiffness, which is the maximum melt strength (draw-down force), and the melt stress are comparably low at a high plasticizer content of 25 wt.-%. This can lead to instabilities in the foam network during foaming and foam collapse can occur due to low cell wall stiffness. This problem can be overcome by keeping the plasticizer content below ca. 20 wt.-%. With this reduced content, melt strength of 15 to 20 cN can be achieved while melt extensibility remains at a high level of 400 to 600 mm $s^{-1}$. These values are excellent for an externally plasticized polymer and agree well with results for non-plasticized foam polymers such as PE, PP, and PS [13, 48, 53].

As a conclusion, the obtained results show that the properties studied are closely linked together and strongly influenced by the plasticizer content as well as by the type of plasticizer used. From these results, it can be concluded that the lower the glass transition of the amorphous externally plasticized CA compounds while the thermal stability remains constant, the wider the window for thermoplastic processing and the easier it will be to adjust the melt rheology of CA to the requirements for the foam extrusion process. The significantly enlarged melt processing range means at the same time a considerable broadening of the window for the foam extrusion. However, if the plasticizer content is too high, melt stiffness and melt strength may become too low for stable foam extrusion processes. By changing the plasticizer content or by using a more miscible plasticizer, excellent melt processing behavior as well as good thermal and rheological properties of CA can be achieved for foam extrusion. To proof the foam extrusion behavior, preliminary foam extrusion tests were conducted. For these tests, an externally plasticized CA compound with 20 wt.-% of glycerol triacetate (GTA) was produced because of a good balance between melt processing, melt flow, melt extensibility, and melt strength. The compound were prepared by means of a co-rotating twin-screw extruder (TSA EMP 26-40) with an L/D of 40/1. The screw speed was set at 250 rpm and the throughput was fixed at 10 kg $h^{-1}$. The temperature profile ranged from 160 °C at the feeding zone to 205 °C at the die. The compound was dried for 24 h at 70 °C (hot air dryer).

### 3.3. Influence of physical blowing agent on melt flow behavior

As pointed out in the introduction, the blowing agent also strongly influences the melt rheology of the polymer during foam extrusion. Therefore, information on gas loaded CA melts is essential for setting the right foam extrusion parameters such as the temperature of the extruder. It is well known that the PBA can act as a plasticizer or solvent [16-18]. The extent to which a PBA influences the rheology and the thermal properties of the polymer is directly related to its physical properties such as molar mass, solubility and diffusivity in the polymer melt. The influence of blowing agent on the flow behavior of a CA melt plasticized with 20 wt.-% GTA was studied by means of in-line rheometer with high speed camera system. These preliminary investigations were carried out on a 60 mm single screw extruder with an L/D of 40/1 (Barmag Oerlikon Textile GmbH & Co. KG). Carbon dioxide ($CO_2$) was selected as an eco-friendly PBA. The slit die gap of the rheometer was fixed at 5 mm and the throughput of the single screw extruder was kept constant at 11 kg $h^{-1}$.

As seen from Figure 9, pressure decreases continuously over the flow length for the externally plasticized CA melt. When carbon dioxide is added to the externally plasticized CA melt, an overall shift of the pressure curve to lower values is observed. A continuous increase in carbon dioxide content leads to a steady decrease in the pressure curve over flow length. The shape of the pressure curve seems to be independent from the carbon dioxide content. The plasticization effect of carbon dioxide used as PBA is not only seen in the pressure decrease but also confirmed by a decrease in viscosity and increase in volume flow of the externally plasticized CA melt at constant processing conditions (temperature, throughput, slit die gap).

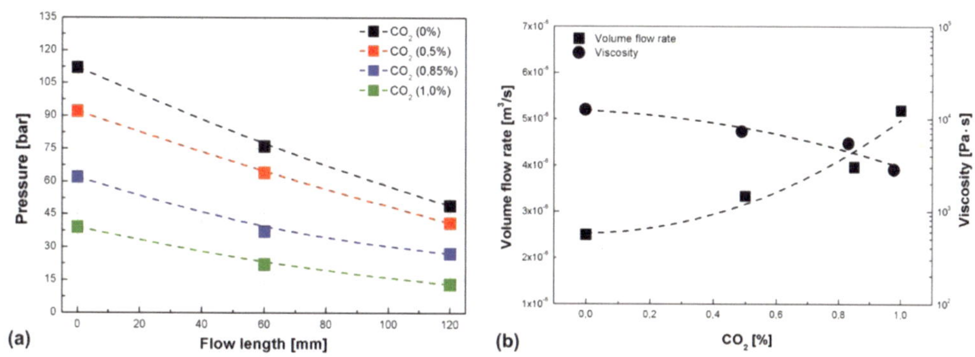


**Figure 9.** (a) Pressure distribution and (b) volume flow rate and viscosity of $CO_2$ loaded CA melt plasticized with 20 wt.-% GTA as a function of flow length and $CO_2$ content at constant rheometer temperature (220 °C) and constant processing parameters.

Since the effective dissolution of a PBA in the polymer melt is mostly pressure controlled, a certain viscosity of the polymer melt is required to achieve sufficient pressure in the extruder for dissolution of the PBA. As seen in Figure 10, due to the constant processing parameters (throughput, slit die gap, rheometer temperature), partial supersaturation and premature separation of carbon dioxide from the externally plasticized CA melt is observed, especially at 1 % $CO_2$ (on weight basis). An explanation for this could be the continuous increase in volume flow and decrease in melt viscosity. As a result, the pressure inside of the extruder is too low to keep carbon dioxide dissolved in the CA melt.

0.0 % $CO_2$ 0.85 % $CO_2$ 1.0 % $CO_2$

**Figure 10.** Influence of $CO_2$ content on supersaturation of $CO_2$ from the CA melt plasticized with 20 wt.-% GTA and prenucleation in the in-line rheometer at constant rheometer temperature (220 °C) and constant processing parameters.

To minimize premature supersaturation of carbon dioxide and to avoid prenucleation in the extruder, the temperature of the extruder and the rheometer can be reduced. This leads to an increase in the viscosity of the gas loaded CA melt. As a result, the pressure in the extruder also increases, keeping the carbon dioxide dissolved in the CA melt and avoiding premature supersaturation in the extruder. This is clearly shown in Figure 11.

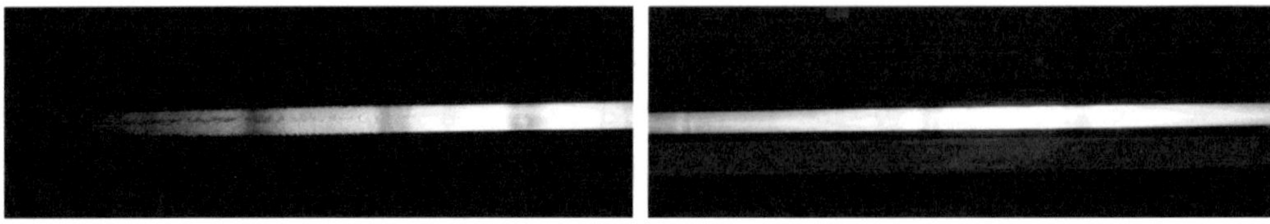

1.0 % $CO_2$ (220 °C) 1.0 % $CO_2$ (205 °C)

**Figure 11.** Effect of rheometer temperature on supersaturation of 1 % $CO_2$ from the CA melt plasticized with 20 wt.-% GTA at constant throughput (11 kg $h^{-1}$) and slit die gap (5 mm).

The results for carbon dioxide used as a PBA show the general complexity between gas loaded polymer melts and the extrusion process. To get more information on the melt flow and viscosity behavior of gas loaded externally plasticized CA melts, further studies have to be conducted using additional blowing agents, for example butane or nitrogen ($N_2$).

## 4. Foam extrusion of externally plasticized cellulose acetate

For the preliminary foam extrusion tests, the same CA compound composition (20 wt.-% GTA) as in subchapter 3.3 was used because of a good balance between melt processing and relevant melt rheology. Extrusion foamed CA rods were produced with different physical blowing agents (PBA). Table 5 shows the PBAs used for these preliminary foam extrusion tests.

| Property | $CO_2$ | $N_2$ | n-butane | trans-1,1,1,3-tetrafluoropropene (HFO 1234ze) |
|---|---|---|---|---|
| Molar mass [g $mol^{-1}$] | 44.0 | 28.0 | 58.1 | 114.0 |
| Boiling point [°C] | -78.5 | -195.79 | -0.5 | -19.0 |
| Critical temperature [°C] | 31.0 | -146.9 | 149.9 | 382.51 |
| Critical pressure [MPa] | 7.38 | 3.4 | 3.8 | 3.63 |
| Vapor pressure (25 °C) [MPa] | 6.43 | - | 0.24 | 0.49 |
| GWP | 1 | - | 3 | 6 |

**Table 5.** Physical blowing agents used and some short characteritics.

Additionally, a 5 % talc masterbatch based on the externally plasticized CA compound was added at 0.2, 0.6 and 0.8 wt.-% to investigate the influence of this nucleating agent on the foam morphology. The talc used has platelet geometry with a diameter $d_{0.5}$ of 2 μm. The physical foam extrusion tests were carried out on a 60 mm single screw extruder with an L/D of 40/1 (Barmag Oerlikon Textile GmbH & Co. KG). The extruder is equipped with a mixing screw optimized for the foam extrusion process. The last 11 D of L before the die are temperature controlled in order to cool the polymer melt. The blowing agent was compressed and injected into the extruder barrel through a pressure hole at 16 D of L using a metering system equipped with a diaphragm pump. By means of a static mixer, the PBA is dispersed in the melt.

As seen from Figure 12, the blowing agent not only influences the melt rheology of CA but also the foam extrusion and expansion behavior at the die.

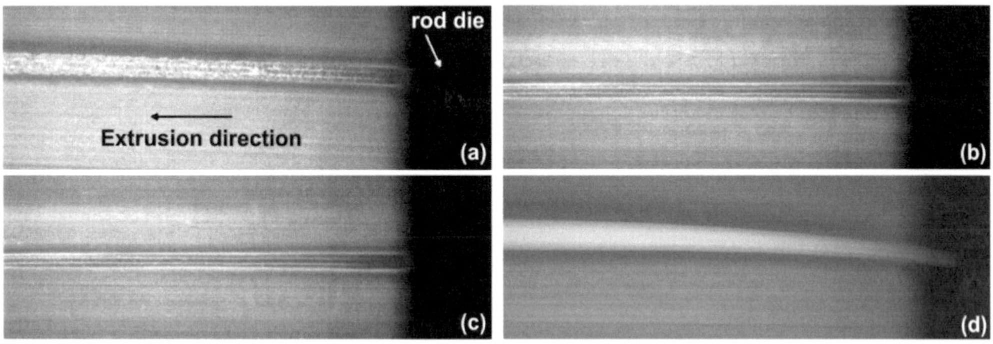


**Figure 12.** Influence of blowing agent and nucleating agent content on foam extrusion process of CA plasticized with 20 wt.-% GTA: (a) 1 % $CO_2$ + 0 wt.-% talc, (b) and (c) 2 % HFO 1234ze + 0 wt.-% talc and (d) 2 % HFO 1234ze + 0.6 wt.-% talc.

Carbon dioxide shows rapid expansion at the die in comparison to HFO 1234ze. An explanation for this could be the higher diffusivity and permeability of carbon dioxide due to its lower molecular size when compared to longer chain PBAs such as HFO 1234ze. Blowing agents having a high diffusivity will be phased out in a shorter time [54]. Additionally, the solubility of carbon dioxide is restricted in most polymer melts when compared to other conventional blowing agents such as hydrofluorocarbons [54]. High pressure in the extruder is required to keep carbon dioxide dissolved in the externally plasticized CA melt. The high pressure gradient at the die in combination with the high vapor pressure and high diffusion rate of carbon dioxide causes a strong expansion process at the die. Not only the PBA but also the addition of nucleating agents, for example talc, affects the foaming behavior at the die. By adding talc to HFO 1234ze, the foaming process of externally plasticized CA is significantly improved that can be seen in Figure 12 (c) and (d). The expansion directly starts at the die due to better nucleation (more and faster nucleation) in conjunction with stable cell growth as both processes start simultaneously.

The investigation of the rod diameter ($D$) and the foam density as a function of PBA content and talc content verify the results got from high speed camera images. As seen in Figure 13, a steady increase in $D$ is observed with increasing concentration of PBA. Generally, the higher the PBA concentration, the more PBA is dissolved in the polymer melt that can cause stronger foaming. The type of PBA and its solubility in the externally plasticized CA melt influences the intensity of the slope. The addition of talc used as a nucleating agent shows a selective influence. For butane no significant influence is observed whereas for HFO 1234ze the addition of talc (0.2 wt.-%) significantly improves the foam expansion process. This is in good agreement with the visual results obtained with the high speed camera system.

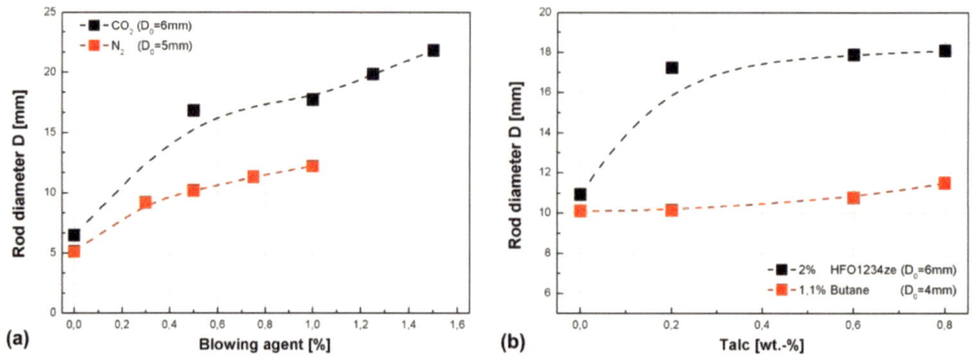


**Figure 13.** (a) Influence of PBA content on rod diameter (*D*) of CA rods foamed with $CO_2$ and $N_2$ and (b) rod diameter (*D*) of CA rods foamed with butane and HFO 1234ze as a function of talc content at constant PBA content.

From Figure 13 one may also conclude that an increase in nucleating agent content from 0.2 to 0.8 wt.-% does not affect the rod diameter considerably. An explanation for this could be the heterogeneous nucleation process in presence of talc. Therefore, the cell nucleation predominates the cell growth. Contrary, homogeneous nucleation occurs in absence of talc. Consequently, cell nucleation rate is limited and the following cell growth process predominates. The nucleating agent is therefore more important for controlling the nucleation rate, the cell density, the foam morphology, and the stabilization of the foam network. It is one parameter that significantly affects the foam density and foam ratio, this is the ratio between the density of the foamed and the unfoamed polymer. As expected, increasing talc content leads to a steady decrease in foam density and continuous increase in the foam ratio, as shown in Table 6. These observations agree well with results from literature [55, 56]. However, too high talc contents can lead to agglomeration effects of the nucleated cells [57, 58] or can cause cell collapse [59]. Therefore, an increase in foam density and a decrease in foam ratio may occur at too high talc contents [60].

The influences of PBA and talc on the foaming process, the foam density, and the foam ratio are supported by investigations of the foam morphology using optical microscopy (OM). Figure 14 shows selected microscopy images of extrusion foamed CA rods.

Butane as well as nitrogen causes a coarse inhomogeneous morphology with a broad cell size distribution and large partially opened cells to some extent. This can be explained by premature phase separation (supersaturation) and cell coalescence due to the poor solubility of butane and nitrogen in the externally plasticized CA melt. *Li* [61] found that less soluble blowing agents tend to diffuse out more rapidly than the more soluble one. Consequently a smaller amount of these blowing agents is dissolved in the polymer melt for foaming. By comparison, 1 % $CO_2$ with 0.6 wt.-% talc shows uniform closed cell morphology with homogeneous cell distribution and thin cell walls. These investigations agree well with the detected foam density and foam ratio. Scanning electron microscopy (SEM) images are of further evidence of the previous results. Blowing agents which show limited solubility in the externally plasticized CA melt such as butane or nitrogen lead to heterogeneous foam morphologies with large and partially opened cells, which is seen in Figure 15 (b). These large cells can act as voids. As a result, final properties such as mechanical performance of these foams may be poor when compared to foams which have a fine and uniform morphology.

| Blowing agent | Content [%] | Talc content [%] | Density [kg m$^{-3}$] | Foam ratio [-] |
|---|---|---|---|---|
| - | - | - | 1310 | 1.0 |
| $CO_2$ | 0.5 | 0.0 | 262 | 5.0 |
| | 1.0 | | 178 | 7.4 |
| | 1.25 | | 140 | 9.4 |
| | 1.5 | | 105 | 12.5 |
| | 1.0 | 0.2 | 162 | 8.1 |
| | | 0.6 | 152 | 8.6 |
| | | 0.8 | 124 | 10.6 |
| Butane | 1.1 | 0.0 | 351 | 3.7 |
| | | 0.2 | 332 | 3.9 |
| | | 0.6 | 322 | 4.1 |
| | | 0.8 | 271 | 4.8 |
| | 1.3 | 0.8 | 213 | 6.2 |

**Table 6.** Density and foam ratio of extruded CA foam rods as a function of PBA and talc.

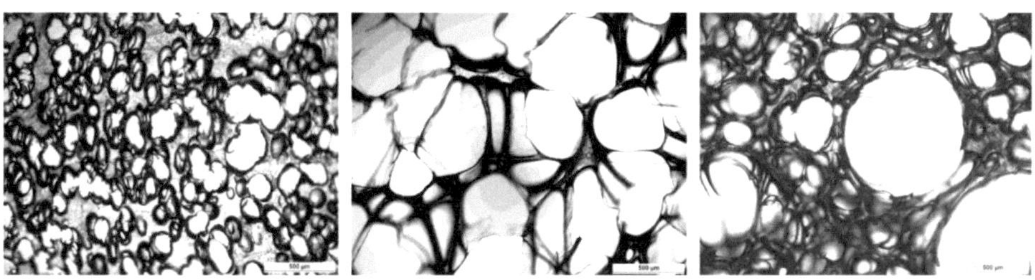

0.5 % $N_2$ + 0.6 wt.-% talc     1 % $CO_2$ + 0.6 wt.-% talc     1.1 % butane + 0.6 wt.-% talc

**Figure 14.** Influence of blowing agent on the foam morphology of selected CA foam rods at constant talc content (0.6 wt.-%) [OM, transmitted light, magnification 50x].

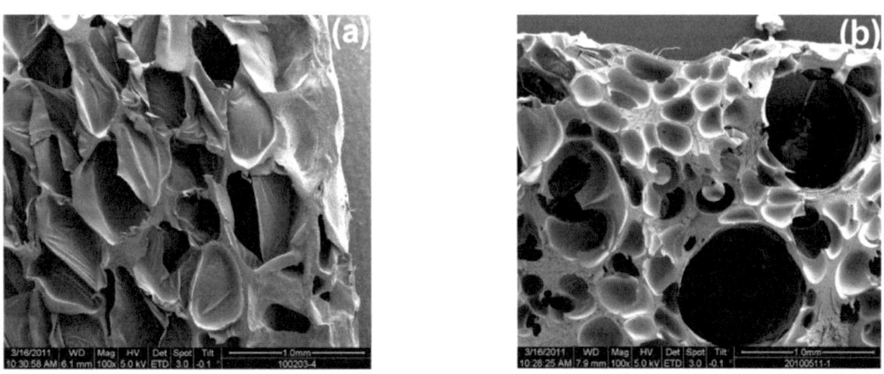


**Figure 15.** Influence of blowing agent on the foam morphology of selected CA foam rods at constant talc content (0.8 wt.-%) for (a) 1 % $CO_2$ and (b) 1.1 % butane (SEM, magnification 100x).

The limited solubility of butane and nitrogen can be explained by their nonpolar character. As seen from Table 7, only the dispersion part of the *Hansen* solubility parameter is accessible for dissolving butane and nitrogen in the polar externally plasticized CA melt. In contrast, carbon dioxide is a more polar or hydrogen bonding gas. Therefore, the polar part and hydrogen bonding part are also available for dissolution of carbon dioxide in the CA melt.

| PBA | $\delta$ [(MPa)$^{0.5}$] | | | |
|---|---|---|---|---|
| | $\delta_t$ | $\delta_h$ | $\delta_d$ | $\delta_p$ |
| $CO_2$ | 17.4 | 5.8 | 15.6 | 5.2 |
| $N_2$ | 11.9 | 0.0 | 11.9 | 0.0 |
| Butane | 14.1 | 0.0 | 14.1 | 0.0 |

**Table 7.** *Hansen* solubility parameter of selected physical blowing agents [43].

Micro-CT measurements of selected CA foam rods are shown in Figure 16. These images confirm the results from OM and SEM. For butane, cell coalescence and partially opened cell structures are evident, especially in the center of the sample cross-section. Similar results were obtained for nitrogen. Due to the limited solubility in the externally plasticized CA melt, premature supersaturation of butane and nitrogen from the melt occurs. Thus, coalescence primarily in the flow center is observed due to this being the area of the lowest flow resistance. Conversely, carbon dioxide causes fine cell morphology with homogeneous cell size distribution and high cell density across the sample.

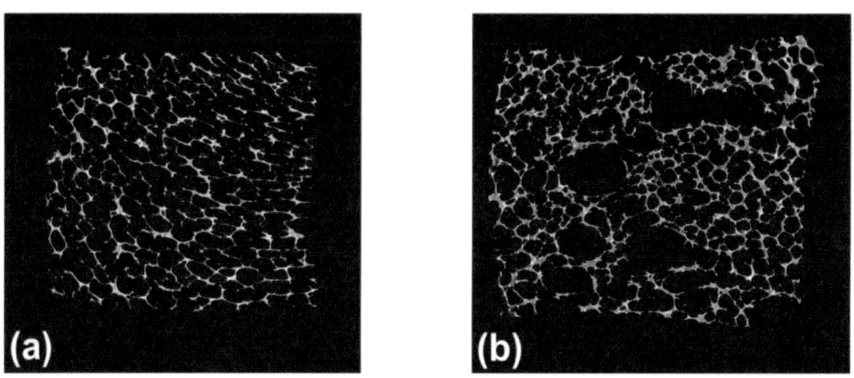


**Figure 16.** Influence of blowing agent on the foam morphology of selected CA foam rods at constant talc content (0.8 wt.-%) for (a) 1 % $CO_2$ and (b) 1.1 % butane (Micro-CT, 2D cross-sections).

To obtain further information about the influence of the blowing agents and the nucleating agent, cell size and cell density ($N_f$) were studied. Cell density ($N_f$) in cells cm$^{-3}$ was calculated according to Eq. (6) [62]

$$N_f = \left(\frac{N}{A}\right)^{\frac{3}{2}} \tag{6}$$

where $N$ is the number of cells and $A$ is the area in $cm^2$. As expected from literature [55, 60], the addition of a nucleating agent such as talc leads to a considerable decrease in cell size (Figure 17). This is in good agreement with the obtained results for the foam density and foam ratio. From Figure 17, one may conclude that no tremendous influence of the PBA concentration on the cell size was found at a talc content of 0.8 wt.-% within the range of PBA concentration studied. Similar results were found in [55] for PP foamed with carbon dioxide above 0.8 wt.-% talc content. It can be assumed that heterogeneous cell nucleation predominates in presence of talc and therefore cell nucleation rate is controlled by the high talc content regardless of the increase in concentration of the PBA used.

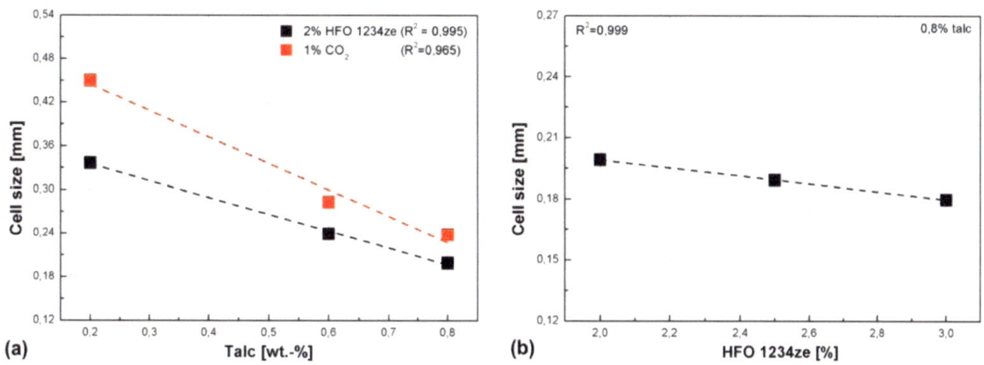


**Figure 17.** (a) Cell size of extruded CA rods foamed with 1 % $CO_2$ and 2 % HFO 1234ze as a function of talc content and (b) as a function of HFO 1234ze concentration at 0.8 wt.-%talc.

As the cell size continuously decreases with increasing talc content, cell density basically increases at the same time due to higher cell nucleation rate. This is shown in Figure 18. For HFO 1234ze, which shows good solubility in the externally plasticized CA melt, an exponential increase in cell density is observed with continuous increase in talc content due to an increase in heterogeneous cell nucleation rate. The higher cell density is directly seen in the finer foam morphology. In contrast, no excessive increase in cell density is observed for butane even at 0.8 wt.-% talc. This agrees well with the previous results obtained for the foam density, the foam ratio, and the foam morphology. Due to the limited solubility of butane, premature supersaturation of butane from the externally plasticized CA melt occurs. As a result, only small amount of butane remains dissolved in the CA melt for foaming at the die. This is directly seen in almost unchanged foam morphologies of the extruded CA rods foamed with 1.1 % butane at 0.2 wt.-% talc and 0.8 wt.-% talc respectively.

Figure 19 (a) shows the relationship between cell size and cell wall thickness. Completely different results are obtained for the selected PBAs, which might be attributed to their solubility in the externally plasticized CA melt. For butane, which shows limited solubility in externally plasticized CA melt, only a slight change in average cell size within the selected talc content is observed. The cell wall thickness of the extruded CA rods foamed with butane is also significantly higher when compared to HFO 1234ze, even at high talc content of 0.8 wt.-%. Similar results were obtained for nitrogen. The limited solubility of butane and nitrogen leads to premature supersaturation. As a result, insufficient amount of these blowing agents are dissolved in the externally plasticized CA melt for cell nucleation and cell growth. Contrary, HFO 1234ze shows good solubility in the externally plasticized CA melt. Consequently, a significant decrease in cell size with increasing talc content is observed. Furthermore, the cell wall thickness decreased continuously with decreasing cell size. This is in good agreement with the literature [63]. Figure 19 (b) shows the relationship between cell size and cell density in a LOG-LOG plot. A linear relationship between cell size and cell density is observed. The values of the slopes obtained by linear fit are close to values obtained by *Ito* [63] for polycarbonate-based nanocomposite foams.

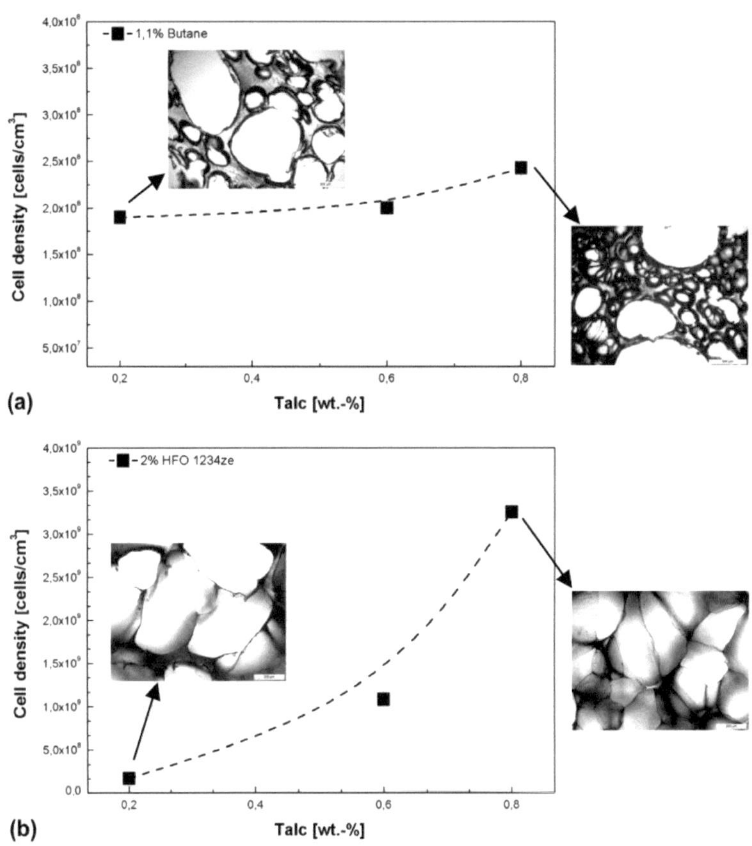


**Figure 18.** Influence of talc content on cell density ($N_f$) of extruded CA rods foamed (a) with 1.1 % butane and (b) with 2 % HFO1234ze.

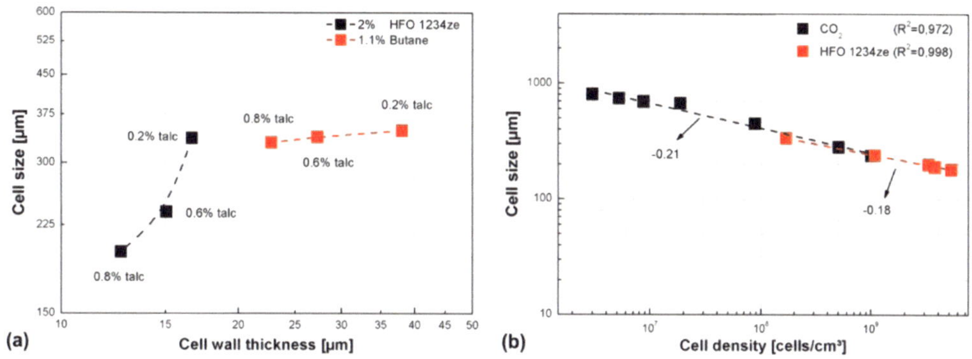


**Figure 19.** (a) Cell size in dependence of cell wall thickness as a function of talc content and PBA type and (b) cell size in dependence of cell density for extruded CA rods foamed with $CO_2$ and HFO 1234ze.

As a conclusion, the preliminary foam extrusion tests showed that externally plasticized CA is a promising biopolymer for foam extrusion technologies. By choosing an appropriate plasticizer type, the optimal plasticizer content, and a suitable blowing agent, highly substituted CA exhibits good foam extrusion behavior. Properties of CA foam rods such as foam density, foam ratio or cell density are in range with other polymers foamed with similar low PBA concentrations and similar talc contents [55]. Further improvements of the foam extrusion of externally plasticized CA can be achieved by either increasing the amount of PBA or by using higher talc content. The use of CBA/PBA mixtures as well as the use of nano-scaled particles as nucleating agents may lead to additional improvements in the foam extrusion behavior and final foam properties. *Park* [64] showed that the addition of nanoclays to the foam process produces finer foam morphologies with higher cell densities and foam ratios due to accelerated cell nucleation and suppressed cell coalescence.

## 5. Conclusions

In this chapter the use of externally plasticized CA for foam extrusion with physical blowing agents was discussed. Properties relevant to the foamability of externally plasticized CA were presented. The influence of plasticizer type, its content, and its miscibility with CA was studied with respect to the requirements for foam extrusion. An increase of plasticizer content leads to a steady decrease in glass transition temperature while the thermal stability of CA remains almost constant. Glass transition temperature of externally plasticized CA decreases about 100 °C when 25 wt.-% of a highly miscible plasticizer, for example TEC or GDA, is added. The melt processing range of the amorphous compounds increases significantly and thermoplasticity is improved over a wide temperature range. Not only the melt processing but also the melt rheology, the melt strength, and the melt extensibility are considerably improved. The polydispersity of CA as well as the strong intra- and intermolecular interactions between the free OH-groups promote high melt strength and high melt extensibility. Melt strength and melt extensibility are in range of conventional foam polymers such as PS or branched PP. If the plasticizer content is too high, melt strength and melt stiffness of the CA compounds are too low with respect to foam network stabilization. This can be overcome by reducing the plasticizer content. When choosing a more compatible plasticizer, similar results can be achieved even at lower plasticizer content. This is very important with regard to the prevention of extensive plasticizer evaporation during melt processing, which is often combined with drastic changes in the properties required for the foam extrusion process. The use of highly miscible plasticizers also minimizes the well known problem of plasticizer migration. Acetate-based plasticizers such as GDA, GTA, and EGDA as well as citrate-based plasticizers such as TEC and ATEC showed excellent miscibility with CA. Approximately 20 wt.-% of these plasticizers were found to be a good concentration in order to achieve a balanced property profile and to fulfill the specifications for foam extrusion. Physical foam extrusion tests of CA externally plasticized with 20 wt.-% GTA were conducted to proof the usability of thermoplastic CA for foam extrusion. Different PBAs, namely carbon dioxide, nitrogen, butane, and trans-1,1,1,3-tetrafluoropropene (HFO 1234ze), were studied. Talc was added as a nucleating agent. It was found, that the type of PBA and its concentration used as well as the nucleating agent content have tremendous effects on the foam extrusion behavior, the foam morphology, and the final foam properties such as foam density, cell size, and cell density. When butane and nitrogen are used as blowing agents only limited foam extrusion performance is achieved. An explanation for this could be the limited solubility of these blowing agents in the externally plasticized CA melt. By comparison, carbon dioxide and HFO 1234ze seems to be suitable blowing agents for externally plasticized CA. Both, carbon dioxide and HFO 1234ze, lead to good foam extrusion behavior and excellent expansion at the die. When talc is added as a nucleating agent this significantly improves the foam morphologies and the final foam properties. An increase in talc content leads to finer and more homogeneous foam morphologies, higher cell densities, smaller cells, and thinner cell walls. As a conclusion, externally plasticized CA foamed with carbon dioxide or HFO 1234ze in presence of talc shows excellent foam extrusion performance. Foam morphologies, cell densities, and cell sizes are comparable to conventional polymers such as PP, PS or polyesters that are foamed with similar low blowing agent content and comparable nucleating agent content. With respect to its property profile, externally plasticized CA is a promising bio-based polymer for foam extrusion and for replacing conventional oil-based polymers, especially PS, in certain foam applications.

## Author details

Stefan Zepnik
*Fraunhofer Institute for Environmental, Safety, and Energy Technology UMSICHT, Oberhausen, Germany*
*Center of Engineering Sciences, Martin Luther University, Germany*

Tilo Hildebrand
*Institute of Plastics Processing (IKV), RWTH Aachen University, Aachen, Germany*

Stephan Kabasci and Thomas Wodke
*Fraunhofer Institute for Environmental, Safety, and Energy Technology UMSICHT, Oberhausen Germany*

Hans-Joachim Radusch
*Center of Engineering Sciences, Martin Luther University, Germany*

## Acknowledgement

This work was funded by the Federal Ministry of Food, Agriculture and Consumer Protection BMELV and the Agency for Renewable Resources FNR.

## 6. References

[1] Eaves D. (2004) Handbook of Polymer Foams. Shawbury: Smithers Rapra Technology Limited. 304 p.
[2] Zhang Q., Xanthos M. (2004) Material Properties Affecting Extrusion Foaming. In: Lee S.-T., Ramesh N.S., editors. Polymeric Foams: Mechanisms and Materials. Boca Raton: CRC Press. pp. 111-138.
[3] Lee S.-T. (2000) Introduction. In: Lee S.-T., editor. Foam Extrusion: Principles and Practice. Boca Raton: CRC Press. pp. 1-14.
[4] Gendron R. (2005) Rheological Behavior Relevant to Extrusion Foaming. In: Gendron R., editor. Thermoplastic Foam Processing: Principles and Development. Boca Raton: CRC Press. pp. 56-116.
[5] Gendron R., Daigneault L.E. (2000) Rheology of Thermoplastic Foam Extrusion Process. In: Lee S.-T., editor. Foam Extrusion: Principles and Practice. Boca Raton: CRC Press. pp. 35-80.
[6] Liao R., Yu W., Zhou C. (2010) Rheological Control in Foaming Polymeric Materials: Part 1. Amorphous Polymers. Polymer. 51: 568-580.
[7] Stange J. (2006) Einfluss rheologischer Eigenschaften auf das Schäumverhalten von Polypropylenen unterschiedlicher molekularer Struktur. PhD Thesis. 201 p.
[8] Gunkel F., Spörrer A.N.J., Lim G.T., Bangarusampath D.S., Altstädt A. (2008) Understanding Melt Rheology and Foamability of Polypropylene-Based TPO Blends. J. Cell. Plast. 44: 307-325.
[9] Yamaguchi M. (2004) Melt Elasticity of Polyolefins: Impact of Elastic Properties on Foam Processing. In: Lee S.-T., Ramesh N.S., editors. Polymeric Foams: Mechanisms and Materials. Boca Raton: CRC Press. pp. 19-72.
[10] Stange J., Münstedt H. (2006) Effect of Long-chain Branching on the Foaming of Polypropylene with Azodicarbonamide. J. Cell. Plast. 42: 445-467.
[11] McCallum T.J., Kontopoulou M., Park C.B., Muliawan E.B., Hatzikiriakos S.G. (2007) The Rheological and Physical Properties of Linear and Branched Polypropylene Blends. Polym. Eng. Sci. 47: 1133-1140.
[12] He C., Costeux S., Wood-Adams P., Delay J.M. (2003) Molecular Structure of High Melt Strength Polypropylene and its Application to Polymer Design. Polymer. 44: 7181-7188.
[13] Muke S., Ivanov I., Kao N., Bhattacharya S.N. (2001) The Melt Extensibility of Polypropylene. Polym. Int. 50: 515-523.
[14] Coccorullo I., Di Maio L., Montesano S., Incarnato L. (2009) Theoretical and Experimental Study of Foaming Process with Chain Extended Recycled PET. eXPRESS Polymer Letters. 3: 84-96.

[15] Wagner M.H., Kheirandish S., Koyama K., Nishioka A., Minegishi A., Takahashi T. (2004) Modeling Strain Hardening of Polydisperse Polystyrene Melts by Molecular Stress Function Theory. Rheol. Acta. 44: 235-243.

[16] Gendron R., Daigneault L.E., Caron L.M. (1999) Rheological Behavior of Mixtures of Polystyrene with HCFC 142b and HFC 134a. J. Cell. Plast. 35: 221-246.

[17] Gendron R., Champagne M.F. (2004) Effect of Physical Foaming Agents on the Viscosity of Various Polyolefin Resins. J. Cell. Plast. 40: 131-143.

[18] Champagne M.F. (2011) Foaming Polystyrene Using Blends of HFC: Solubility and Processing Behaviour. Proceedings of 13$^{th}$ International Conference on Blowing Agents and Foaming Processes. Düsseldorf. Paper 12.

[19] Lee S.-T., Park C.B., Ramesh N.S. (2007) Polymeric Foams: Science and Technology. Boca Raton: CRC Press. 220 p.

[20] Mihai M., Huneault M.A., Favis B.D. (2010) Rheology and Extrusion Foaming of Chain-Branched Poly(lactic acid). Polym. Eng. Sci. 50: 629-642.

[21] Garancher J.P., Parker K., Shah S., Weat S., Fernyhough A. (2011) Industrial Trials of E-PLA Foams. bioplastics MAGAZINE. 6: 40-41.

[22] Tai H. (2012) Batch Foaming of Amorphous Poly (DL-Lactic Acid) and Poly (Lactic Acid-co-Glycolic Acid) with Supercritical Carbon Dioxide: $CO_2$ Solubility, Intermolecular Interaction, Rheology and Morphology. In: De Vincente J., editor. Rheology. Rijeka: InTech. pp. 133-148. Available: www.intechopen.com.

[23] Pilla S. (2011) Biodegradable PLA/PBAT Foams. bioplastics MAGAZINE. 6: 36-38.

[24] Zhang J.-F., Sun X. (2007) Biodegradable Foams of Poly(lactic acid)/Starch. I. Extrusion Condition and Cellular Size Distribution. J. Appl. Polym. Sci. 106: 857-862.

[25] Nabar Y., Narayan R., Schindler M. (2006) Twin-Screw Extrusion Production and Characterization of Starch Foam Products for Use in Cushioning and Insulation Applications. Polym. Eng. Sci. 46: 438-451.

[26] Fang Q., Milford A.H. (2001) Preparation and Characterization of Biodegradable Copolyester-Starch based Foams. Bioresource Technology. 78: 115-122.

[27] Schützenberger P. (1865) Action de l'acide acétique anhydre sur la cellulose, l'amidon, les sucres, la marmite et ses congénères, les glucosides et certaines matières colorantes végétales. C. R. Acad. Sci. 61: 485-486.

[28] Edgar K.J. (2004) Cellulose Esters, Organic. Vol. 9. In: Mark H.F., editor. Encyclopedia of Polymer Science and Technology. Weinheim: Wiley-VCH. pp. 129-158.

[29] Müller F., Leuschke C. (1992) Organische Celluloseester/Thermoplastische Formmassen. In: Bottenbruch L., editor. Technische Thermoplaste: Polycarbonate, Polyacetale, Polyester, Celluloseester. Munich: Hanser. pp. 396-457.

[30] Zugenmaier P. (2004) Characteristics of Cellulose Acetates. In: Rustemeyer P., editor. Macromolecular Symposia – Special Issue: Cellulose Acetates: Properties and Applications. Macromol. Symp. 208: 81-166.

[31] Mohanty A.K., Wibowo A., Misra M., Drzal L.T. (2003) Development of Renewable Resource-Based Cellulose Acetate Bioplastic: Effect of Process Engineering on the Performance of Cellulosic Plastics. Polym. Eng. Sci. 43: 1151-1161.

[32] Zepnik S., Berdel K., Hildebrand T., Kabasci S., Radusch H.-J., van Lück F., Wodke T. (2011) Foam (Sheet) Extrusion of Externally Plasticized Cellulose Acetate. Proceedings of 13th International Conference on Blowing Agents and Foaming Processes. Düsseldorf. Paper 7.
[33] Zepnik S., Berdel K., Hildebrand T., Kabasci S., Radusch H.-J., van Lück F., Wodke T. (2011) Influence of Physical Blowing Agent and Talc Content on the Foam Extrusion Behaviour and Foam Morphology of Externally Plasticized Cellulose Acetate. Proceedings of 3rd International Conference on Biofoams. Capri. pp. 262-270.
[34] Zepnik S., Kabasci S., Radusch H.-J., Wodke T. (2012) Influence of External Plasticization on Rheological and Thermal Properties of Cellulose Acetate with Respect to Its Foamability, J. Mater. Sci. Eng. A. 2: 152-163.
[35] Howell C.J., Trott D.W., Riley J.L. (1979) Process for Extruding Plasticized Open Cell Foamed Cellulose Acetate Filters. US Patent 4 180 536.
[36] Mori H., Yoshida M., Matsui M., Nakanishi M. (2001) Biogradable Cellulose Acetate Foam and Process for its Production. US Patent 6 221 924.
[37] Jackson W.J., Darnell W.R. (1985) Process for Foaming Cellulose Acetate Rod. US Patent 4 507 256.
[38] Redl A., Morel M.H., Bonicel J., Guilbert S., Vergnes B. (1999) Rheological Properties of Gluten Plasticized with Glycerol: Dependence on Temperature, Glycerol Content and Mixing Conditions. Rheol. Acta. 38: 311-320.
[39] Liu D., Tian H., Zhang L. (2007) Inluence of Different Amides as Plasticizer on the Properties of Soy Protein Plastics. J. Appl. Polym. Sci. 106: 130-137.
[40] Wypych G. (2004) Handbook of Plasticizers. Toronto: ChemTec Publishing. 687 p.
[41] Blasi P., Schoubben A., Giovagnoli S., Perioli L., Ricci M., Rossi C. (2010) Ketoprofen Poly(lactide-co-glycolide) Physical Interaction. AAPS Pharm. Sci. Tech. 8: E1-E8.
[42] Guo J.-H. (1993) Effects of Plasticizers on Water Permeation and Mechanical Properties of Cellulose Acetate: Antiplasticization in Slightly Plasticized Polymer Film. Drug Dev. Ind. Pharm. 19: 1541-1555.
[43] Hansen C.M. (2007) Hansen Solubility Parameters: A User's Handbook. Boca Raton: CRC Press. 544 p.
[44] Barnes H.A. (2000) A Handbook of Elementary Rheology. Aberystwyth: Cambrian Printers. 210 p.
[45] Spitael P., Macosko C.W. (2004) Strain Hardening in Polypropylenes and Its Role in Extrusion Foaming. Polym. Eng. Sci. 44: 2090-2100.
[46] Cheng S., Phillips E. (2006) Rheological Studies on Radiation Modified Polyethylene Resins. Proceedings of Society of Plastics Engineers (SPE) ANTEC Conference. Charlotte.
[47] Zwynenburg J. (2008) Predicting Polyolefin Foamability Using Melt Rheology. Available: http://www.testplastic.com/pdfs/foams-2008-jim-zwynenburg.pdf.
[48] Bernnat A. (2001) Polymer Melt Rheology and the Rheotens Test. PhD Thesis. 119 p.
[49] Koestner R.J., Mollon C.T., Schunk T.C., Gamble W.J. (2009) Method of Manufacture of a Polymeric Film with Anti-Blocking Properties. US Patent 7 597 956.

[50] Okunev P.A., Dorofeev S.P., Nikolskii K.S., Kubaenko T.M., Kulikova E.S., Tsokolaev R.B. (1972) Gel-Penetrating Chromatography Method of Investigating the Influence of the Preparation Conditions on the Molecular Weight Distribution and Properties of Cellulose Acetate. Fibre Chemistry. 3: 166-168.

[51] Minoshima W., White J.L., Spruiell J.E. (1980) Experimental Investigation of the Influence of Molecular Weight Distribution on the Rheological Properties of Polypropylene Melts. Polym. Eng. Sci. 20: 1166-1176.

[52] Minegishi A., Nishioka A., Takahashi T., Masubuchi Y., Takimoto J.-I., Koyama K. (2001) Uniaxial Elongational Viscosity of PS/a Small Amount of UHMW-PS Blends. Rheol. Acta. 40: 329-338.

[53] Muke S., Ivanov I., Kao N., Bhattacharya S.N. (2001) Extensional Rheology of Polypropylene Melts from the Rheotens Test. J. Non-Newtonian Fluid Mech. 101: 77-93.

[54] Vachon C. (2005) Research on Alternative Blowing Agents. In: Gendron R., editor. Thermoplastic Foam Processing: Principles and Development. Boca Raton: CRC Press. pp. 152-204.

[55] Kaewmesri W., Lee P.C., Park C.B., Pumchusak J. (2006) Effects of $CO_2$ and Talc Contents on Foaming Behvaior of Recyclable High-melt-strength PP. J. Cell. Plast. 42: 405-427.

[56] Pushpadass H.A., Babu G.S., Weber R.W., Milford H.A. (2008) Extrusion of Starch-based Loose-fill Packaging Foams: Effects of Temperature, Moisture and Talc on Physical Properties. Packag. Technol. Sci. 21: 171-183.

[57] Micheali W., Schuhmacher H. (2006) The Effect of Talcum Particle Diameter on the Structure of PE Foam Sheets. Proceedings of 8th International Conference on Blowing Agents and Foaming Processes. Munich. Paper 13.

[58] Jung U.P. (2007) Investigation of Foaming Behaviour of Thermoplastic Polyolefin (TPO) Blend. Bachelor Thesis. 40 p.

[59] Heinz R. (2002) Prozessoptimierung bei der Extrusion thermoplastischer Schäume mit $CO_2$ als Treibmittel (Process Optimization for the Extrusion of Thermoplastic Foams Using $CO_2$ as a Blowing Agent). PhD Thesis. 120 p.

[60] Naguib H.E., Park C.B., Lee P.C. (2003) Effect of Talc Content on the Volume Expansion Ratio of Extuded PP Foams. J. Cell. Plast. 39: 499-511.

[61] Li G., Leung S.N., Wang J., Park C.B. (2006) Solubilities of Blowing Agent Blends. Proceedings of AIChE Annual Meeting. San Francisco. Paper # 75377.

[62] Yao J., Berzegari M.R., Rodrigue D. (2010) Polyethylene Foams Produced Under a Temperature Gradient with Expancel and Blends Thereof. Proceedings of 12th International Conference on Blowing Agents and Foaming Processes. Cologne. Paper 17.

[63] Ito Y., Yamashita M., Okamoto M. (2006) Foam Processing and Cellular Structure of Polycarbonate-Based Nanocomposites. Macromol. Mater. Eng. 291: 773-783.

[64] Park C., Zhai W. (2010) Improving the Foaming Behaviour of Linear Polypropylene-Based TPO by Introducing Nanoclay. Proceedings of 12th International Conference on Blowing Agents and Foaming Processes. Cologne. Paper 16.

Chapter 3

# Production of Biofuels from Cellulose of Woody Biomass

Pedram Fatehi

Additional information is available at the end of the chapter

http://dx.doi.org/10.5772/50740

## 1. Introduction

Today, fossil resources supply approximately 86% of energy and 96% of organic chemicals used in the world [1]. The continuous depletion of petroleum fuel is one of the prime concerns these days. Other concerns associated with the large-scale utilization of fossil fuels are availability, global warming and uneven geographic distribution [2,3]. Also, the global population is expected to increase by approximately 3 billion people by 2050, which substantially increases the need for fuels. One estimate indicated that the world energy consumption would increase by 35% over the next 20 years in order to meet the growing demand of industrialized countries and the rapid development of emerging economies [2]. As the fossil fuels are depleting in the coming years, new technologies should be developed in order to produce fuels from renewable biomass resources [4].

Currently, biomass accounts for 9.8% of the world's primary energy use annually, among which 30% is used in modern forms (e.g. liquid biofuel and steam), and 70% is used traditionally (combustion for domestic heating with the energy density of 15-20 MJ/kg) [2,4,5]. Biofuel is a type of fuel whose energy is derived from biomass. It includes solid biofuels such as wood pellets, wood cube and wood puck; liquid biofuels such as ethanol and butanol; and gaseous biofuel such as hydrogen. The world liquid biofuel production would increase from 1.9 million of barrels per day in 2010 to 5.9 million barrel per day in 2030 [2]. In the United States, expectation for the production of biofuels is 136 billion litter mandated by 2022, of which 61 million liter are made from cellulosic materials [6].

As described earlier, biomass has directly been used for producing heat and electricity. The power generation engines were designed so that they directly used biomass as energy source during World War II. However, this direct utilization has several major problems: if used as biofuel, the uneven geographical distribution of biomass necessitate its transportation, the bulky nature of biomass (low heat value) leads to costly and complicated transportation systems; engines utilizing biomass usually possess a poor efficiency and high environmental impact (e.g. $CO_2$ emission). However, according to the first law of thermodynamics, any biological or chemical conversion of biomass to biofuel will consume energy, thus a part of the energy stored in biomass will be lost during the conversion and the biofuel (product) will ultimately have a lower energy than will biomass (raw material). There are several advantages for the conversion of biomass to biofuel that outweighs this deficiency: there is a large demand for liquid fuel (e.g. ethanol) in the transportation sector; the biofuel production process is more environmentally friendly (i.e. less $CO_2$ emission); the digested materials from biorefinery can be readily used as an excellent and sustainable fertilizer for cultivation and crops (a true recyclable process); the energy density (MJ/kg) of biofuel will be higher than that of biomass, and the common problem of combustion, e.g. fly ash disposal and super heater corrosion, can be eliminated via biofuel production [2,7,8].

One of the challenges of biofuel production is the difficulties in increasing the bulk density of the resource, while preserving its energy content [9]. The future biofuel should 1) have a high energy density on a mass; 2) be produced at yields near the stoichiometric maximum from a given biomass feed; 3) be compatible with existing fuel distribution infrastructures; 4) have a minimum impact on environment; and 5) not affect the global food supplies [9-11].

The first generation biofuels are presently produced from sugars, starches and vegetable oils, but these products have several issues: 1) their availability is limited by soil fertility and per-hectare yield; and 2) their contribution to savings of $CO_2$ emissions and fossil energy consumption are limited by high energy input for their cultivations and conversions [12-14]. However, lignocellulosic biomass seems to be more attractive because 1) it is the most widespread renewable source available on earth (overall chemical energy stored in biomass by plants is approximately 6-7 times of total human energy consumption annually [15]); 2) it is locally available in many countries; and more importantly 3) it does not compete with food or food industries [12]. However, the conversion of woody biomass to fermentable sugars is more difficult than that of agro-based biomass because of the presence of more hemicelluloses (not easily fermentable) and lignin as well as more condensed and crystallized structure of cellulose in woody biomass [9].

Current technologies to produce biofuel from cellulosic materials involve gasification, pyrolysis/liquefaction and hydrolysis of biomass. This book chapter excludes 1) the studies on the gasification and pyrolysis/liquefaction of biomass (as an intact raw material) for biofuel production; and 2) the studies on the production of fuel additives, e.g. levulinic acid and furfural [16] and the studies on the production of biodiesel [17-22]. Instead, recent advancements and challenges associated with the production of ethanol, butanol, hydrogen and new furan-based biofuel from cellulosic biomass will be discussed.

In order to produce biofuel from cellulose of biomass, the lignocellulosic biomass should be first dissembled to facilitate the isolation of cellulose from other constituents, i.e. lignin and hemicelluloses. Subsequently, cellulose macromolecules should be depolymerized, as depolymerization significantly improves the chemical and biological conversions of cellulose to biofuel. Then, the depolymerized cellulose, i.e. glucose, should be converted to biofuel via biological treatments, and finally the biofuel should be purified. Additionally, biomass should be deoxygenated as the presence of oxygen reduces the heat content of molecules and creates high polarity, which impairs its blending with existing fossil fuels [12].

## 2. Pretreatment of biomass

The complex structure of lignocellulosic biomass makes its utilization in biofuel production difficult. Lignocellulosic raw materials are generally composed of 40-50% cellulose, 25-35% hemicelluloses and 15-20% lignin [8]. A pretreatment stage is necessary to dissociate the plant cell wall in order to improve the accessibility of chemicals and/or microorganisms to cellulose for possible conversions [23]. The pretreatment processes target the removal of lignin, which improves the digestibility of cellulose in the following hydrolysis process [24]. Table 1 lists various pretreatment processes of woody biomass conducted in the past in order to improve the performance of fermentation processes in producing biofuels.

| Pretreatment type | Example |
|---|---|
| Physical pretreatment | Ball milling, Irradiation |
| Physicochemical pretreatment | Steam explosion, hot water pretreatment |
| Chemical pretreatment | Acid, alkali, solvent |
| Biological | Fungi |
| Enzyme | Cellulase |

**Table 1.** Various pretreatment processes of lignocellulosic feedstock [14]

Physical pretreatment consists of mechanical disruption of lignocelluloses, which is an environmentally friendly process. This process increases the surface area of biomass and decreases the crystallinity of cellulose, but it does not cause an expensive mass loss [25]. Irradiation using gamma rays, electron rays and microwaves are other physical methods to beak the structure of lignocelluloses. Microwave irradiation has been applied in many fields including food drying chemical synthesis and extraction [14].

Physicochemical pretreatment is another approach to separate the lignocelluloses of woody materials. Hydrothermal treatment, such as hot water pretreatment and steam explosion, is a suitable method particularly prior to enzyme hydrolysis. Hot water pretreatment process is conducted under pressure at an elevated temperature of 230-240 °C for 15 min to maintain water in liquid form, which produces less inhibitory compounds (e.g. furfural) compared to steam explosion method [26,27]. However, the viscosity of the spent liquor produced in this method is rather high, which makes its handling process challenging. The steam explosion is practically applied in industry via steaming biomass at an elevated temperature, e.g. 170 °C [28-30]. To limit the production of inhibitors, process conditions should be precisely adjusted. Steam explosion has different subcategories, such as ammonia fiber explosion and acid-explosion, in which acid or ammonia is also added to the system during the steaming process.

The chemical pretreatment of cellulosic biomass includes oxidizing hydrolysis, acid or alkaline hydrolysis and solvent extraction. In this context, lime treatment (e.g. treatment of woody biomass at 180 °C with lime solutions) has been considered as an effective chemical pretreatment method, because of its low cost and wide use in agro- and wood-based processes [4,27,31-34]. Pretreatment with dilute acid at intermediate temperatures (e.g. 160 °C) is usually considered the most cost-effective method to loosen the cell wall matrix via degrading the hemicelluloses of biomass [27,35].

Biological pretreatment, such as fungal, is milder in its operational conditions than physical or chemical pretreatment. The oxidative biodegradation of lignin by white-rot fungi has been widely studied in the past [36]. The main advantages of biological pretreatment are low energy input, no chemical requirement, mild environmental conditions and environmentally friendly working manner [25]. However, the biological pretreatment processes usually need a long retention time and have a low yield. They are also sensitive to the process conditions such as temperature and pH. Enzymatic pretreatment of biomass has been comprehensively studied in the past and will be discussed in a separate section.

## 3. Hydrolysis

Generally, microorganisms have poor cellulose/biomass digestibility and a limited efficiency in producing biofuels. Hydrolysis has been commonly applied in industrial scales to improve the efficiency of microorganisms in producing biofuels prior to fermentation, which relies on the decomposition of polysaccharides to monosaccharides [37].

### 3.1. Enzyme hydrolysis

In this process, enzyme is used for decomposing polysaccharides into monosaccharides. Microorganisms that usually produce enzyme are fungal species, such as *hypocerea jecornia, trichoderma reesei,* and bacteria species, such as *clostridium thermocellum, cellulomonas flavigena*. Three main enzymes hydrolyzing cellulose to glucose are *cellulase, 1-4- β-D-endoglucanase and 1-4- β-D-cellobiohydrolase* [38]. Process parameters, e.g. pH, temperature, time, significantly affect the performance of enzyme hydrolysis. Furthermore, porosity and crystallinity and the lignin content of biomass seem to significantly influence the efficiency of this process [25].

Equation 1 describes the enzymatic hydrolysis model of cellulose to glucose [39]:

$$X = 1 - \left[\frac{K_e + e_0}{K_e(K_a e_0 t + 1) + e_0}\right]^b \tag{1}$$

where $e_0$ is the initial enzyme concentration (g/l), X is conversion efficiency (<1), t is the reaction time (h), $k_a$ is the enzyme deactivation constant (g/l.h), $b=k_e k_a k_1$ is the fitted constant (dimensionless), $k_e$ is the adsorption equilibrium constant (g/l), and $k_1$ is the rate constant of sugar formation (1/h). When $e_0 \rightarrow \infty$ , X converges to a constant at constant time according to equation 2:

$$X = 1 - \left[\frac{1}{K_e K_a t + 1}\right]^b \quad (2)$$

and when $t \rightarrow \infty$, X converges to a maximum conversion of 1.

Figure 1 shows the results of one study on the enzyme hydrolysis of pre-steamed corn Stover at 50 °C and pH 4.8, the conversion of celluloses to reducing sugar was 60% after 48h. Fitting the experimental data of Figure 1 into equation 1 resulted in $k_e$, $k_a$, $k_1$, and b of 0.9975 (g/l), 0.9837 (l/g.h), 0.2843 (1/h), and 0.2897, respectively [39].

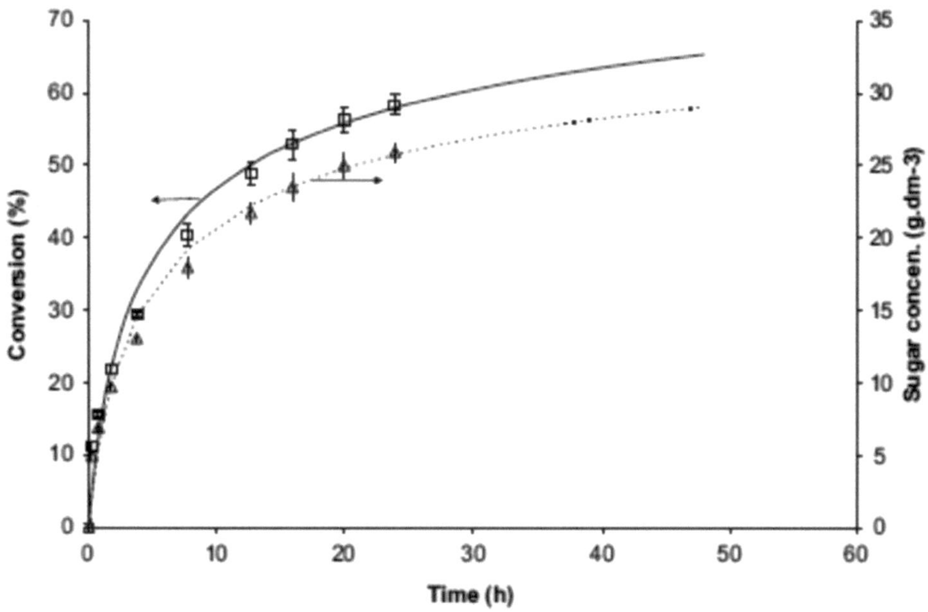


**Figure 1.** The conversion of cellulose and the concentration of reducing sugars as a function of time in the enzymatic pretreatment. □, Δ experimental points; solid and dash lines denote the model (Eq. 1) values [39].

Enzymatic hydrolysis has several advantages compared with acid hydrolysis such as a lower equipment cost and higher glucose yield without sugar degrading products or by-products [24].

## 3.2. Acid hydrolysis

This method has been extensively applied to convert oligomeric sugars to monomeric sugars in the past [40-43]. It has a short process time (i.e. less than 1 h) and produces a higher sugar yield (>85%), but operates at a relatively high temperature, e.g. >120 °C (compared to enzyme hydrolysis). However, acid hydrolysis may result in the production of undesirable by-products that should be eliminated prior to fermentation processes. These detoxification processes are generally costly and complicated [25,40-42].

## 4. Detoxification process

Natural occurring and process-induced compounds may retard the fermentation processes for producing biofuels, which complicates the fermentation process [42-46]. These inhibitors are phenols, acetic acid, furfural, metal ions and lignosulfonates, and can chemically or biologically be removed from hydrolysates prior to fermentation processes.

Boiling and overliming have been extensively used in different studies to reduce the concentration of the inhibitors [44-46]. Boiling was proposed to reduce the concentration of volatile components, e.g. furfural, and overliming was proposed to create some insoluble inorganic salts that adsorb inhibitors [47]. Alternatively, the inhibitors can be removed by employing the concept of adsorption and flocculation phenomena. In this regard, commercial adsorbents, e.g. activated carbons, fillers, e.g. calcium carbonate or lime, or ion exchange resins may be suitable choices. The adsorption/flocculation processes could effectively remove these inhibitors and research in improving the performance of this process is on-going [28,30,48-50].

Alternatively, the inhibitors can be removed from hydrolysates via biological treatments. For example, a mutant yeast, *S. cerevisiae* YGSCD 308.3, was applied to reduce the acetic acid content of the ammonia-based hardwood hydrolysate in one study [51]. This yeast grows on acetic acid, but not on xylose, glucose, mannose and fructose. This detoxification process resulted in an ethanol production (fermentation was conducted at a temperature of 30 °C, and 300 rpm for 24 h) with a 73% yield of that of the maximum theoretical value in a laboratory scale. However, the presence of acetic acid did not allow ethanol production in the control sample of this study [51].

## 5. Ethanol production

As of January 2008, 136 ethanol plants in the US produced 7.5 billion gallons of ethanol annually, and this number is expected to increase by 13.3 billion gallons/year via constructing additional 62 plants [4]. The current biofuel market is largely dominated by ethanol (90% of the world biofuel production) [2]. However, most of ethanol produced today comes from starch (as in maize grains) or sucrose (from sugarcane), i.e. first generation biofuel. Lignocellulosic biomass, on the other hand, represents more abundant feedstock for ethanol production, i.e. the second generation biofuel [4]. However, the cost of producing lignocellulosic ethanol could be almost double of that of corn-derived ethanol [2]. In this context, the US department of Energy (DOE) has committed over $ 1 billion dollars toward a realization of a 2012 goal of making lignocellulosic ethanol at a competitive cost of $1.33 per gallon [52].

The production of ethanol from biomass has been criticized in the past since a large amount of $CO_2$ is produced (released) as the by-product of this fermentation process. However, one study showed that, if softwood (unspecified species and fermentation conditions) biomass were considered as raw material, and ethanol were produced from cellulose and hemicelluloses and FT-diesel from lignin, this integrated process could lead to 54% of mass conversion efficiency, 67% of carbon conservation efficiency, but more interestingly, 88% of energy conversion efficiency [12]. In other words, the majority of mass, but just 12% of energy, would be lost in ethanol production process. Consequently, ethanol production process would definitely improve the energy density of biomass, which is a critical factor of biofuel. This analysis depicted that it was extremely crucial to widen the assessments beyond the sole mass balance to obtain a complete understanding of biofuel production [12].

## 5.1. Process alternatives

The biological processes of ethanol production usually involve hydrolysis, which breaks cellulose to glucose, and fermentation to convert glucose to ethanol. Ethanol can be produced via separate hydrolysis and fermentation (SHF) or simultaneous saccharification and fermentation (SSF). The SHF facilitates the operation of hydrolysis and fermentation processes under separate optimized conditions. Previous studies on pre-steamed agro-based raw materials (corn Stover) revealed that the yield (ethanol produced per unit mass of dried feedstock g/g) of SHF was higher than that of SSF [53]. Also, SHF has a faster hydrolysis rate than does SSF under optimized conditions [11,39]. In SSF process, cellulose is hydrolyzed to glucose by cellulase, while yeast spontaneously ferments glucose to ethanol [54]. The SSF usually has a higher productivity (ethanol produced per unit mass of dried feedstock per unit time, g/g.h) than does SHF, since SSF has a shorter operating time [39]. Other advantages of SSF over SHF include less inhibition of enzymes and a longer time of enzymatic hydrolysis. Usually, fermentation temperature and the presence of ethanol are not at optimized conditions for cellulase activities in the SSF process, which leads to poor enzyme performance. To overcome this difficulty, several approaches were followed in the past, which are demonstrated as follows:

1. Enzyme immobilization onto a solid support was proposed to improve the enzyme activity at non-optimal conditions (e.g. SSF). In one study, dissolved cellulase (500 μl, in 10 mM acetate buffer, pH 4.8 with 0.01 % (w/v) sodium azide) was immobilized on Aerosol OX-50 silica (40 nm average diameter particle, 1.4 $m^2$ surface area), and the SSF was carried out using this coated silica [54]. The results showed that the ethanol yields of the SSF process containing immobilized enzyme were 2.3 and 2.1 times higher than that of the SSF process containing enzyme in solutions at pH 4.8 and 5.3, respectively. The higher ethanol yield of the immobilized enzyme process was likely due to higher glucose yields as a result of increased enzymatic stability at the non-optimal enzyme conditions required for the SSF [54]. However, this process results in a higher ethanol yield under the optimum conditions of fermentation rather than that of enzyme hydrolysis. This is because 1) the optimum conditions for the immobilized enzyme reaction may not coincide with those for the enzyme in solutions; 2) although the same mass of enzyme can be used in the solution and immobilized systems, it does not mean that the same amount of enzyme is available to participate in the hydrolysis reactions; and 3) enzyme immobilization results in a random orientation of adsorbed enzyme on the silica substrates, and this leads to some inactive enzyme due to buried or inaccessible active sites [54].

2. Semi-simultaneous saccharification and fermentation (SSSF) is another approach to produce ethanol. In this method, the hydrolysis process is applied under optimized conditions, which breaks down cellulose to glucose, and subsequently fermentation is conducted on the product of hydrolysis without removing the hydrolystate. In other words, it is an operating mode between the SHF and SSF, thus it has the advantages of both SHF and SSF (a higher productivity and yield). In one study on microcrystalline cellulose (Avicel PH101), three alternative processes were carried out to produce ethanol: 24h hydrolysis+48h SSF (called SSSF), 72 h SSF, and 24 h hydrolysis+48h fermentation (SHF) [39]. The hydrolysis process was conducted under the conditions of 50 °C, pH 4.8 and 2 g Novozymes enzyme, while the fermentation was performed under the conditions of 36 °C, pH 4.8 and 300 rpm using *S. cereviasiae*. The results showed that the optimal ethanol productivity for SSSF, SSF and SHF were 0.222, 0.194, and 0.176 g/l.h, respectively. The corresponding maximum ethanol concentration was 16, 14 and 12.6 g/l with equivalent theoretical ethanol yields of 70.5%, 61.8%, and 56.1%, respectively [39].

## 5.2. Theory

The theoretical ethanol yield (Y) can be calculated using equation 3 [39]:

$$Y = \frac{0.9E}{0.511C_0}100\% \tag{3}$$

where E is the final ethanol concentration (g/l) and $C_0$ is the initial cellulose concentration (g/l). Also, the fermentation efficiency ($e_f$), from sugar to ethanol, can be determined using equation 4:

$$e_f = \frac{E}{0.511G_h}100\% \tag{4}$$

where $G_h$ is the glucose concentration after hydrolysis (g/l). When glucan in cellulose is completely converted into glucose ($C_0$=0.9 $G_h$). The theoretical yield of SSF yield is equal to fermentation efficiency [39].

## 5.3. Microorganism for ethanol production

Ethanol production via fermentation has been the subject of several research activities. In this regard, many yeasts and bacteria have been introduced or modified and their ethanol production efficiencies have been assessed.

### *5.3.1. Bacteria*

*Escherichia coli* can consume hexoses and pentoses for ethanol production. However, *E. coli* is less rebust against several factors including pH, salt concentration and temperature. It also exhibits a low ethanol (<35 g/l) and butanol tolerance (< 20g/l) [51]. *Z. mobilis* has also been applied in ethanol production, has a tolerance of up to 120 g/l ethanol, and its ethanol production yield approaches 97% of the maximum theoretical value under optimized conditions. However, it can only ferment glucose, fructose and sucrose (hexoses), and it has a low tolerance to acetic acid [55,56]. *Clostridia* have also been applied in ethanol production under optimized temperature of 35-37 °C at a pH range of 6 and 9, which can ferment hexoses and pentoses [57]. *Corynebacterium glutamicum* has also been applied for ethanol production, but it cannot ferment pentoses, unlike *E. coli.* [58].

### 5.3.2. *Yeast*

Hexoses can effectively be converted to ethanol with a high yield (0.4-0.51 g ethanol/ g glucose) and a high productivity (up to 1 g/l.h) by *Saccharomyces cerevisiae* or re-combinant *S. cerevisiae* [53,55]. *S. cerevisiae* is the best known microorganism used for fermenting glucose to ethanol, but it cannot ferment pentoses. In this respect, the fermentation rate of xylose is 3-12 times lower than that of glucose by *S. cerevisiae* [45]. *Kluyveromyces* is another thermotolorant species with up to 98% of the maximum theoretical ethanol yield at a temperature above 40 °C, but it cannot accommodate pentose [59]. *Hansenula polymorpha* is another thermotolorant species with an optimized temperature of 37 °C, and its fermentation operation is possible up to a temperature of 48 °C. *Kluyveromyces* and *H. polymorpha* are suitable for SSF processes [60]. *P. stipitis* is another naturally fermenting pentose and hexose. It can be used for SSF set-up even though its optimized growth temperature is around 30 °C [53,55].

## 5.4. Ethanol recovery

The ethanol production is coincided with yeast cell growth in fermentation, hence the yeast is a by-product of the process. Pure yeast is a value-added product of the process and can inevitably decrease the net cost of the process [53]. Currently, centrifugation is applied to separate yeast cells from ethanol, which is an expensive process with a high energy demand. Yeast cell immobilization technologies using inert carries or chemicals have also been applied in ethanol industry for separating ethanol from cells [53]. Alternatively, yeast flocculation process was introduced in the 1980s and commercialized in 2005 in China and comprehensively used in brewing industry. It involves lectin-like proteins and selectively binds the mannose resides of cell wall of adjacent yeast cell. In this flocculation process, calcium ions are needed and the flocculation occurs spontaneously [61-63]. Upon formation, the flocs would either sediment (large yeast) or rise to the surface (ale yeasts). This flocculation process has the advantages of 1) allowing greater yeast cell biomass concentrations because of no inert carrier; 2) being a simpler and more economically competitive process and 3) fostering the yeast cell viability because the continuous renewal of cells resulting from breaking up of relatively large flocs. The yeast flocs can be purged from the fermenter maintaining the biomass concentration inside the fermenter at specified levels [64]. Different configurations of immobilization process are available for industry: airlift, single packed column, two-stage packed column and $CO_2$ suspended bed [64].

## 5.5. SPORL and dilute acid ethanol production processes

Sulfite pretreatment to overcome recalcitrance of lignocellulose (SPORL) has been developed by the U.S. Forest Service, Forest Products Laboratory and the University of

Wisconsin-Madison as a promising biorefinery process to produce ethanol from cellulosic materials [32,65,66]. The SPORL process has a short chemical pretreatment at a high temperature to remove recalcitrance of the substrate without a significant delignification, and a disk refiner to increase the surface area, which is necessary for the sugar dissolution in the latter stages for fermentation. The potential products of the SPORL process are ethanol, hemicellulosic sugars and lignosulfonate [67]. Figure 2 shows the process diagram of the SPORL process. As can be seen, the wood chips are pretreated with bisulfite and/or sulfuric acid at a temperature of 160-190 °C, a pH of 2-5, and liquid/wood ratio of 2-3 for 10-30 min. The bisulfite charge is 1-3% and 6-9% for hardwood and softwood species, respectively, and acid ranges between 1 and 2% [68]. The solid substrates of the chemical treatment are then fibrilized using a mechanical disk refiner. The chemical pretreatment has a direct effect on the refining stage of the SPORL process. Subsequently, the solid substrate is enzymatically hydrolyzed, fermented, and distilled to produce ethanol [67]. Hemicelluloses are also isolated from the spent liquor and fermented to ethanol, while lignosulfonate is a by-product of this process. Softwoods species have usually poor digestibility in enzymatic saccharification [65]. The SPORL process is particularly effective in improving the enzymatic saccharification efficiency of softwoods.

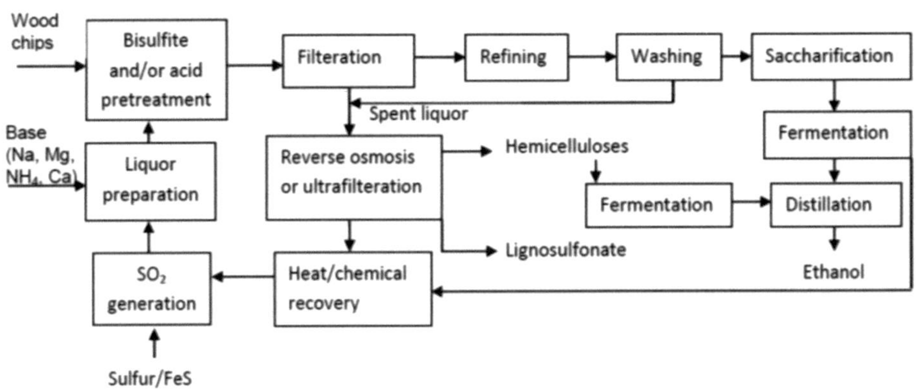


**Figure 2.** Process flow diagram of the SPORL or dilute acid treatment [67].

Alternatively, ethanol can be produced by pretreating biomass with sulfuric acid, and subsequent mechanical pulping (refining) of the pretreated biomass. This process is called dilute acid process. However, a high temperature and a very low pH of this process impart serious equipment corrosion problems [70].

It was reported that the SPORL pretreatment was more efficient than the dilute acid pretreatment for recovering sugars from the solid substrate and spent liquor. Inhibitors were more in the spent liquor of the dilute acid pretreatment, but the concentration of lignin was higher in the spent liquor of the SPORL pretreatment. Due to the lower inhibitors presented in the spent liquor, the ethanol production would be facilitated with the SPORL system. The ethanol production from the substrate and spent liquor of pine species under various SPORL pretreatment conditions at 180 °C for 25 min and then fermenting via using *S. cerevisiae* (ATCC 200062) at 32 °C for 72 h at 100 rpm are listed in Table 2 [70].

| Acid charge, % | Bisulfate charge, % | Initial pH | Substrate ethanol[1] | Substrate ethanol efficiency[2] | Hydrosate ethanol[1] | Hydrolysate ethanol efficiency[2] | Total ethanol yield |
|---|---|---|---|---|---|---|---|
| 2.21 | 4 | 1.8 | 136.7 | 73.7 | 52.8 | 65.1 | 189.5(49.2) |
| 2.21 | 8 | 1.9 | 209.4 | 83.8 | 66.9 | 94.2 | 276.3(71.7) |
| 1.4 | 8 | 2.3 | 193.2 | 76.4 | 36.3 | 70.4 | 229.5(59.6) |
| 0 | 8 | 4.2 | 165.6 | 69.6 | 1.7 | 11.3 | 166.7(43.3) |

[1]:l/ton wood; [2]:percentage of the theoretical yield

**Table 2.** Ethanol production in the SPORL system of pine species under various conditions [70].

As can be seen, by increasing the bisulfite concentration from 4% to 8%, the ethanol production and efficiency were increased for both the substrate and spent liquor, resulting in 21% increase in the total ethanol production. It is also noticeable that the ethanol production from the substrate or spent liquor was reduced by reducing the acid charge (increasing pH) of the pretreatment.

## 6. Butanol production

Although ethanol has been considered as a promising biofuel, it has several drawbacks: the heat value of ethanol is 27 MJ/kg, while that of FT-diesel is 42.7 MJ/kg implying that ethanol contain a much lower energy density compared with diesel fuel used in automobile industry [2,12]; it has also a high water solubility that prevents it from being an ideal biofuel, and high concentration ethanol blends can cause corrosion of some metallic components in tanks and deterioration of rubbers and plastic used in car engines [2]. Consequently, the incentives for obtaining a better alternative biofuel are high.

The industrial synthesis of biobutanol was commenced during 1912-1914 by acetone-butanol- ethanol (ABE) fermentation of molasses and cereal grains using *clostridium acetobutylicum* [71]. However, butanol was primarily used as a solvent for the production of other chemicals prior to 2005. Butanol has similar energy density and polarity to those of gasoline [71]. It has an adequate blending ability with gasoline and compatibility to combustion engines [72]. It can be shipped through existing fuel pipelines, whereas ethanol must be transported via rail and truck [73,74]. It has octane-improving power and low volatility (six times less than the volatility of ethanol). These novel properties made butanol a promising biofuel [75]. The economics of butanol fermentation is favorable even with the present technology [76]. However, the capital cost of butanol fermentation is presently higher, but its production cost is less, than that of petrochemical process to produce butanol [77]. Despite steadily growing production, its market remains tight and its price high due to low investment and hefty demand in coming years [76].

### 6.1. Butanol fermentation process

To produce butanol via fermentation as the second generation biofuel, cellulose should be initially converted to glucose, as most of the butanol-fermenting microorganisms can digest glucose and not cellulose. Similar to ethanol production, the enzymatic hydrolysis of polysaccharides to monosaccharides and then fermentation to biosolvents were carried out in the past [78]. The drawbacks of this process is its energy intensity, which makes its commercialization costly [78,79].

The production of butanol from cellulose mainly relies on the application of *clostridia* species (*C. acetobutylicum, C. beijerinckii, C. pasteurianum*). *Clostridia* acetone-butanol-ethanol (ABE) fermentation from carbohydrates used to be the largest biotechnological process second to yeast ethanol fermentation and the largest process ever conducted under sterile conditions [72,76]. *Clostridia* species have several advantages: their thermophilic nature permits their utilization at a high temperature of 60 °C (facilitates sterilization process), they can digest pentoses, and more interestingly they can ferment cellulose, which implies that cellulose hydrolysis to glucose and glucose fermentation can be proceeded spontaneously. Butanol can only be generated if the cells enter sporulation process, but this process discourages cell growth. Thus, balancing cell growth and sporulation timing is a key factor influencing the carbon flow towards cell growth and butanol generation [74]. However, these species have some disadvantages including the low solvent resistance, comparative difficulty in genetic modification, and increased energetic demand for cellulase production under anaerobic environments [80].

In the ABE process, the coproduction of acetone, butanol and ethanol causes a poor selectivity with respect to butanol production. The theoretical mass and energy yields of ABE fermentation are 37% and 94%, respectively [81]. It was reported that the substrate costs account for 60% of the total production cost, and the butanol production will not be feasible if the fermentation yield is less than 25% [80]. The best results reported for the ABE process were 8.2 g/l acetone, 2.2 g/l ethanol and 17.6 g/l butanol [82].

### 6.2. Production improvement

The production of butanol faces some challenges: 1) selection of sustainable biomass 2) low production rate, 3) constrains executed on butanol inhibition and 4) high product recovery costs [72,83]. Different strategies can be performed to improve the production of butanol via fermentation as described in the following sections.

#### *6.2.1. Eliminating inhibitors*

The phenolic compounds of cellulosic materials inhibit the butanol fermentation process (similar that of ethanol fermentation process) [78]. In one study, the majority of inhibitory compounds were phenolic compounds, while furfural and hydroxymethyl furfural (HMF) did not affect the cell growth and ABE fermentation [84]. It was reported that, during the initial growth phase of cells, other by-products were also produced such as acetic acid and butyric acid, which inhibited the butanol production [84]. In one study, the presence of 1 g/l ferulic acid and vanillin acid reduced the cell growth by 70% and 56%, respectively [84]. The choice of detoxification method would depend on the compositions of hydrolysates and the species of fermenting microorganism [84]. Previously, lime treatment [85], evaporation, adsorption using ion exchange resins and activated carbon were used prior to fermentation as detoxification methods for butanol production [84]. However, the detoxification efficiency depends on the chemical structure of inhibitors [84]. Other inhibitors of the ABE process are substrate inhibition, salt concentration, presence of dead cells, low water activity, $O_2$ diffusion, macromolecules accumulation and nutrient deficiency [86].

### *6.2.2. Removal of butanol*

*Clostridia* species are known to be solventogenic in producing acetone, butanol, and ethanol, but are still subjected to negative inhibition by their own products [72,74]. The hurdles are being resolved using genetic engineering techniques, metabolic engineering strategies and integrated continuous fermentation processes [72]. In this context, one study showed that the presence of butanol at a concentration of higher than 7.4 g/l impaired the cell growth [87,88]. In this case, butanol may penetrate the cytoplasmic membrane and disrupt several physicochemical characteristics of the cells [72]. On the contrary, the *Clostridium* BOH3 species showed the advantage of high resistance against butanol (up to 16 g/l) [74].

In the literature, the simultaneous production and the removal of butanol was carried out in order to maintain a low butanol concentration in the fermentation medium (broth) using liquid-liquid extraction, adsorption, perstraction, reverse osmosis, pre-evaporation and gas stripping [89], among which gas stripping has received great attentions. Gas stripping offers several advantages including feasibility and simplicity of the process, reduction in butanol concentration without affecting culture, concentration of nutrients and reaction intermediates [86]. Furthermore, a membrane separation with super critical extraction may be a feasible method in butanol removal in future [72].

### *6.2.3. Modifying microorganism*

One alternative to increase the butanol production was reported to be the genetic engineering for strain improvement with insertion of butanol producing gene of *Clostridia* in high butanol tolerant organisms. In this context, the genetic manipulating of *Clostridia* was reported to decrease its sensitivity to the presence of butanol, which eventually increased the butanol concentration up to 17.8 g/l in the fermentation medium [87]. Also, research on aerobic producing butanol using genetically engineered organics like, *E. coli*, are being attempted [72]. The strain improvement is effective in improving the yield, but has marginal influence on the economics.

## 6.3. Process configuration

The fermentation of glucose to butanol can be conducted in batch or continuous process. Generally, continuous butanol processes, (free cells, immobilized cells, and cell recycling) are more economical over batch processes. Other advantages are the reduction in sterilization and inoculation times and the superior productivity. In free cell continuous fermentation, cells are free to move within the fermentation broth due to agitation or air lifting. This maintains the microbial cells and nutrients in the suspension and helps in promoting mass transfer [72]. In immobilized cell fermentation, cells are stationary, which have the advantages of the long survival time of cells in solventogenic phase. In one study, immobilized cell fermentation using *C. acetobutylicum* produced 20% higher butanol yield than did free cell fermentation [90]. However, the scale up process using immobilized technology seems to have technical problems due to excessive cell growth in the packed beds and blockage [85]. Recycling technique using a membrane technology (e.g. filter) after the fermentation has also been attempted in the literature and the results showed 10 times higher cell concentration and 6 times higher butanol production compared with conventional continuous process using *C. saccharoperbutylacetonicum* [85]. On the contrary, batch process has a less capital cost and contamination problem (sterilization) in fermentation, and is more flexible.

## 7. Hydrogen production

Hydrogen has been considered as one of the major energy carriers in future due to its high energy content and its capability to overcome the air pollution and global warming [91]. Renewable carbohydrates (e.g. cellulose) are suitable raw materials for hydrogen production, i.e. second generation biofuel, because they are less expensive than other hydrogen carriers, e.g. hydrocarbons, biodiesel, methanol and ethanol [3,37]. Glucose is widely used substrate for hydrogen production [37]. Today, the production cost of hydrogen from renewable biomass is appealing ($ 60/dry ton or $ 3.6 per GJ) [15]. The most important energy application of hydrogen is transportation, especially for light-duty vehicles [3]. However, the large scale implementation of the hydrogen economy has some obstacles: sustainable hydrogen production, high density hydrogen storage, hydrogen distribution infrastructure, fuel cell cost and life time, and safety concerns [3]. Thus, new technologies and strategies should be developed to make the hydrogen production more economically attractive and industrially feasible [14].

### 7.1. Hydrogen production processes

Table 3 lists various pathways to produce hydrogen from glucose [3]. As can be seen, the theoretical and practical yields of hydrogen production via chemical catalytic reactions, i.e. gasification, pyrolysis and hydrolysis (accompanied by aqueous phase reforming (APR)), of glucose are in a similar range. The gasification is conducted at a high temperature of 1000 K in the presence of oxygen and water. Pyrolysis is carried out at a high temperature but in the absence of oxygen. The main advantage of APR over gasification is the lower extent of undesirable decomposition reaction [3]. The APR is carried out at a lower temperature (400-550 K) and a medium pressure (50-70 bar) [3]. The water medium of this process promotes the occurrence of the hydrogen production reaction. Although the chemical catalytic reaction seem to have considerably higher hydrogen production yields, their prerequisite high energy input and poor selectivity toward hydrogen production are barriers in their industrial implementations.

However, the biological conversion of cellulose to hydrogen is performed at much lower temperatures, which implies that the energy input of this process is much lower than that of chemical catalytic processes. In contrast, the theoretical and practical yields of hydrogen production via this method are rather low (see dark anaerobic fermentation in Table 3) [30,92,93]. In principle, up to 12 mol of hydrogen can be produced per mole of glucose and water via fermentation. However, natural microorganisms produce as much as 4 mol of hydrogen/ mol of glucose along with 1 mol of acetate [94]. The current yield of hydrogen production in cellulose fermentation ranges from 1 to 2 mol $H_2$/mole of hexose sugar [95]. To increase the hydrogen yield, a bioelectrochemically-assisted microbial fuel reactor was reported to convert 2 mol of acetate to up to 8 mol of hydrogen with the aid of electricity [96]. In this respect, the overall production yield of hydrogen increased to 9 mol per mol of glucose [96]. However, this process also needs electricity input and has not been studied in large scales.

| Method | Theoretical yield, % | Practical yield, % | Energy efficiency, % |
|---|---|---|---|
| Dark fermentation (DF) | 4 | 1-3.2 | 10-30 |
| DF+ electricity-assisted microbial fuel cell | 12 | 9 | 75 |
| Ethanol fermentation/partial oxidation reforming | 10 | 9 | 60 |
| Gasification | 12 | 2-8 | 35-50 |
| Pyrolysis | 12 | 2.5-8 | 30-50 |
| Hydrolysis+ aqueous reforming | 12 | 6-8 | 30-50 |
| Synthetic pathway biotransformation | 12 | 12 | 122 |

**Table 3.** Methods for converting glucose to hydrogen as biofuel [3].

Synthetic pathway biotransformation is a new biocatalytic technology based on the application of enzymes to convert cellulose to hydrogen. This process is much simpler than biological treatments. The biotransformation of carbohydrates to hydrogen by cell-free synthetic pathway (enzymes) has numerous advantages: high production yield (12 $H_2$/glucose unit), 100% selectivity, high energy conversion efficiency (122% based on combustion energy), high purity hydrogen generated, mild reaction conditions, low cost bioreactor and no toxicity hazards. In one study using 2 mM cellubiose concentration as the substrate, the overall yields of hydrogen and $CO_2$ were 11.2 and 5.64 mol per mol of anhydroglucose unit corresponding to 93% and 94% of the theoretical yields, respectively [97]. Furthermore, these enzymatic reactions are reversible, thus the removal of products favors the unidirectional reaction towards the desired products. One study on the biotransformation of starch and water revealed that these reactions were spontaneous and endothermic ($\Delta G_0$= -49.8 kJ/mol, $\Delta H_0$=+598 kJ/mol). Thus, these reactions are driven by entropy gain rather than enthalpy loss [3]. Such entropy-driven reactions can generate more output chemical energy in the form of hydrogen than input energy in the form of polysaccharides [97]. This is very interesting as most of the chemical conversions of cellulose/glucose to biofuel are exothermic. In other words, the output biofuel has less energy than do raw materials, i.e. cellulose/glucose. However, this method currently has a low hydrogen production rate [3]. Research in this area is on-going and no concrete conclusion has been made yet.

## 7.2. Biological production of hydrogen

The biological production of hydrogen has received great attentions due to its potential as an inexhaustible, low-cost and renewable source of clean energy [91]. The photosynthetic production of hydrogen basically uses $CO_2$ and water via direct photolysis. The yield of this process is high, but there is a shortage of photobioreactors on large scales, which constrains its investigation in laboratory scales [98]. Compared to photosynthesis, anaerobic hydrogen production process is more feasible (less expensive), has higher rates and thus has been widely studied [99]. This process seems to be closer to be implemented in commercial scales.

Several breakthroughs have been made in understanding the fundamentals of hydrogen production including the isolation of microbial strains with a high hydrogen production capacity and the optimization of the microbial fermentation process [100,101]. The performance of anaerobic fermentation depends on a number of factors, e.g. temperature, pH, alkalinity oxidation-reduction potential, particle size of lignocellulosic materials, substrate content and inoculum source [91,102]. It was reported that the optimized pH for producing hydrogen in a dark fermentation is between 5.5 and 6.7 [102]. However, there are contradictory reports about the influence of ethanol on hydrogen production in dark fermentation processes in that some studies reported a possible competition between ethanol and hydrogen production [103-105], while others reported a high hydrogen production accompanied by a high ethanol production [106,107]. In another study, the addition of ethanol to the growth medium at the initiation of the fermentation process resulted in 54% $H_2$ and 25% acetate increases, respectively, using *C. Thermocellum* bacteria [108].

In addition to hydrogen, organic acids are produced in dark fermentation. To achieve adequate overall energy efficiency, the energy stored in the organic acids produced in the dark fermentation processes should also be utilized. This can be conducted via combining the concepts of dark fermentation and anaerobic digestion [109].

On the contrary, dark fermentation has some drawbacks: 1) poor yield per substrate in the conversion of biomass to $H_2$ by microbial fermentation; 2) sensitivity to end-product accumulation; and 3) high environmental and economic costs of biomass production to generate fermentable substrate [13,110-112].

## 7.3. Microorganisms to produce hydrogen

Thermophilic cellulosic bacteria promote their greater operating temperature to produce hydrogen, which facilitate biomass pretreatment, maximize enzymatic reaction rates, and favor the equilibrium point of $H_2$ in direction of $H_2$ production [13,14]. *Clostridium thermocellum* and *caldicellulosiruptor saccharolyticus* have been the main focus of hydrogen production research [95]. *Caldicellulosiruptor saccharolyticus* is a gram-positive and extremely thermophilic, has been reported to produce hydrogen from cellulose even at a high temperature of 70 °C, and is the most promising candidate for large scales hydrogen production [98,113]. Hydrogen has been reported to be produced from pentoses via using thermophile *Caldicellulosiruptor saccharolytricus* with a high yield of 334.7 ml $H_2$/g of sugar accounting for 67% of maximum theoretical yield of 497.6 ml $H_2$/g of sugar [37,114,115]. Furthermore, the optimum pH in the fermentation of cellulose to produce hydrogen via *C. Thermocellum* was between 7 and 7.2 [116].

In the literature, some mesophilic bacteria have been investigated for hydrogen production including *Clostridium cellulolyticum, Clostidium ellulovorans, Clostridium phytofermentans* and *Clostridium termitidis*. However, their low operating temperature (32-40 °C) introduces additional operating units to the hydrogen production process and increases the risks in process operations, which constrains the practical implementation of mesophilic bacteria in large industrial scales [3].

## 7.4. Fermentation culture

Although the pure culture usually provides a high production efficiency, it is difficult to apply in industrial scales. In fact, the preparation of pure cultures in indutrial scales is difficult and expensive, which will definitely affect the production cost of hydrogen. Instead, hydrogen can be produced with mixed cultures. The mixed culture has been claimed to be more effective in substrate conversion than pure culture [95].

However, mixed culture may encounter the drawbacks of the competition of substrates with non-hydrogen producing microbial population as well as the consumption of produced hydrogen by hydrogen consuming bacteria. One alternative to address this difficulty is to pretreat the mixed culture with base, heat and/or an anaerobic condition, which eliminates/inhibits the non-producing/consuming hydrogen bacteria [34,113,117]. Heat pretreatment is commonly used in anaerobic fermentation in order to improve the hydrogen production by activating spore-forming *clostridium* and inhibiting hydrogen consuming non-spore-forming bacteria. However, the heat pretreatment might inactivate hydrogen producing bacteria existing in natural feedstock [34,37]. The elucidation of anaerobic activated sludge microbial community utilizing monosaccharides will be an important foundation towards the industrialization of hydrogen production. In one study, a mixed microbial culture was obtained via the anaerobic activation of sludge in a continuous stirred-tank reactor (29 days of acclimatization) [118]. In this study, glucose had the highest specific hydrogen production rate (358 mL/g.g of mixed liquid volatile suspended solid) and conversion rate (82 mL/g glucose) among glucose, fructose, galactose and arabinose under the fermentation conditions of 35 °C and pH of 5 [34,118].

## 7.5. Improving production efficiency

### *7.5.1. Dark fermentation*

The presence of hydrogen in fermentation reactors seems to affect the performance of dark fermentation process. It was reported that sparling the bioreactor with nitrogen would bring the concentration of hydrogen at a low level in the fermentation. Alternatively, using hollow fiber silicone rubber membrane effectively reduced biogas partial pressure in the fermentation resulting 10% improvement in the rate and 15% increase in the yield of hydrogen production [119]. To improve the hydrogen production, some strategies were carried out in the literature involving bioprocess engineering and bioreactor design that maintains a neutral pH during fermentation and ensures the rapid removal of hydrogen and $CO_2$ from the aqueous phase [95].

### *7.5.2. Enzymatic treatment*

Enzymes applied in the biotransformation of hydrogen are expensive. A combination of enzyme immobilization and thermo stable enzymes was reported to increase the life time of enzyme used for hydrogen production [120]. In one study, changing the process parameters (e.g. temperature, enzyme concentration, substrate concentration and metabolic channeling) in enzymatic reaction were reported to affect the hydrogen production rates [3]. A conservative estimation reported that the hydrogen production rates could increase to 23.6 $H_2$/l/h using a high-cell density [121]. The over-expression of enzymes catalyzing reactions towards the desired product is another method to increase the hydrogen production. The heterologous gene expression of pyruvate decarboxylate and alcohol dehydrogenase from *Zymomonas mobilis* within *C. cellulolyticum* was shown to increase the hydrogen yield by 75%, while that of acetate and ethanol increased by 93% and 53%, respectively [122].

### 7.6. Process configuration

Similar to ethanol production, hydrogen production can be conducted in a one- or two-stage process. The simultaneous saccharification and fermentation process, i.e. one-stage process, is less expensive and more commercially feasible, but may be less efficient since the preferred conditions for cellulose degradation and dark fermentation could be significantly different [14]. An important parameter in this process is to choose a microorganism, such as *thermococcus kodakaensis*, *clostridium thermolacticum* and *clostridium thermocellum*, that has the capability to produce cellulolytic enzymes and hydrogen simultaneously [14,123]. In this process, the production rate and efficiency are limited by enzymatic saccharification. The two-stage hydrogen production process, on the other hand, is performed via cellulose hydrolysis in one stage and fermentation of the hexose in another stage [124]. This process might be more effective in terms of hydrogen yield, but it is more complicated.

## 8. Production of furan-based biofuel

Recently, a new second generation biofuel was introduced via the chemical conversion of cellulose. In this approach, hydroxymethyl furfural (HMF) is initially produced from cellulose, and HMF is subsequently converted to 2,5-dimethylfuran (DMF). DMF has an energy content of 31.5 MJ/l, which is similar to that of gasoline (35 MJ/l) and 40% greater than that of ethanol (23 MJ/l) [1]. DMF (bp 92-94 °C) is less volatile than ethanol (bp 78 °C), and is immiscible with water, which makes it an appealing liquid biofuel [1].

Figure 3 shows the process block diagram of DMF production from fructose. This process contains two main parts: 1) HMF production and its purification and 2) DMF production from HMF and its downstream purification.

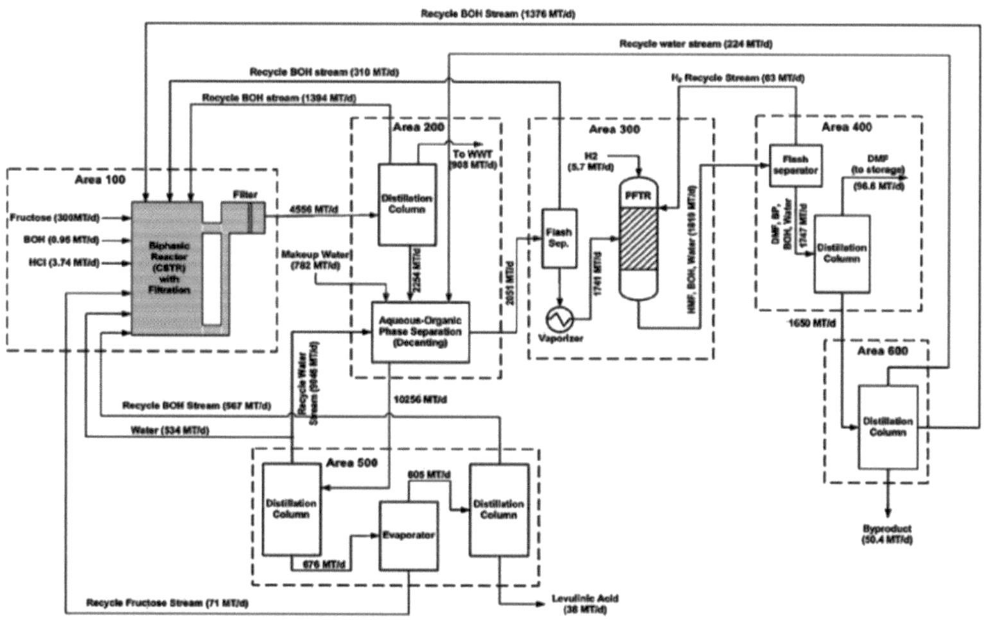


**Figure 3.** DMF production from fructose [6].

Two-solvent catalyst system of butanol/water (at 180 °C) was proposed for producing HMF from fructose [1]. In this process, the conversion of glucose/fructose takes place in the aqueous phase (with HMF yield of 83%), in which HCl acts as a catalyst to convert fructose to HMF. Also, NaCl is added to the system and enhances the transportation of HMF from aqueous phase to organic phase (butanol), which prevents HMF from further degradation in the aqueous phase. Afterwards, the HMF is purified via various distillation/separation units and butanol is recycled to the HMF production reactor [1].

In another study, methyl isobutyl ketone (MIBK)/water system (in the present of $TiO_2$ and 250 °C) resulted in 35% HMF yield from glucose [125,126]. In this respect, ionic liquids, e.g.1-ethyl-3-methyl-imidazolium chloride ionic (in the presence of LiCl, CuCl or CrCl) showed a higher yield (>65%) at a temperature of 160 °C [127,128]. In another research, dimethyl sulfoxide (DMSO) was used as solvent to produce HMF from glucose with HMF yield of 53% [129]. However, these solvents are usually toxic and difficult to prepare, and their separation process from HMF is challenging [129]. These drawbacks will most likely limits their application in producing HMF in laboratory scales.

Alternatively, the hydrothermal conversion of cellulose to HMF (275-300 °C for less than 30 min) in homogenous systems in the presence of sulphuric acid resulted in a 20% HMF yield. However, this system under a neutral condition produced a lower yield (<16%), but a product (HMF) with a higher purity (50-60%) [130]. In another research, the yield of HMF from fructose was 65% in the presence of phosphoric acid, which was conducted at 228 °C for less than 5 min [1]. The most promising and industrially attractive method to produce HMF was the acetic acid/water mixture (10 g/l acid at 200 °C for 20 min), which resulted in the production of HMF from fructose with 60% yield [131]. This process resulted in 53% HMF yield under the same conditions, but without acetic acid [131]. Alternatively, the hot compressed water treatment of cellulose at a temperature of 523 K for 5 min (pressure 2.5 MPa under nitrogen) produced 15% HMF, and the addition of 10% (wt.) $TiO_2$ to the system increased the yield to 28% [1]. This method is more feasible than other homogenous catalytic methods, as the catalyst in this system can be easily separated and thus recycled to the system [132]. Furthermore, the addition of acetone to water/$TiO_2$ system induced a higher HMF yield (35%). This process was conducted under atmospheric condition, which is more industrially attractive, but the presence of acetone in the system brings recycling issues and uncertainties at large scale applications [1].

The produced HMF in Figure 3 will be fed into a PFTR reactor, in which $H_2$ is added (in the presence of chromium II or copper-ruthenium-carbon catalysts) that is necessary for the conversion of HMF to DMF [6]. Finally, DMF is purified using several distillation/separation processes, and the organic solvent (butanol) is recycled to the system (Figure 3).

However, this process has several drawbacks: 1) it uses hydrogen which is a biofuel and currently expensive; 2) it uses butanol (another biofuel) as a solvent, which brings difficulties to its large scale implementations; 3) it uses NaCl to enhance the extraction of HMF from aqueous phase to organic butanol. NaCl introduces uncertainties in the downstream processes and its removal is cost intensive; 4) as described above, expensive chromium II or copper-ruthenium-carbon was used as a catalyst in the PFTR reactor [6]. It was reported that the production cost of DMF using this process is approximately 2 $/l, which is not presently economical. The process optimization seemed to lower the production cost to approximately 1 $/l, which is still high and not competitive with other appealing biofuels [6].

Alternatively, cellulose can be converted to 5-chloromethylfurfural (CMF) via heating cellulose in concentrated HCl at the presence of LiCl. Subsequently, the products should be extracted with 1,2-dichloroethane. This process yielded 71% CMF in one study [133]. The CMF was then converted to DMF and 5-ethoxymethylfurfural. This process yielded 84% isolated DMF, but the presence of LiCl is considered as one drawback of this system [133]. Alternatively, the CMF was converted to ethoxymethylfurfural (EMF) via stirring in ethanol solution. EMF has a boiling point of 235 °C, and energy density that is similar to energy density of gasoline and 40% higher than that of ethanol [133]. The EMF can also be used a biofuel, but this process seem to be complicated and faces with several technological challenges.

The most important parameters affecting the production cost of DMF via the aforementioned process are feedstock cost, production yield, by-product prices, catalyst cost, and total purchased equipment cost [6]. The productions of DMF and CMF are not feasible using current technologies. Although these chemicals were produced at laboratory scales, their commercialization is under serious questions as their production processes require various expensive solvents that are not easily recycled. The separation and purification of final products from solvents are also costly.

## 9. Conclusions

The pretreatment of woody materials is an important step for producing biofuels via fermentation. The physicochemical pretreatment is the most promising approach to dissemble cellulosic biomass. Enzymatic hydrolysis is very selective in terms of decomposing cellulose chains, but its process conditions are not very industrially attractive. However, acid hydrolysis is fast with a high yield, but it produces some by-products. These by-products are inhibitors of downstream fermentation processes and hence should be removed from the process prior to fermentation. Adsorption and evaporation have been the most successful approaches to eliminate these inhibitors in detoxification stages prior to fermentation. *S. cerevisiae, C. acetobutylicum* and *C. thermocellum* are the most promising microorganisms for ethanol, butanol and hydrogen productions, respectively. Presently, the major challenges in the production of these biofuels are the rather low production yield of biofuels and the sensitivity of microorganisms to the presence of inhibitors and biofuels in the fermentation media. The productions of ethanol and butanol from woody biomass are close to be commercialized, but hydrogen production is still facing with difficulties in increasing the production rate. Although 2,5-dimethylfuran (DMF) has appealing properties and can be potentially used as biofuel, its production with present technologies is not economical and more industrially attractive processes should be developed in order to have a commercialized DMF process.

## Author details

Pedram Fatehi
*Chemical Engineering Department, Lakehead University, Thunder Bay, ON, Canada*

## Acknowledgement

The author would like to thank NSERC Canada for providing funds for this review research.

## 10. References

[1] Binder JB, Raines RT. Simple Chemical Transformation of Lignocellulosic Biomass into Furrans for Fuels and Chemicals. Journal of American Chemical Society 2009; 131: 1979-1985.

[2] Serrano-Ruiz JC, West RM, Dumesic J. Catalytic Conversion of Renewable Biomass Resources to Fuels and Chemicals. Chemical and Biomolecular Engineering-Annual Review 2010; 1: 79-100.

[3] Zhang Y-HP. Renewable Carbohydrates Are a Potential High-Density Hydrogen Carrier. International Journal of Hydrogen Energy 2010; 35: 10334-10342.
[4] Sivakumar G, Vali DR, Xu J, Burner DM, Jr JOL, Ge X, Weathers PJ. Bioethanol and Biodiesel: Alternative Liquid Fuels for Future Generations. Engineering and Life Science 2010; 10(1): 8-18.
[5] Visser EM, Filho DO, Martins MA, Steward BL. Bioethanol Production Potential from Brazilian Biodiesel Co-products. Biomass and Bioenergy 2011; 35: 489-494.
[6] Kazi FK, Patel AD, Serrano-Ruiz JC, Dumesic JA, Anex RP. Techno-Economic Analysis of Dimethylfuran (DMF) and Hydroxymethylfurfural (HMF) Production from Pure Fructose in Catalytic Processes. Chemical Engineering Journal 2011; 169: 329-338.
[7] Alvandi S, Agrawal A. Experimental Study of Combustion of Hydrogen Syngas/Methane Fuel Mixtures in a Porous Burner. International Journal of Hydrogen Energy 2008; 33(4): 1407-1415.
[8] Kaparaju P, Serrano M, Thomsen AB, Kongjan P, Angelidaki I. Bioethanol, Biohydrogen and Biogas Production from Wheat Straw in a Biorefinery Concept. Bioresource Technology 2009; 100: 2562-2568.
[9] Fischer CR, Klein-Marcuschamer D, Stephanopoulos G. Selection and Optimization of Microbial Hosts for Biofuels Production. Metabolic Engineering 2008; 10: 295-304.
[10] Johnson JMF, Coleman MD, Gesch R, Jaradat A, Mitchell R, Reicosky D, Wilhelm WW. Biomass-bioenergy Crops in the United States: A Changing Paradigm. American Journal of Plant Science and Biotechnology 2007; 1: 1-28.
[11] Weber C, Farwick A, Benisch F, Brat D, Dietz H, Subtil T, Boles E. Trends and Challenges in the Microbial Production of Lignocellulosic Bioalcohol Fuels. Applied Microbiology and Biotechnology 2010; 87: 1303-1315.
[12] Cherubini F, Stromman AH. Production of Biofuels and Biochemicals from Lignocellulosic Biomass: Estimation of Maximum Theoretical Yield and Efficiencies Using Matrix Algebra. Energy and Fuels 2010; 24: 2657-2666.
[13] Jones PR. Improving Fermentative Biomass-Derived $H_2$-Production by Engineering Microbial Metabolism. International Journal of Hydrogen Energy 2008; 33: 5122-5130.
[14] Cheng CL, Lo YC, Lee KS, Duu JL, Lin CY, Chang JS. Biohydrogen Production from Lignocellulosic Feedstock. Bioresource Technology 2011; 102: 8514-8523.
[15] Zhang Y-HP. A Sweet Out-Of-The-Box Solution to the Hydrogen Economy: Is the Sugar-Powered Car Science Fiction? Energy and Environmental Science 2009; 2: 272-282.
[16] Dautzenberg G, Gerhardt M, Kamm B. Bio-based Fuels and Fuel Additives from Lignocellulose Feedstock via the Production of Levulinic Acid and Furfural. Holzforschung 2011; 65: 439-451.
[17] Koda R, Takao N, Shinji H, Sriappareddy T, Kazunori N, Tsutomu N, Chiaki O, Hideki F, Kondo A. Ethanolysis of Rapeseed Oil to Produce Biodiesel Fuel Catalyzed by *Fusarium Heterosporum* Lipase-Expressing Fungus Immobilized Whole-Cell Biocatalysts. Journal of Molecular Catalysis B: Enzymatic 2010: 66(1-2): 101-104.
[18] André A, Diamantopoulou P, Philippoussis A,Sarris D, Komaitis M, Papanikolaou S. Biotechnological Conversions of Bio-Diesel Derived Waste Glycerol into Added-Value

Compounds by Higher Fungi: Production of Biomass, Single Cell Oil and Oxalic Acid. Industrial Crops and Products 2010; 31(2): 407-416.

[19] Venkata Subhash G, Venkata Mohan S. Biodiesel Production from Isolated Oleaginous Fungi *Aspergillus sp.* Using Corncob Waste Liquor as a Substrate. Bioresource Technology 2011; 102(19): 9286-9290.

[20] Meng X, Yang J, Xu X, Zhang L, Nie Q, Xian M. Biodiesel Production from *Oleaginous* Microorganisms. Renewable Energy 2009; 34(1): 1-5.

[21] Arai S, Nakashima K, Tanino T, Ogino C, Kondo A, Fukuda H. Production of Biodiesel Fuel from Soybean Oil Catalyzed by Fungus Whole-Cell Biocatalysts in Ionic Liquids. Enzyme and Microbial Technology 2010; 46: 51-55.

[22] Xia C, Zhang J, Zhang W, Hu B. A New Cultivation Method for Microbial Oil Production: Cell Pelletization and Lipid Accumulation by *Mucor Circinelloides*. Biotechnology for Biofuels 2011; 4: 15-24.

[23] Fan Y, Zhang Y, Zhang S, Hou H, Ren B. Efficient Conversion of Wheat Straw Wastes into Biohydrogen Gas by Cow Dung Compost. Bioresource Technology 2006; 97(3): 500-505.

[24] Cara C, Moya M, Ballesteros I, Negro MJ, Gonzalez A, Ruiz E. Influence of Solid Loading on Enzymatic Hydrolysis of Steam Exploded or Liquid Hot Water Pretreated Olive Tree Biomass. Process Biochemistry 2007; 42: 1003-1009.

[25] Lin ZX, Huang H, Zhang HM, Zhang L, Yan LS, Chen JW. Ball Milling Pretreatment of Corn Stover for Enhancing the Efficiency of Enzymatic Hydrolysis. Applied Biochemistry and Biotechnology 2010; 162: 1872-1880.

[26] Negro MJ, Manzanares P, Oliva JM, Ballesteros I, Ballesteros M. Changes in Various Physical/Chemical Parameters of Pinus Pinaster Wood after Steam Explosion Pretreatment. Biomass and Bioenergy 2003; 25: 301-308.

[27] Cheng JJ, Timilsina GR. Status and Barriers of Advanced Biofuel Technologies: A Review. Renewable Energy 2011; 36: 3541-3549.

[28] Shi H, Fatehi P, Xiao H, Ni Y. A Combined Acidification/PEO Flocculation Process to Improve the Lignin Removal from the Pre-Hydrolysis Liquor of Kraft-Based Dissolving Pulp Production Process. Bioresource Technology 2011; 102: 5177-5182.

[29] Liu Z, Fatehi P, Jahan MS, Ni Y. Separation of Lignocellulosic Materials by Combined Processes of Prehydrolysis and Ethanol Extraction. Bioresource Technology 2011; 102: 1264-1269.

[30] Fatehi P, Shen J, Ni Y. Treatment of Pre-Hydrolysis Liquor of Kraft-Based Dissolving Pulp Process with Surfactant and Calcium Oxide. Journal of Science and Technology for Forest Products and Processes 2011; 1(3): 31-37.

[31] Shen J, Fatehi P, Soleymani P, Ni Y. A Process to Utilize the Lignocelluloses of Pre-Hydrolysis Liquor in the Lime Kiln of Kraft-Based Dissolving Pulp Production Process. Bioresource Technology 2011; 102: 10035-10039.

[32] Fatehi P, Ni Y. Integrate Forest Biorefinery-Sulfite Pulping. In: Zhu J, Zhang X, Pan X. (ed.) Sustainable Production of Fuels, Chemicals, and Fibers from Forest Biomass. American Chemical Society Symposium Series 2011; 1067: p409–441.

[33] Fatehi P, Ni Y. Integrated Forest Biorefinery- Prehydrolysis/Dissolving Pulping Process. In: Zhu J, Zhang X, Pan X. (ed.) Sustainable Production of Fuels, Chemicals, and Fibers from Forest Biomass. American Chemical Society Symposium Series 2011; 1067: p475–506.

[34] Chairattanamanokorn P, Penthamakeerati P, Reungsang A, Lo YC, Lu WB, Chang JS. Production of Biohydrogen from Hydrolyzed Bagasse with Thermally Preheated Sludge. International Journal of Hydrogen Energy 2009; 34: 7612-7617.

[35] Gomez LD, Stelle-King CG, McQueen-Mason SJ. Sustainable Liquid Biofuels from Biomass: the Writing's on the Walls. New Phytologist 2008; 178: 473-485.

[36] Taniguchi M, Suzuki H, Watanabe D, Sakai K, Hoshino K, Tanak T. Evaluation of Pretreatment with *Pleurotus Ostreatus* for Enzymatic Hydrolysis of Rice Straw. Journal of Bioscience and Bioenergy 2005; 100(6): 637-643.

[37] Lin CY, Hung WC. Enhancement of Fermentative Hydrogen/Ethanol Production from Cellulose Using Mixed Anaerobic Cultures. International Journal of Hydrogen Energy 2008; 33: 3660-3667.

[38] Harnpicharnchai P, Champreda V, Sornlake W, Euwilaichitr LA. A Thermotolerant Beta-glucosidase Isolated from an Endophytic Fungi, *Periconia sp.,* with a Possible Use for Biomass Conversion to Sugars. Protein Expression Purification Journal 2009; 67:61-69.

[39] Shen J, Agblevor FA. Modeling Semi-Simultaneous Saccharification and Fermentation of Ethanol Production from Cellulose. Biomass and Bioenergy 2010; 34: 1098-1107.

[40] Saeed A, Fatehi P, Ni Y, van Heiningen A. Impact of Furfural on the Sugar Analysis of Pre-Hydrolysis Liquor of Kraft-Based Dissolving Pulp Production Process Using the HPAEC Technique. BioResources 2011; 6(2): 1707-1718.

[41] Liu Z, Fatehi P, Ni Y. A Proposed Process for Utilizing the Hemicelluloses of Pre-Hydrolysis Liquor in Papermaking. Bioresource Technology 2011; 102: 9613-9618.

[42] Liu X, Fatehi P, Ni Y. Adsorption of Lignocellulosic Materials Dissolved in Hydrolysis Liquor of Kraft-Based Dissolving Pulp Production Process on Polymer-Modified Activated Carbons. Journal of Science and Technology for Forest Products and Processes 2011; 1(1): 46-54.

[43] Leschinsky M, Zuckerstatter G, Weber HK, Patt R, Sixta H. Effect of Autohydrolyis of *Eucalyptus Globulus* Wood on Lignin Structure. Part 2: Effect of Autohydrolysis Intensity. Holzforschung 2008; 62: 653-658.

[44] Helle S, Cameron D, Lam J, White B, Duff S. Effect of Inhibitory Compounds Found in Biomass Hydrolysates on Growth and Xylose Fermentation by a Genetically Engineered Strain of *S. Cerevisiae*. Enzyme and Microbial Technology 2003; 33: 786-792.

[45] Helle, SS, Murry A, Lam J, Cameron DR, Duff SJB. Xylose Fermentation by Genetically Modified *Saccharomyces Cerevisiae* 259ST in Spent Sulfite Liquor. Bioresource Technology 2004; 92: 163-171.

[46] Helle SS, Lin T, Duff SJB. Optimization of Spent Sulfite Liquor Fermentation. Enzyme and Microbial Technology 2008; 42: 259-264.

[47] Nigam JN. Ethanol Production from Hardwood Spent Sulfite Liquor Using an Adapted Strain of *Pichia Stipitis*. Journal of Industrial Microbiology and Biotechnology 2001; 26:145-150.

[48] Saeed A, Fatehi P, Ni Y. Chitosan as a Flocculant for Pre-Hydrolysis Liquor of Kraft-Based Dissolving Pulp Production Process. Carbohydrate Polymer 2011; 86: 1630-1636.

[49] Liu X, Fatehi P, Ni Y. Adsorption of Lignocellulosic Materials Dissolved in Pre-Hydrolysis Liquor of Kraft-Based Dissolving Pulp Process on Oxidized Activated Carbons. Industrial Engineering and Chemistry Research 2011; 50: 11706-11711.

[50] Shen J, Fatehi P, Soleimani P, Ni Y. Lime Treatment of Pre-hydrolysis Liquor from the Kraft-based Dissolving Pulp Production Process. Industrial Engineering and Chemistry Research 2012; 51: 662-667.

[51] Schneider H. Selective Removal of Acetic Acid from Hardwood-Spent Sulfite Liquor Using a Mutant Yeast. Enzyme and Microbial Technology 1996; 19: 94-98.

[52] Waltz E. Cellulosic Ethanol Booms despite Unproven Business Models. Natural Biotechnology 2008; 26: 8-9.

[53] Ohgren K, Bura R, Lesnicki G, Saddler J, Zacchi G. A Comparison between Simultaneous Saccharification and Fermentation and Separate Hydrolysis and Fermentation Using Steam-Pretreated Corn Stover. Process Biochemistry 2007; 42(5): 834-839.

[54] Lupoi JS, Smith EA. Evaluation of Nanoparticle-Immobilized Cellulase for Improved Ethanol Yield in Simultaneous Saccharification and Fermentation Reactions. Biotechnology and Bioengineering 2011; 108(12): 2835-2843.

[55] Knoshaug EP, Zhang M. Butanol Tolerance in a Selection of Microorganisms. Applied Biochemistry and Biotechnology 2009; 153: 13-20.

[56] Rogers PLK, Lee J, Skotnicki ML, Tribe DE. Ethanol Production by *Zymononas mobilis*. Advanced Biochemical Engineering 1982; 22: 37-84.

[57] Warnick TA, Methe BA, Leschine SB. *Clostridium Phytofermentans sp*. nov., a Celluloytic Mesophile from Forest Soil. International Journal of System and Evolution Microbiology 2002; 52: 1155-1160.

[58] Kawaguchi H, Sasaki M, Vertes AA, Inui M, Yukawa H. Identification and Functional Analysis of the Gene Cluster for L-arabinose Utilization in *Corynebacterium Glutaicum*. Applied Environmental Microbiology 2009; 75(11): 3419-3429.

[59] Banat IM, Nigam P, Marchant R. Isolation of Thermotolorant, Fermentative Yeasts Growing at 52 °C and Producing Ethanol at 45 °C and 50 °C. World Journal of Microbiology and Biotechnology 1992; 8: 259-263.

[60] Ryabova OB, Chmil OM, Sibirny AA. Xylose and Cellobiose Fermentation to Ethanol by the Thermotolorant Methylotrophic Yeast *Hansenula Polymorpha*. FEMS Yeast Research 2003; 4: 157-164.

[61] Verstrepen KJ, Derdelelinckx G, Verachtert H, Delvaux FR. Yeast Flocculation: What Brewers Should Know. Applied Microbiology and Biotechnology 2003; 61: 197-205.

[62] Xu TJ, Zhao Q, Bai FW. Continuous Ethanol Production Using Self-Flocculating Yeast in a Cascade of Fermentors. Enzyme and Microbial Technology 2005; 37: 634-640.

[63] Bai FW, Anderson WA, Moo-Young M. Ethanol Fermentation Technologies from Sugar and Starch Feedstocks. Biotechnology Advances 2008; 26: 89-105.

[64] Ge XM, Zhang, L, Bai FW. Impact of Temperature, pH, Divalent Cations, Sugars, and Ethanol on the Flocculating of SPSC01. Enzyme and Microbial Technology 2006; 39: 783-787.

[65] Zhu JY, Pan XJ, Wang GS, Gleisner R. Sulfite Pretreatment (SPORL) for Robust Enzymatic Saccharification of Spruce and Red Pine. Bioresource Technology 2009; 100: 2411-2418.

[66] Wang GS, Pan XJ, Zhu JY, Gleisner R, Rockwood D. Sulfite Pretreatment to Overcome Recalcitrance of Lignocellulose (SPORL) for Robust Enzymatic Saccharification of Hardwoods. Biotechnology Progress 2009; 25(4): 1086-1093.

[67] Zhu W, Zhu JY, Gleisner R, Pan XJ. On Energy Consumption for Size-Reduction and Yields from Subsequent Enzymatic Saccharification of Pretreated Lodgepole Pine. Bioresource Technology 2010; 101: 2782-2792.

[68] Zhu JY, Pan XJ. Woody Biomass Pretreatment for Cellulosic Ethanol Production: Technology and Energy Consumption Evaluation. Bioresource Technology 2010; 101: 4992-5002.

[69] Shuai L, Yang Q, Zhu JY, Lu FC, Weimer PJ, Ralph J, Pan XJ. Comparative Study of SPORL and Dilute-Acid Pretreatments of Spruce for Cellulosic Ethanol Production. Bioresource Technology 2010; 101: 3106-3114.

[70] Zhu JY, Zhu W, O'Bryan P, Dien BS, Tian S, Gleisner R, Pan XJ. Ethanol Production from SPORL-Pretreated Lodgepole Pine: Preliminary Evaluation of Mass Balance and Process Energy Efficiency. Applied Microbiology and Biotechnology 2010; 86: 1355-1365.

[71] Durre P. Biobutanol: An Attractive Biofuel. Biotechnology Journal 2007; 2: 1525-1534.

[72] Kumar M, Gayen K. Developments in Biobutanol Production: New Insights. Applied Energy 2011; 88: 1999-2012.

[73] Schwarz WH, Gapes R. Butanol-Rediscovering a Renewable Fuel. BioWorld Europe 2006; 1:16-19.

[74] Bramono SE, Lam YS, Ong SL, He J. A Mesophilic *Clostridium* Species That Produces Butanol from Monosaccharides and Hydrogen from Polysaccharides. Bioresource Technology 2011; 102: 9558-9563.

[75] Keis S, Shaheen R, Jones TD. Emended Description of *Clostridium Acetobutylicum* and *Clostridium Beijerinckii* and Descriptions of *Clostridium Saccharoperbutylacetonicum sp.* nov. and *Clostridium Saccharobutylicum sp.* nov. International Journal of System Evolution and Microbiology 2011; 51: 2095-2103.

[76] Zverlov VV, Berezina O, Velikodvorskaya GA, Schwarz WH. Bacterial Acetone and Butanol Production by Industrial Fermentation in the Soviet Union: Use of Hydrolyzed Agricultural Waste for Biorefinery. Applied Microbiology and Biotechnology 2006; 71: 587-597.

[77] Marlatt JA, Datta R. Acetone-Butanol Fermentation Process Development and Economic Evaluation. Biotechnology Progress 1986; 2(1): 23-28.

[78] Ezeji T, Qureshi N, Blaschek HP. Butanol Production from Agriculture Residues: Impact of Degradation Products on *Clostridium Beijerinckii* Growth and Butanol Fermentation. Biotechnology and Bioengineering 2007; 97(6):1460-1469.
[79] Ezeji T, Blaschek HP. Fermentation of Dried Distillers's Grains and Solubles (DDGS) Hydrolysates to Solvents and Value-Added Products by Solventogenic *Clostridia*. Bioresource Technology 2008; 99(12): 5232-5242.
[80] Demain AL, Newcomb M, Wu JHD. Cellulose, *Clostridia* and Ethanol. Microbiology and Molecular Biology Review 2005; 69: 124-154.
[81] Qureshi N, Blaschek HP. ABE Production from Corn: a Recent Economic Evaluation. Journal of Industrial Microbiology and Biotechnology 2001; 27: 292-297.
[82] Nair RV, Green EM, Watson DE, Bennett GN, Papoutsakis ET. Regulation of the Sol Locus Genes for Butanol and Acetone Formation in *Clostridium Acetobutylicum* ATCC 824 by A Putative Transcriptional Repressor. Journal of Bacteriology 1999; 181: 319-330.
[83] Tashiro Y, Takeda K, Kobayashi G, Sonomoto K, Ishizaki A, Yoshino S. High Butanol Production by *Clostridium Saccharoperbutylactonicum* N1-4 in Fed-Batch Culture with pH-Stat Continuous Butyric Acid and Glucose Feeding Method. Journal of Bioscience and Bioengineering 2004; 98(4): 263-268.
[84] Cho DH, Lee YJ, Um Y, Sang BI, Kim YH. Detoxification of Model Phenolic Compounds in Lignocellulosic Hydrolysates with Peroxidase for Butanol Production from *Clostridium Beijerinckii*. Applied Microbiology and Biotechnology 2009; 83: 1035-1043.
[85] Qureshi N, Lai LL, Blaschek HP. Scale-up of a High Productivity Continuous Biofilm Reactor to Produce Butanol by Adsorbed Cells of *Clostridium Beijerinckii*. Food Bioproduct Processing 2004; 82: 164-173.
[86] Maddox IS, Qureshi N, Thomson KR. Production of Acetone-Butanol-Ethanol from Concentrated Substrates Using *Clostridium Acetobutylicum* in an Integrated Fermentation-Product Removal Process. Process Biochemistry 1999; 30(3): 209-215.
[87] Lee SY, Park JH, Jang SH, Nielsen LK, Kim, J, Jung KS. Fermentative Butanol Production by *Clostridia*. Biotechnology and Bioengineering 2008; 101(2): 209-228.
[88] Liu S, Qureshi N. How Microbes Tolerate Ethanol and Butanol. New Biotechnology 2009; 26: 117-121.
[89] Zheng YN, Li LZ, Xian M, Ma YJ, Yang JM, Xu X, He DZ. Problems with the Microbial Production of Butanol. Journal of Industrial Microbiology and Biotechnology 2009; 36: 1127-1138.
[90] Huang WC, Ramey DE, Yang ST. Continuous Production of Butanol by *Clostridium Acetobutylicum* Immobilized in a Fibrous Bed Bioreactor. Applied Biochemistry and Biotechnology 2004; 115: 887-898.
[91] Yuan X, Shi X, Zhang P, Wei Y, Guo R, Wang L. Anaerobic Biohydrogen Production from Wheat Straw stalk by Mixed Microflora: Kinetic Model and Particle Size Influence. Bioresource Technology 2011; 102: 9007-9012.
[92] Adams MWW, Stiefel EL. Biological Hydrogen Production: Not So Elementary. Science 1998; 282: 1842-1843.
[93] Hallenbeck PC, Benemann JR. Biological Hydrogen Production: Fundamentals and Limiting Processes. International Journal of Hydrogen Energy 2002; 27: 1185-1193.

[94] Thauer K, Jungermann K, Decker K. Energy Conservation in Chemotrophic Anaerobic Bacteria. Bacteriology Review 1977; 41: 100-180.
[95] Levin BL, Carere CR, Cicek N, Sparling R. Challenges for Biohydrogen Production via Direct Lignocellulose Fermentation. International Journal of Hydrogen Energy 2009; 34: 7390-7403.
[96] Logan BE, Regan JM. Microbial Fuel Cells-Challenges and Applications. Environmental Science and Technology 2006; 40: 5172-5180.
[97] Ye X, Wang Y, Hopkins RC, Adams MWW, Evans BR, Mielenz JR, Zhang YH. Spontaneous High-Yield Production of Hydrogen from Cellulosic Materials and Water Catalyzed by Enzyme Cocktails. Chemistry and Sustainability 2009; 2:149-152.
[98] Herbel Z, Rakhely G, Bagi Z, Ivanova G, Acs N, Etelka K, Kovacs L. Exploitation of the Extremely Thermophilic *Caldicellulosiruptor Saccharolyticus* in Hydrogen and Biogas Production from Biomasses. Environmental Technology 2010; 31(8-9): 1017-1024.
[99] Fang HHP, Zhang T, Liu H. Biohydrogen Production from Starch in Wastewater under Thermophilic Conditions. Journal of Environment and Management 2003; 69: 149-156.
[100] Ren NQ, Wang BZ, Ma F. Hydrogen Bio-production of Carbohydrate by Anaerobic Activated Sludge Process. In: Proceeding of Water Environmental Federation and 68th Annual Conference Expo, Miami Beach, Florida, Oct 21-25 1995, 145-153.
[101] Rachman MA, Furutani Y, Nakashimada Y, Kakizono T, Nishio N. Enhanced Hydrogen Production in Altered Mixed Acid Fermentation of Glucose by Enterobacter Aerogenes. Journal of Fermentation Bioengineering 1997; 83: 358-363.
[102] Li JZ, Ren NQ. The Operational Controlling Strategy about the Optimal Fermentation Type of Acidogenic Phase. China Environmental Science 1998; 18(5): 398-402.
[103] Li C, Fang HHP. Fermentative Hydrogen Production from Wastewater and Solid Wastes by Mixed Cultures. Critical Review in Environmental Science and Technology 2007; 37: 1-39.
[104] Hawkes FR, Dinsdale R, Hawkes DL, Hussy L. Sustainable Fermentative Hydrogen Production: Challenges for Process Optimization. International Journal of Hydrogen Energy 2002; 27: 1335-1347.
[105] Khanal SK, Chen WH, Li L, Sung S. Biological Hydrogen Production: Effect of pH and Intermediated Products. International Journal of Hydrogen Energy 2005; 29: 1123-1131.
[106] Ren N, Li J, Li B, Wang Y, Liu S. Biohydrogen Production from Molasses by Anaerobic Fermentation with a Pilot Scale Bioreactor System. International Journal of Hydrogen Energy 2006; 31: 2147-2157.
[107] Koshinen PEP, Lay CH, Beck SR, Tolvanen AH, Kaksonen AH, Orlygsson J, Lin CY, Puhakka JA. Bioprospecting Thermophilic Microorganisms from Icelandic Hot Springs for Sustainable Energy ($H_2$ and/or Ethanol) Production. Energy and Fuels 2008; 22: 124-140.
[108] Rydzak T, Levin DB, Cicek N, Sparling R. Growth Phase Dependent Enzyme Profile of Pyruvate Catabolism and End-Product Formation in *Clostridium Thermocellum* ATCC 27405. Journal of Biotechnology 2011; 140: 169-175.
[109] Ljunggren M, Zacchi G. Techno-Economic Analysis of a Two-step Biological Process Producing Hydrogen and Methane. Bioresource Technology 2010; 101: 7780-7788.

[110] Hill J, Nelson e, Tilman D, Polasky S, Tiffany D. Environmental, Economic, and Energetic Costs and Benefits of Biodiesel and Ethanol Biofuels. Proceeding of National Academic Science USA 2006; 103: 11206-11210.

[111] Angenent LT, Karim K, Al-dahhan MH, Wrenn BA, Domiguez-Espinosa R. Production of Bioenergy and Biochemicals from Industrial and Agricultural Wastewater. Trends in Biotechnology 2004; 22: 477-485.

[112] De Vrije T, Mars AE, Budde MA, Lai MH, Dijkema C, de Warrd P, Claassen PA. Glycolytic Pathway and Hydrogen Yield Studies of the Extreme Thermophile *Caldicellulosiruptor Saccharolyticus*. Applied Microbiology and Biotechnology 2007; 74: 1358-1367.

[113] Panagiotopoulos IA, Bakker RR, de Vrije T, Koukios EG, Claassen PAM. Pretreatment of Sweet Sorghum Bagasse for Hydrogen Production by *Caldicellulosiruptor Saccharolyticus*. International Journal of Hydrogen Energy 2010; 35: 7738-7747.

[114] Kadar Z, de Vrije T, van Noorden G, Budde M, Szengyel Z, Reczey K, Claassen P. Yield from Glucose, Xylose, and Paper Sludge Hydrolysate during Hydrogen Production by the Extreme Thermophile *Caldaicellulosriruptur Saccharolyticus*. Applied Biochemistry and Biotechnology 2004; 114(1): 497-508.

[115] Lo Y, Chen W, Hung C, Chen S, Chang J. Dark $H_2$ Fermentation from Sucrose and Xylose Using $H_2$-Producing Indigenous Bacteria: Feasibility and Kinetic Studies. Water Research 2008; 42(4-5): 827-842.

[116] Islam R, Cicek N, Sparling R, Levin DB. Influence of Initial Cellulose Concentration on the Carbon Flow Distribution during Batch Fermentation by *Clostridium Thermocellum* ATCC 27405. Applied Microbiology and Biotechnology 2009; 82: 141-148.

[117] Modan SV, Bhaskar YV, Krishna PM, Rao NC, Babu VL, Sarna PN. Biohydrogen Production from Chemical Wastewater as Substrate by Selectively Enriched Anaerobic Mixed *Consortia*: Influence of Fermentation pH and Substrate Composition. International Journal of Hydrogen Energy 2007; 32: 2286-2295.

[118] Li J, Ren N, Li B, Qin Z, He J. Anaerobic Biohydrogen Production from Monosaccharides by a Mixed Microbial Community Culture. Bioresource Technology 2008; 99: 6528-6537.

[119] Liang TM, Cheng SS, Wu KL. Behaviour Study on Hydrogen Fermentation Reactor Installed with Silicone Rubber Membrane. International Journal of Hydrogen Energy 2002; 27: 1157-1165.

[120] Myung S, Wang YR, Zhang Y-HP. Fructose-1,6 Bisphosphatase from a Hyper-Thermophilic Bacterium Thermotoga Maritime: Characterization, Metabolite Stability and its Implications. Process Biochemistry 2010; 45(12): 1882-1887.

[121] Yoshida A, Nishimura T, Kawaguchi H, Inui M, Yukawa H. Enhanced Hydrogen Production from Formic Acid by Formate Hydrogen Lyase-Overexpressing *Escherichia Coli* Strains. Applied Environmental Microbiology 2005; 71: 6762-6768.

[122] Guedon E, Desvaux M, Petitdemanage H. Improvement of Cellulolytic Properties of *Clostridium Cellulolyticum* by Metabolic Engineering. Applied Environmental Microbiology 2002; 68: 53-58.

[123] Kanai T, Imanka H, Nakajima A, Uwamori K, Omori Y, Fukui T, Atomi H, Imanaka T. Continuous Hydrogen Production by the Hyperthermophilic Archaeon, *Thermococcus Kadakaraensis* KOD1. Journal of Biotechnology 2005; 116: 271-282.

[124] Mosier N, wyman C, Dale B, Elander R, Lee YY, Holtzapple M, Ladisch M. Features of Promising Technologies for Pretreatment of Lignocellulosic Biomass. Bioresource Technology 2005; 96: 673-686.

[125] McNeff CV, Nowlan DT, McNeff LC, Yan B, Fedie RL. Continuous Production of 5-hydroxymethylfurfural from Simple and Complex Carbohydrates. Applied Catalyst A: General 2010; 384: 65-69.

[126] Su Y, Brown HM, Huang X, Zhou XD, Amonette JE, Zhang ZC. Single-Step Conversion of Cellulose to 5-Hydroxymethylfurfural (HMF), A Versatile Platform Chemical. Applied Catalyst A: General 2009; 361: 117-122.

[127] Wang P, Yu H, Zhan S, Wang S. Catalytic Hydrolysis of Lignocellulosic Biomass into 5-Hydroxymethylfurfural in Ionic Liquid. Bioresource Technology 2011; 102: 4179-4183.

[128] Tao F, Song H, Chou L. Catalytic Conversion of Cellulose to Chemicals and Ionic Liquid. Carbohydrate Research 2011; 346: 58-63.

[129] Chheda JN, Roman-Leshkov Y, Dumesic JA. Production of 5-HydroxymethylFurfural and Furfural by Dehydration of Biomass-Derived Mono- and Poly-Saccharides. Green Chemistry 2007; 9: 342-350.

[130] Yin S, Pan Y, Tan Z. Hydrothermal Conversion of Cellulose to 5-Hydroxymethyl Furfural. International Journal of Green Energy 2011; 8: 234-247.

[131] Li Y, Lu X, Yuan L, Liu X. Fructose Decomposition Kinetics in Organic acid-Enriched High Temperature Liquid Water. Biomass and Bioenergy 2009; 33: 1182-1187.

[132] Chareonlimkun A, Champreda V, Shotiprul A, Laosiripojana N. Reactions of C5 and C6-Sugars and Lignocellulose under Hot Compressed Water (HCW) in the Presence of Heterogeneous Acid Catalysts. Fuel 2010; 89: 2873-2880.

[133] Mascal M, Nikitin EB. Direct, High-Yield Conversion of Cellulose into Biofuel. Angewandte Chemie (International Edition) 2008; 47: 7924-7926.

Chapter 4

# Fungal Biodegradation of Agro-Industrial Waste

Shereen A. Soliman, Yahia A. El-Zawahry and Abdou A. El-Mougith

Additional information is available at the end of the chapter

http://dx.doi.org/10.5772/56464

## 1. Introduction

Fungi possess an efficient hydrolytic system capable to convert lignocellulosic material to essential metabolites for growth. Usually, these fungi secrete enzymes, including cellulases (cellobiohydrolases, endoglucanases), hemicellulases (xylanases) and β-glycosidases. In terms of enzyme novelty, interest is focused on not only finding enzymes which could break down lignocellulose much more rapidly, but also enzymes which could withstand pH, temperature and inhibitory agents. Mutant strains of *Trichoderma reesei* have been selected that produce extracellular cellulases up to 35 g/l [1,2]. It has been suggested that increasing the specific enzyme activity is the most likely approach to improving the commercial prospects of lignocellulose hydrolysis [3].

Lignocellulose consists of lignin, hemicellulose and cellulose [4,5]. The major components of lignocellulosic biomass are cellulose (C6 sugars), hemicellulose (C6 and C5 sugars) and lignins (polyphenols) [6]. Large amounts of lignocellulosic "waste" are generated through agricultural practices, paper-pulp industries, and can pose an environmental pollution problem. Lignocellulosic waste is often disposed of by biomass burning, which is not restricted to developing countries alone, but is considered a global phenomenon [7]. However, plant biomass considered as "waste" can potentially be converted into various different value added products as illustrated in (Fig 1)including biofuels, chemicals, animal feed, textile and laundry, pulp and paper [8,9,10,11]. Production of ethanol and other alternative fuels from lignocellulosic biomass can reduce urban air pollution, decrease the release of carbon dioxide in the atmosphere, and provide new markets for agricultural wastes [12].

One effective approach to reduce the cost of enzyme production is to replace pure cellulose with relatively cheaper substrates, such as lignocellulosic materials. There are reports of successful attempts to produce cellulases on lignocellulosic materials [13]. Environmental pollution is a worldwide threat to public health. Biological degradation, for both economic and ecological reasons, has become an increasingly popular treatment of agricultural and industrial waste [14]. Continuous accumulation of industrial wastes poses a serious environmental problem [15,14,16].

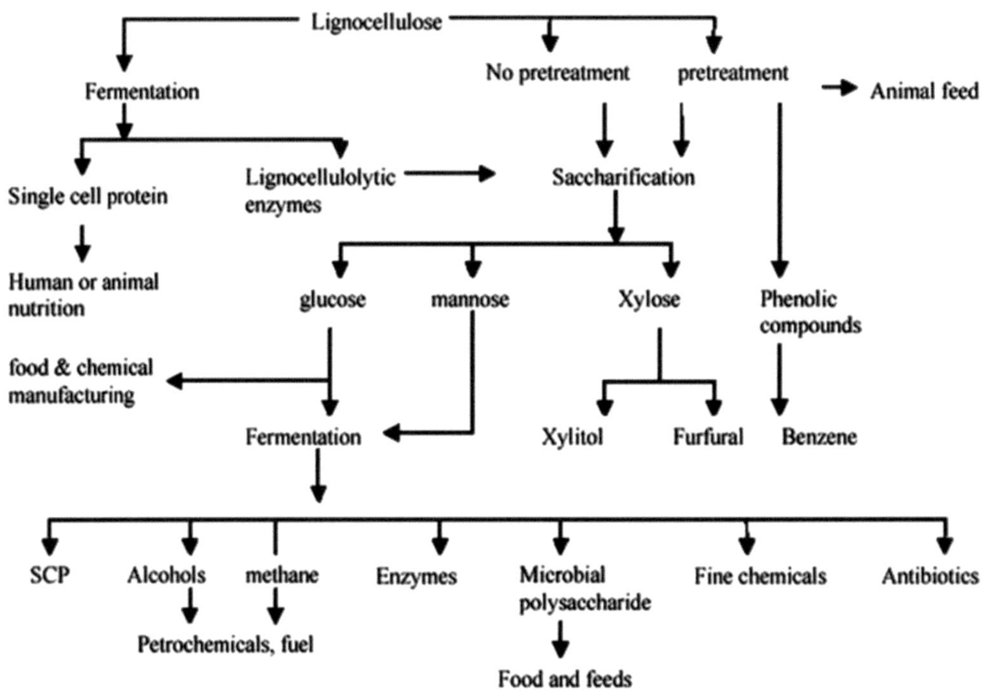


**Figure 1.** Lignocellulose bioconversion into value-added by-products Howard *et al.*, (2003)

## 1.1. Cellulose

Cellulose is composed of insoluble, linear chains of glucose units linked by β-1, 4-glucosidic bonds (Fig 2). It is composed of highly crystalline regions and (non-crystalline) regions forming a structure generally resistant to enzymatic hydrolysis, especially the crystalline regions [17]. The cellulose fibers are usually embedded in an amorphous matrix of hemicellulose and lignin [18]. The presence of lignin in the biomass lowers the biodegradability both of the cellulose and hemicellulose [19]. Numerous pretreatment methods including biological methods have been developed for separation of lignocellulosic to cellulose, hemicellulose, and lignin [6]. Biological methods based on the enzymology have been suggested [20,21]. According to [22] the main problem in cellulose hydrolysis is the secondary and tertiary structures, not the primary linkage structure.

Chemically, cellulose $(C_6H_{10}O_5)_n$ molecules are linear glucans [23,24,18]. Partial hydrolysis of cellulose produces a range of oligosaccharides including cellobiose, cellotriose, and cellotetrose [25,26]. Intra-molecular hydrogen bonds between hydroxyl groups on the same cellulose chains produce the high viscosity, and rigidity associated withcellulose polymers. The hydroxyl groups at the ends of each cellulose chain have different chemical properties. At one end of the cellulose chain the number one carbon contains an aldehyde hydrate group with reducing activity, while at the other end the number four carbon is an alcoholic hydroxyl with non-reducing activity. In native cellulose the cellulose chains are oriented in a parallel super molecular structure (Fig 2) in which inter-molecular hydrogen bonding between contiguous cellulose molecules, results in a sheet-like structure in the cellulose fibers [6].

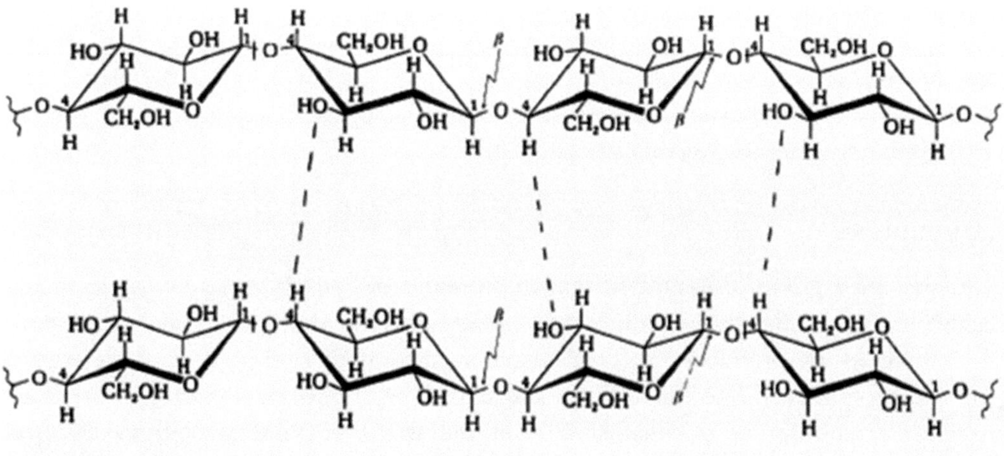

**Figure 2.** Structure of cellulose

The hydrogen bond strength is 25 KJ/mol, which is almost one hundred times stronger than Van der Waals forces (about 0.15KJ/mol), but less than one-tenth the strength of the O-H covalent bond (460 KJ/mol) [27]. Both inter and intra chain hydrogen bonds, which result in a sufficiently packed structure that prevents penetration, not only by enzymes, but also by small molecules, such as water. However, some regions, e.g. non-crystalline regions, permit penetration by larger molecules, including cellulases [28]. Obviously, reduction of crystallinity of cellulose and removal of lignin and hemicellulose are important goals for any pretreatment process [29].

## 1.2. Cellulolytic substrates under study (Wood dust).

Cellulose and hemicellulose are macromolecules from different sugars. The composition and percentages of these polymers vary from one plant species to another. Moreover, the composition within a single plant varies with age, stage of growth, and other conditions [30,5]. Cellulose makes up about 45% of the dry weight of wood.

Hemicellulose is a complex carbohydrate polymer and makes up 25–30% of total wood dry weight. It consists of D-xylose, D-mannose, D-galactose, D-glucose, L-arabinose, 4-O-methyl-glucuronic, D-galacturonic and D-glucuronic acids. Sugars are linked together by ß-1,4- and occasionally ß-1,3-glycosidic bonds [31]. Another major difference is the degree of polymerization, that hemicellulose has branches with short lateral chains consisting of different sugars [12].

Wood is an essential material for man. It is a material source for pulp and paper production, manure in the agricultural sector and fuel in the energy sector. Wood debris and by-product of wood processing pollutes the environment [32]. Wood wastes and their disposal have environmental concern worldwide especially when these wastes are biodegradable to useful goods [33]. DuPont, a company with environmental waste management, stated that 'waste manufacturing may be a product looking for a market [34]. The use of biological degradation have greater advantages over the use of chemical degradation because biotechnological synthesized products are less toxic and environmentally friend [35]. Sawdust as a lignocellulosic material can undergo enzymatic degradation to produce protein, glucose, and subsequently ethanol [33].

## 1.3. Cellulases

Cellulases are a group of hydrolytic enzymes capable of hydrolysing cellulose to smaller sugar components like glucose [36]. In industry, these celluloytic enzymes have found novel applications in the production of fermentable sugars, ethanol, organic acids, detergents, pulp and paper industry, textile industry and animal feed [37,38,39, 40, 41,16]. Today, these enzymes account for approximately 20% of the world enzyme market, mostly from Filamentous fungi, in particular *Trichoderma* and *Aspergillus* [42]. Cellulases can be divided into three major enzyme activity classes [43,44,45]; endoglucanses or endo1,4β-glucanase (EC 3.2.1.4), cellobiohydrolase (EC 3.2.1.91) and β-glucosidase (D-glucoside glucohydrolase) (EC 3.2.1.21) [46,47,48].

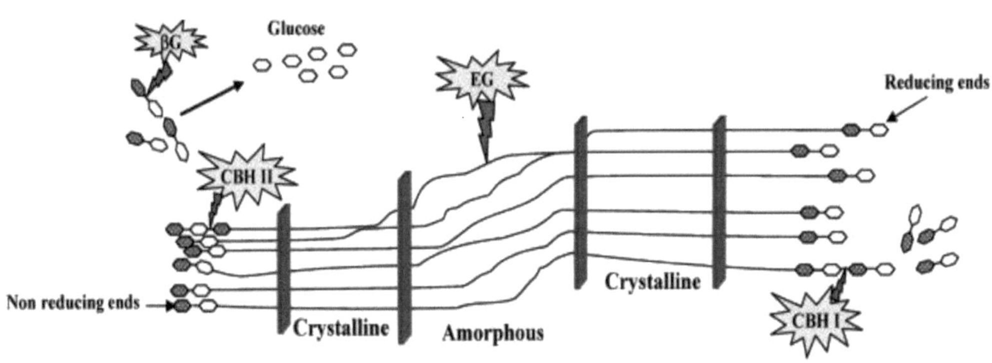


**Figure 3.** Different type of cellulase and their mode of action on cellulose

Extensive research efforts have been focused on lowering the cost of enzyme production, such as: (1) Screening for organisms with novel enzymes; (2) Improvement of existing industrial organisms and enzyme engineering; (3) factors such as substrate, culturing conditions and recycling of enzymes [49]. Strain improvement, choice of substrate and culture conditions in relation to improvement of fungal cellulolytic enzyme production were also studied [50]. Other investigators have looked for cheaper substrates [51,52,48].

## 1.4. Cellulase structure

Most fungal cellulases consist of two domains, i.e. a larger catalytic domain and a smaller cellulose binding domain (CBD) Table (1). These domains are joined by a glycosylated linker peptide [53,54]. The catalytic domain contains an active site [55]. The presence of CBD is essential to the degradation of solid crystalline cellulose [56,57]. Many studies have shown that removal of the CBD typically results in a decrease of about 50-80 % of the activity of fungal cellulases [58,59]. The biochemical role of a CBD is to keep the enzyme catalytic unit close to the substrate surface [60].

| Enzyme | Family | Amino acid residues | Molecular mass kDa | Isoelectric point (pI) | Structural organization[b] |
|---|---|---|---|---|---|
| EGI | 7 | 437 | 50–55 | 4.6 | 368 33 36 |
| EGII | 5 | 397 | 48 | 5.5 | 36 34 327 |
| EGIII | 12 | 218 | 25 | 7.4 | 218 |
| EGIV | 61 | 326 | (37)[a] | - | 233 56 37 |
| EGV | 45 | 225 | (23)[a] | 2.8–3 | 166 23 36 |
| EGVI | unknown* | unknown* | 95–105 | 5.6–6.8 | unknown* |
| CBHI | 7 | 497 | 59–68 | 3.5–4.2 | 430 31 36 |
| CBHII | 6 | 447 | 50–58 | 5.1–6.3 | 36 44 365 |

a The molecular mass calculated from amino acid sequence
b ■, the catalytic domain; □, linker region; ■, CBD
* gene not described

**Table 1.** Properties of cellulases from Trichoderma reesei (Saloheimo et al., 1997).

## 1.5. Cellulolytic organisms

A wide variety of microbes including bacteria, fungi and actinomycetes are involved in the decomposition of cellulose, of which fungi have generally been considered to be the main cellulolytic organisms [61,62]. By 1976 more than 14000 fungi active against cellulose and other insoluble fibres had been collected [63,1]. Many fungi are capable of growing on cellulose as the sole carbon source, including those normally found on wood [64]. Fifty two fungal isolates with cellulolytic activities were isolated on Czapex' dox with a filterpaper and CMC agar media [65]. The cellulase systems of the mesophilic fungi *Trichoderma reesei* and *Phanerochaete chrysosporium* are the most thoroughly studied [66]. The production of cellulase by the members of genus *Fusarium* were investigated by several workers [67,68,69]. Filamentous fungi, particularly *Aspergillus* and *Trichoderma* species, are well known as efficient producers of these cellulases [70,71]. *Trichoderma* species are common soil inhabiting fungi with a strong ability to degrade cellulose [72,6]. Trichoderma reesei produce two exo-glucanases (Table1) and at least six endoglucanases and two β-glucosidases [57,54,73]. The filamentous fungus *Trichoderma reesei* has a long history in the production of hydrolytic enzymes, which was widely used in the textile, pulp and paper industries [3,38].

## 1.6. Enzyme applications

### *1.6.1. In pulp and paper*

Novel enzyme technologies can reduce environmental problems, eliminating caustic chemicals for cleaning paper machines, reduce manufacturing costs and create novel high-value products. Today recombinant DNA has allowed the cloning of enzymes modified for temperature and pH stability [74]. Cellulases have been used in many processes in the paper industry [75,76,77]. Endoglucanase treatment of pulps was shown to decrease the viscosity and chain length and increase the reactivity of a pulp made from eucalyptus and acacia [78]. Despite the progress achieved, more effort is needed for lignocellulosic enzymes and/or microorganisms to have significant industrial impact [74].

### *1.6.2. Enzymatic deinking*

One of the greatest challenges in paper recycling is removal of contaminants; some of the most problematic contaminants are polymeric inks and coating. Toners such as those used in laser and xerographic copy machines are thermally fused to the surface of the printed page [79]. Cellulases are particularly effective in facilitating the removal of toners from office waste papers [30]. Kim et al.,(1991) showed that crude cellulases applied to pulps facilitate deinking [80].

### *1.6.3. Textile industry*

Cellulases are important tools in the textile industry. They provide an economical and ecological way to treat cotton and cotton-containing fabrics. Pretreatment with cellulases reduces the pill-formation and increases the durability and softness of the fabric. Currently textile enzymes have a market value equivalent to 12% of the industrial enzyme market [81]. Today a commercial endoglucanase cellulase product, Cellulsoft Ultra L (expressed in genetically modified *Aspergillus*) is available from Novozymes [82].

Endoglucanases have been used to release the microfibrils from the surface of dyed cellulosic fabrics and thus to restore the original colors [83,84]. Trichoderma viride cellulase that removal of surface fibrils from linen fabrics can be accomplished without high weight losses or reduction in tensile strength [85].

## 1.7. Bio-fuel

In the 20th century, the world economy has been dominated by technologies that depend on fossil energy, such as petroleum, coal, or natural gas to produce fuels, chemicals and power

[86]. However, fossil energy sources are not infinite. Global crude oil production is predicted to decline from 25 billion barrels to approximately 5 billion barrels in 2050 [87]. A search for other energy sources is advisable. Biomass is a potential renewable energy source that could replace fossil energy for transportation [86]. Ethanol is used as a chemical feedstock. Brazil produces ethanol from the fermentation of cane juice whereas in the USA corn is used. In the USA, gasoline fuels contain up to 10% ethanol by volume [9] Ethanol blended with gasoline (10:90) reduces carbon monoxide emissions [12]. By December 2011 there were nearly 10 million E85-capable vehicles (85% ethanol) on U.S roads [88]. Production of fuel ethanol from fermenting sugar or crops such as corn has dropped to around 800 thousand barrels per day (kb/d) in the US, according to the International Energy Agency's latest Oil Market Report (OMR) 2012.

### 1.8. Enhancing the feeding value of Agro-industrial by-products

Agro-industrial by-products are wastes, often causing environmental pollution and hazard when left unutilized. As grain production remains insufficient to meet human and animal needs; the alternative is to employ feed ingredients which do not have direct human food value [89]. Changes in the protein and cellulose of agro-industrial by-products after fermentation with *Aspergillus niger, Aspergillus flavus* and *Penicillium sp*. in solid state [89] indicate the possibility of enhancing the feeding value of these by-products. Biological treatment of agricultural residues is a new method for improvement of digestibility [90]. Attempts are being made to improve the digestibility of lignocellulosic feeds by the use of microbial additives [91].

## 2. Materials and methods

### 2.1. Organism, cultivation and growth conditions

Trichoderma sp FJ937359 previously isolated from Sharkia, Egypt, was cultivated on Czapek medium containing (1%) wood dust as a sole carbon source. The culture was incubated for 14 days at 30°C on orbital shaker at 150 rpm. At the end of the incubationtime, residues were removed and the filtrate was centrifuge at 5000 rpm. The resulting clear supernatant was used as the source of crude enzymes [92].

### 2.2. Determination of enzyme activity

Endoglucanase activity was routinely measured according to [93]. One ml appropriate filtrate dilution, was added to 1 ml of 1% carboxymethyl cellulose (CMC) dissolved in 50 mM sodium acetate buffer, pH 5.0. After incubation at 50 °C for 60 min, the reaction was stopped by addition of 3 ml dinitrosalicylic acid. After 10 min in a boiling water bath, the enzymatic hydrolysis of CMC was determined at 540 nm. One unit of CMCase activity was defined as the amount of enzyme that released 1 μmol of reducing sugars as glucose equivalents $min^{-1}$.

## 2.3. Determination of protein concentration

The protein content of the crude enzyme was assayed by Folin-phenol reagent, using bovine serum albumin as a standard [94].

## 2.4. Purification of endoglucanases (CMCases)

Trichoderma sp. Shmosa Tri FJ937359 culture filtrate was used as source of crude enzyme. Proteinwasprecipitated byslow addition of ammonium sulphate until the desirable saturation 80%. The obtained precipitated protein was resuspended in known volumeof 0.1 M citrate phosphate buffer (pH 5.0) and dialyzed via cellophane bags (MWCO 10 KDa) against citrate-phosphate buffer (pH 5.0). A pharmacia sephadex G 100 column (2.5 x 90 cm) was used for purification of the dialyzed enzyme. The solid phase consists of sephadex $G_{100}$ (10 gm) swollen in 0.1 M citrate phosphate buffer (pH 5.0) for 24 hour at 4.0°C, sodiumazide (0.02%) was added to prevent any microbial growth. Fractions were collected and assayed for CMCase activity and protein content.

## 2.5. Estimation of enzymes molecular weights

For molecular weight determination, the enzyme preparations and known molecular weight markers were subjected to electrophoresis (SDS-PAGE) with 10% acrylamide gel; 0.2% CMC was incorporated into the separating gel prior to the addition of ammonium persulphate [95]. After electrophoresis, the gel was stained with Coomassie Blue R dye. For CMCase activity, the gel washed at room temperature in solution A (sodium phosphate buffer, pH 7.2, containing isopropanol 40%), solution B (sodium phosphate buffer, pH 7.2) for 1 h, respectively then solution C (sodium posphate buffer, pH 7.2, containing 5 mM b-mercaptoethanol and 1 mM EDTA) at 4 °C overnight. The gel was then incubated at 37 °C for 4 h, stained with 1% Congo red for 30 min, destained in 1 M NaCl for 15 min [96], clear bands indicated the CMCase activitiy.

## 2.6. Enzyme characteristics

### 2.6.1. *Effect of substrate concentration on CMCase I and II activities*

The (CMCase) activity was carried out for the two enzymes at different concentrations of the substrate (CMC): 0.12, 0.25, 0.5, 0.75, 1.0, 1.5, 2.0, 2.5 and 3.0 % at pH 5.0 using citrate-phosphate buffer. Activity was determined using DNS reagent (3, 5-dinitrosalicylic acid) [93].

### 2.6.2. *Effect of reaction time on CMCase I and II activities*

(CMCase) activity was determined for the two enzymes CMCase I and II at different incubation time of reaction mixture; 15, 30, 60, 90, 120and 150min. Activity was determined as previously described.

### 2.6.3. *Effect of reaction temperature on CMCase I and II activities*

The enzymatic activity of (CMCase) was estimated for CMCase I and II at the following reaction temperatures 20, 30, 40, 50, 60, 70 and 80°C.

### 2.6.4. *Thermal stability of CMCase I and II*

The thermal stability of (CMCase)activity was tested by preheating of enzymes at 40, 50, 60 and 70 °C. Activity was estimated every 30 min.

### 2.6.5. *Effect of pH-value of reaction mixture on CMCase I and II activities*

The pH optima of the (CMCase) enzymes were determined at a pH range from 2 to 9, using citrate–phosphate buffer (0.1 M citric acid, 0.2 M $Na_2HPO_4$, pH 3.0 to 7.0), and Tris buffer (0.08 M Tris, 0.1 M HCl, pH for pH 8 and 9.0. The (CMCase)activity at each pH-value was estimated as previously described.

### 2.6.6. *pH-stability of CMCase I and II*

The pH stability of purified (CMCase)enzymes was determined by preincubation of enzymes at a pH range from 2 to 9for 24 hours [97] then activities were assayed as previously described at each pH value.

### 2.6.7. *Effect of some metallic ions on CMCase I and II activities*

The effect of KCl, NaCl, $CaCl_2$, $CoCl_2$, $CuSO_4.5H_2O$, $ZnSO_4.7H_2O$, $MgSO_4.7H_2O$, as well as EDTA and SDS at a concentration of 20 ug/ml of reaction mixture on CMCase I and II. Activities were investigated at optimum conditions as described before.

### 2.6.8. *Enzyme substrate specificity*

The specificity of the enzymes for their substrate was investigated using different substrates: CMC (control), Chitin, Starch, β. glucan, Xylan and Cellobiose.

# 3. Experimental results

## 3.1. Cellulase production

There is a growing demand for specific, efficient and cheap cellulases. Therefore it is important to gain more information about the production, purification and activity of these enzymes produced by microorganisms. Cellulases yield appear to depend on factors like carbon source, pH value, temperature, aeration, growth time [98]. To establish a successful fermentation process it is necessary to make the environmental and nutritional conditions favorable for the microorganism for over-production of the desired metabolite [99]. Sawdust is reported to be more suitable for cellulase production as it gave the highest yield of enzyme compared to bagasse and corncob [100]. The highest cellulase productivity with sawdust may be due to its very high percentage of cellulose which is the major component of cell walls of wood.

The use of inexpensive biomass resources as substrates can help to reduce cellulase production cost [52]. Research efforts have been undertaken to replace the expensive carbon and nitrogen sources with cheap raw materials in the media to bring down the production cost of cellulases [48]. The use of agro-industrial residues as the basis for cultivation media is a matter of great interest, aiming to decrease the costs of enzyme production and meeting the increase in awareness on energy conservation and recycling. The negative attitude in which wastes are viewed as valueless, has been replaced by a positive view in which wastes are recognized as raw materials of potential value [49,101]. Production of low-cost cellulolytic enzymes using inexpensive growth media has been investigated by different research workers [102,103]. Lignocellulosic materials from crop residues, wood, and wood residues are considered as the least expensive sources of cellulosic substrates [104,105]. In our study wood dust was selected as substrate for CMCase productionby *Trichoderma sp.* (Tri) Shmosa Tri FJ937359. This selection was due to wood dust being widely available, relatively inexpensive and a highly cellulolytic substrate.

### 3.2. Cellulases purification and activity

The crude enzymes obtained from culture filtrate of Trichoderma sp. Shmosa Tri FJ937359 grown on wood dust was subjected to a purification protocol. Results showed in table (2) indicated that specific activity increased to 62.86 U/mg (1.26 fold) by salting out compared with the crude enzyme. Many workers used ammonium sulphate for precipitation of Cellulase [106,107]. The results showed also that there are two peaks I and II for CMCase with specific activities 170.53 and 166.51 U/mg, respectively. The peaks were used for subsequent characterization and properties of enzymes. These purification techniques are comparable to those reported by [108,109,110,111] using sephadex G-100; sephadex G-150; sephadex G -50 and G -200 and superdex-200 HR respectively.

| Purification steps | | Protein (mg/ml) | Total protein (mg) | Total activity (U/ml) | Specific activity (U/ mg) | Recovery (%) | Fold |
|---|---|---|---|---|---|---|---|
| Crude enzyme | | 0.340 | 595 | 29505 | 49.58 | 100 | 1 |
| Precipitate with $(NH4)_2SO_4$ | | 0.237 | 47.4 | 2980 | 62.86 | 10.09 | 1.26 |
| Gel filtration | Peak I | 0.132 | 3.3 | 562.75 | 170.53 | 1.90 | 3.43 |
| Sephadex $G_{100}$ | Peak II | 0.152 | 3.8 | 632.75 | 166.51 | 2.14 | 3.35 |

**Table 2.** Purification profile (CMCases) produced by Trichoderma sp. grown on wood dust

### 3.3. Enzymes molecular weights

The molecular weights of CMCases were estimated using the technique of sodium dodecyl sulfate polyacrylamide gel electrophoresis (SDS-PAGE). The results showed that two clear bands appeared (I, II) when stained with 1% Congo red and destained in 1 M NaCl (zymogram method), indicating the presence of CMCase activity (Fig 4). The results showed also that the molecular weights of the 2 enzymes I, II were, 58 KDa and 34 KDa, respectively Fig (4). These findings are in agreement with other investigators that reported the presence of different isoenzymes with different molecular weights of CMCase produced by various microorganisms [112,113,114]. Our results are also in comparable to these obtained by Holt and Hartman (1994), they use a zymogram method to detect endoglucanases from Trichoderma reesei [115]. The molecular weight of EG endoglucanase purified from the culture filtrate of Trichoderma sp. C-4, was 51 kDa [106]. The EGs of the mesophilic fungi *Trichoderma reesei* and *Phanerochaete chrysosporium* have molecular weight ranges from 25 to 50 kDa [12]. However carboxymethyl cellulase purified from Trichoderma viride, was examined by (SDS-PAGE) and the molecular weight was 66kDa [116]. Two endoglucanases were purified to homogeneity (EG-III and EG-IV), from the culture filtrate of a mutant strain Trichoderma sp. M7, the molecular weights determined to be 49.7 and 47.5 kDa [114]. The difference in the molecular weights of CMCase enzymes may due to the biological aspect of the fungal strains, these aspects include intrinsic genomic traits, identity of the gene encoding enzyme and the proteomic level.

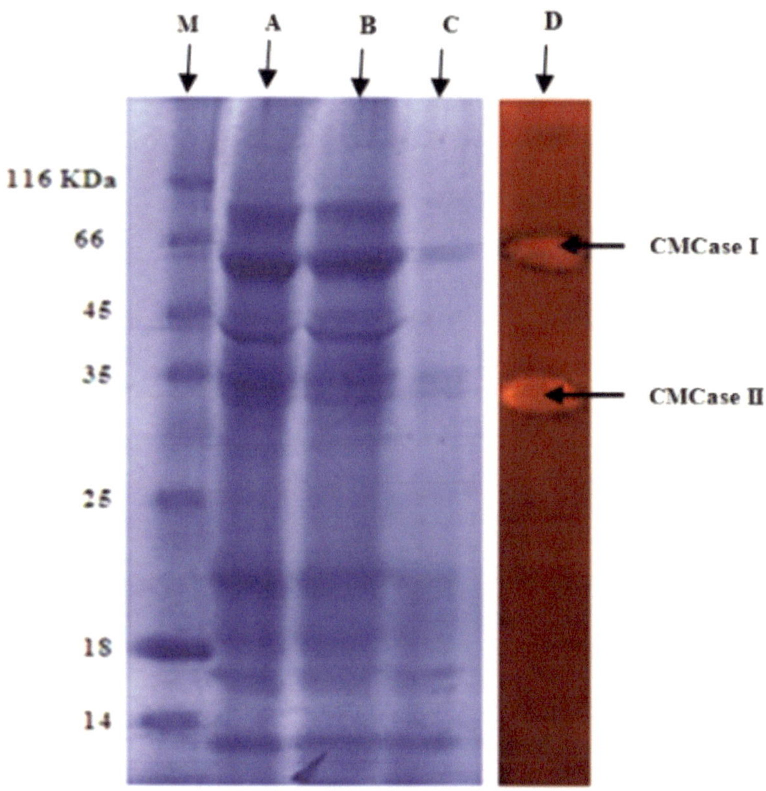


**Figure 4.** SDS-PAGE profile of CMCase (I,II) from Trichoderma sp. (M) Standard protein markers, (A) Crude enzyme preparation, (B) dialyzed enzyme preparation, (C) active fractions of column chromatograph and (D) zymogram stain with Congo red

### 3.4. Effect of substrate concentration on CMCase I and II activities

The results in Fig. (5) illustrate that the optimum specific activities of CMCase I and II were obtained at a substrate concentration of 1% (w/v). Increasing substrate concentration beyond 1 % (w/v) caused a decrease in the specific activities of both enzymes. This is probably because at high substrate concentration, many substrate molecules are around the enzyme molecules, crowding the active site or may be bound to regions which are not the active site. These results are supported by Bakare, et al., (2005) who reported that the activities of cellulases are greatly influenced by the concentration of the substrate [107]. At a fixed enzyme concentration, an increase in the concentration of substrate results in an increase in enzyme activity until a saturation point is reached beyond which enzyme activity decreases. Our results showed also that the Km value of CMCase I and II were 4.0 and 3.1 mg per ml, respectively. Km values denote the amount of substrate needed to achieve half the maximal initial reaction rate [107]. Petrova et al., (2009) purified two endoglucanases EG-III and IV from Trichoderma sp. M7, which exhibited Km's of 2.9 and 3.8 mg/ml, respectively [114].

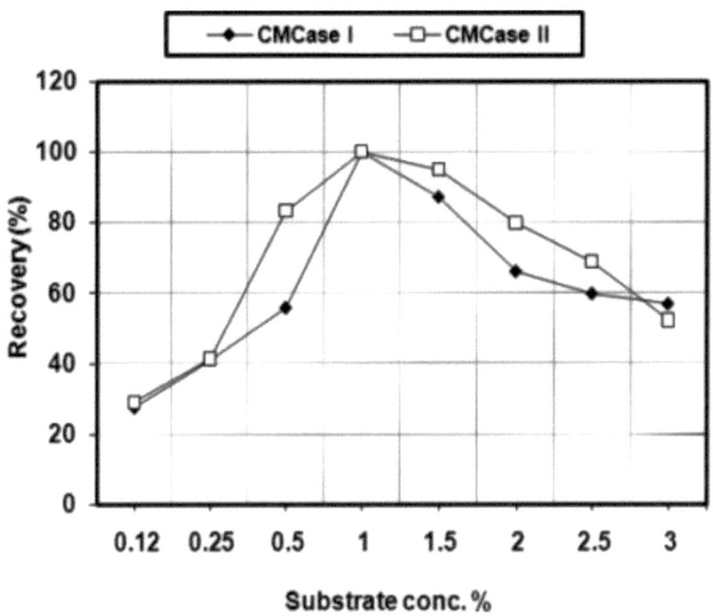


**Figure 5.** Effect of substrate concentrations on the activities of the purified (CMCase I) and (CMCase II

### 3.5. Effect of reaction time on CMcase I and II activities

The results presented in Fig (6) showed that the highest specific activities of CMCase I and CMCase II were recorded after one hour. Our results also indicatethat CMCase I and II retained 25.2 and 42.3 % of their activities respectively, after 3 hours of reaction. Our data is in agreementwith previous studies that recorded the amount of reducing sugar produced under the action of cellulases; for Trichoderma koningii and Aspergillus niger reducing sugars gradually increased with the increase in incubation time and reach the maximum at 60 min [67,117].

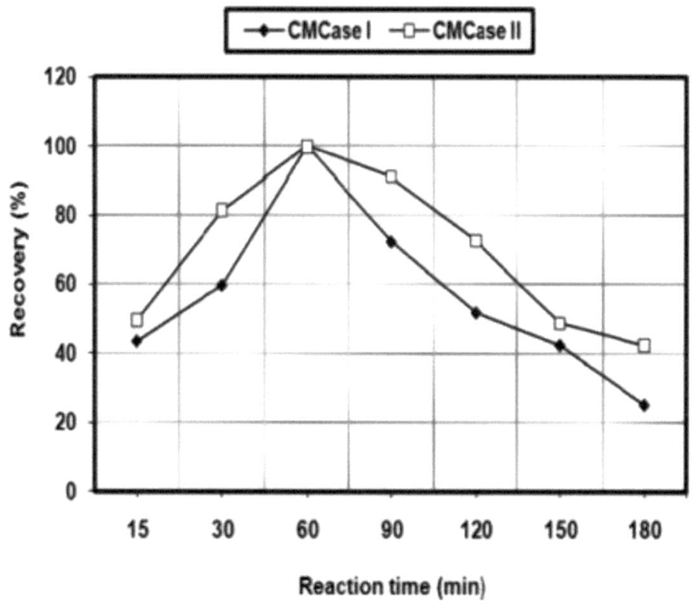


**Figure 6.** Effect of reaction time on the activities of the purified (CMCase I) and (CMCase II).

### 3.6. Effect of reaction temperature and thermal stability on CMcase I and II

Data in Fig (7) illustrates that the specific activities of CMCase enzymes increased with increasing reaction temperature, reaching optimum values at 50°C for both CMCase I and II.

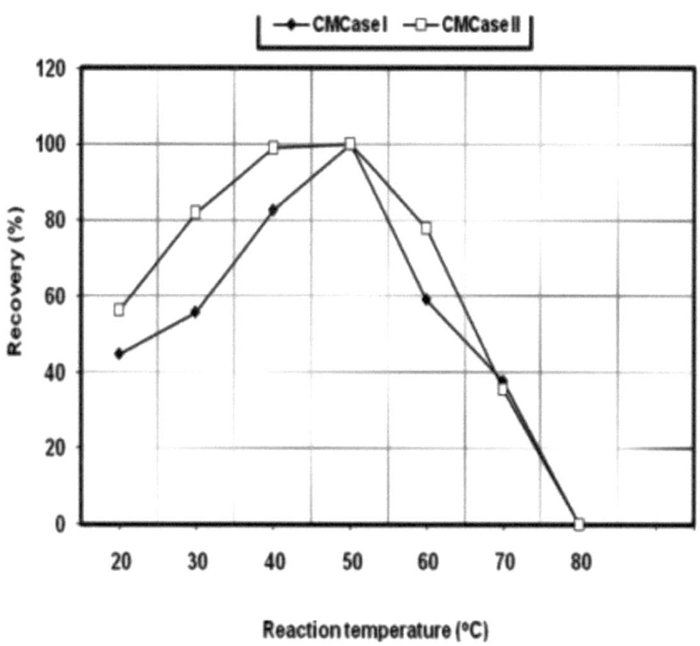


**Figure 7.** Effect of reaction temperature on the activities of the purified (CMCase I) and (CMCase II)

In general, cellulases have high temperature optima when compared with other enzyme systems [108]. Cellulolytic enzyme activities increased with increasing temperatures up to 60° to 65°C for the three enzyme components from *Trichoderma longibrachiatum* [118]. Ülker and Sprey, (1990) purified the low molecular weight endoglucanase from Trichoderma reesei and found the optimal temperature to be 52°C [119]. Petrova et al., (2009) stated that the optimal temperature values for two purified endoglucanases EG-III and EG-IV from Trichoderma sp. M7, were found to be 60°C and 50°C, respectively [114]. Our results show that the thermal inactivation of the purified CMCase I and II enzymes increased with increasing preheating temperature as well as exposure time (Fig 8). The enzymes can withstand 60 min at 50ºC without loss of enzymatic activity. The results also illustrate that CMCase I and II retained 14.0 and 26.5 % of their original activities after 90 min at 70°C. Our findings are in agreement with that reported for CMCase and FPase from Trichoderma sp. A-001 that lost 20–33% of their activities when kept at 60°C for 4 hours before assaying, and a beta-glucosidase that lost 37% of its activity when maintained at 70°C for 4 h [120]. Endoglucanase purified from the culture filtrate of Trichoderma sp. C-4 was assayed at various temperatures ranging between 30°C and 70°C and the optimum temperature was found to be 50°C. The enzyme showed stability at 50°C for 60 min but lost 50% of its maximal activity after 10 min at 60°C [106]

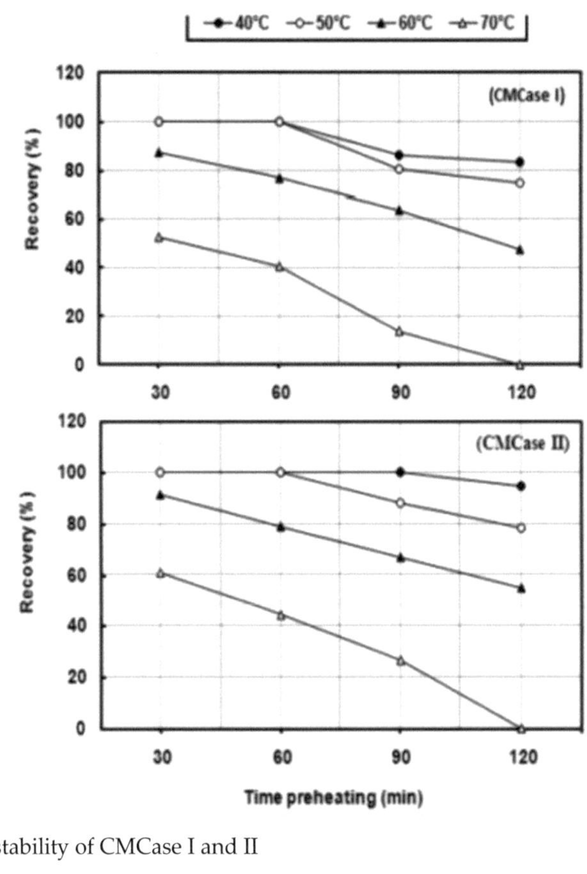


**Figure 8.** Thermal stability of CMCase I and II

The optimum temperature for C1cellulase of *Trichoderma viride* was found to be 40 °C [121]. El-Zawahry and Mostafa (1983) recorded that the optimum temperature for cellulases activities produced by Trichoderma viride ranged from 50 to 60 °C [67].

### 3.7. Effect of PH values of reaction mixture and PH stability on CMCase I and II

Results in Fig. (9) show that the specific activities of CMCase I and II increase gradually with increasing pH of the reaction mixtures,reaching a maximal value at pH 5.0 for both CMCase I and II, consistent with other researchers [106,114]. Kalra et al., (1986) reported that purified enzyme preparations from Trichoderma longibrachiatum showed an optimal pH of 5.0 for CM cellulose [118]. However, Carboxymethyl cellulase purified from Trichoderma viride, showed an optimum pH 4.0 at 50 °C [116]. We also found that both CMCase I and II retained 23.2 and 22.9% of their activity at pH values as high as 9.0, respectively. The results in Fig (10) show that CMCase I was active at room temperature after 24 hrs over a broad pH range (3.0-9.0). On the other hand CMCase II was relatively stable in pH range (4.0-6.0). In general, the pH-stability curves of the enzymes are much broader than the pH-activity curves [108].

Fungal cellulases, in general, are stable at over the pH range 3 – 8 and usually active over the pH range 3.5 to 7, in citrate, phosphate or acetate buffers [112]. The optimum activities of C1 and Cx cellulases produced by *Trichoderma viride* was obtained at pH 5.0 [67]. Catriona et al. (1994) reported that the pH range over which the cellulases were highly active is fairly broad (pH 5.0 – 7.0) [123]. The instability of these enzymes at very low or very high pH values is due to the fact that they are proteins which are generally denatured at extreme pHvalues [124].

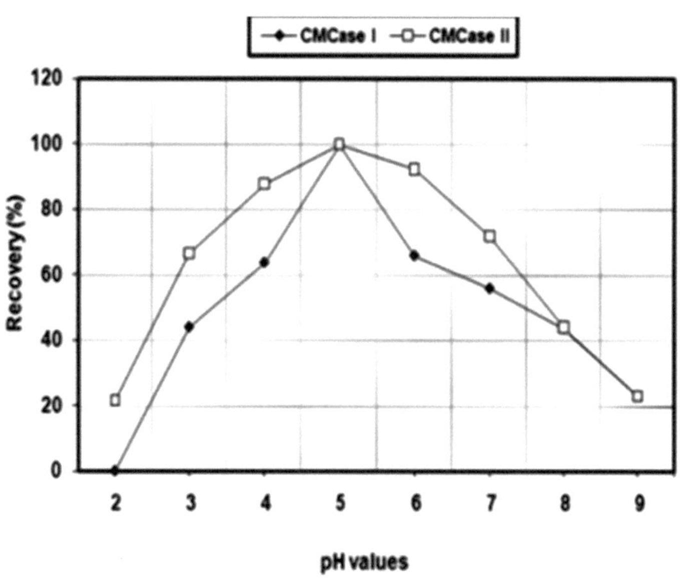


**Figure 9.** Effect of pH values of reaction mixture on the activities of the purified (CMCase I) and (CMCaseII)

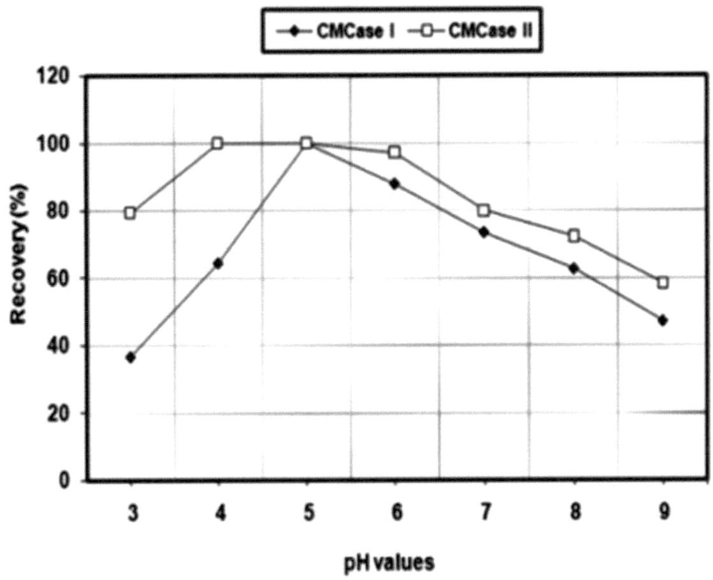


**Figure 10.** pH stability of CMCase I and II

### 3.8. Effect of some metallic ions and chemicals substances on CMCase I and II

The obtained results in Fig. (11) show the inhibitory effect of metallic ions on the activities of CMCase I and II. CMCase I was found to be more sensitive than CMCase II. $Co^{+2}$ was the most inhibitory ion for CMCase I, while $Hg^{+2}$ was the most inhibitory ion for CMCase II. Heavy metal ions such as Hg+2 and Ag+ have significant or complete inhibitory effect on

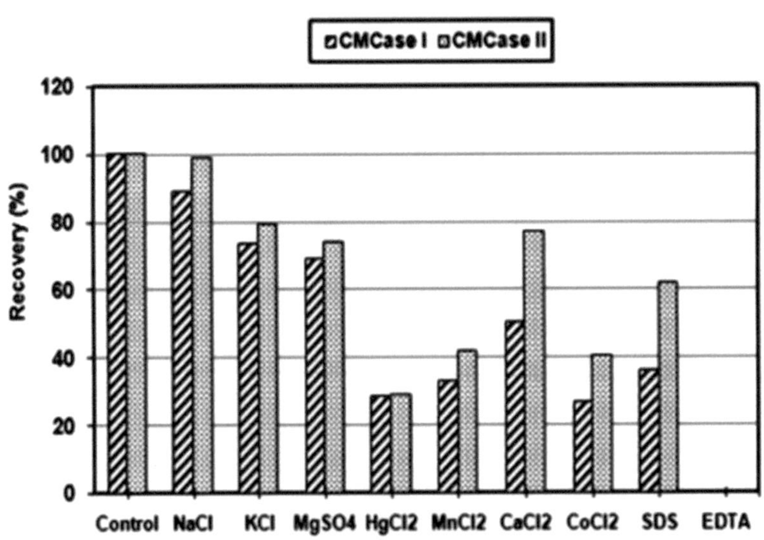


**Figure 11.** Effect of some metallic ions and chemicals on CMCase I and II produced by Trichoderma sp. grown

CMCase purified from Trichoderma viride [116]. Petrova et al., (2009) proved that Mn+2, Cu+2 and Pd+2 strongly inhibited EG-III and EG IVendoglucanases purified from Trichoderma sp [114]. These inhibitory effects may be due to the toxic effect of these ions. Our results also show complete inactivation of the two enzymes in presence of EDTA. EDTA is known as an ionic chelator [125] and its inhibition ability indicates that specific ions might be actively involved in the catalytic reaction of the enzyme [126]. EDTA inhibited enzyme activity at all concentrations while by low concentrations of Na+ and Mg++ stimulated enzyme activity [123].

## 3.9. Substrate specificity of CMCase I and II enzymes

Both enzymes clearly have high hydrolytic activity towards carboxymethyl cellulose CMC,neither showed any hydrolytic activity against chitin, starch and cellobiose (Fig 12). On the other hand both CMCase I and II had relatively low hydrolytic activity towards β glucan and xylan. Ülker and Sprey, (1990) purified endoglucanase from Trichoderma reesei, gave a strong increase in CMC-fluidity but the enzyme had no specificity toward crystalline cellulose (Avicel) or xylan [119]. Purified endoglucanases from Trichoderma sp. EG-IV catalyzed the hydrolysis of Na-CMC whereas EG-III displayed high activity towards xylan [114]. Although *Km* values may serve only denote the amount of substrate needed to achieve half the maximal initial reaction rate [108], *Km* is a measure of the apparent affinity of an enzyme for its substrate [107]. An endo-1,4-β-D-glucanase I was purified from *Bacillus circulans* F-2. It showed a high-level of activity towards carboxymethyl cellulose (CMC) as well as *p*-nitrophenyl-b-D-cellobioside, xylan, Avicel and filter paper [127].

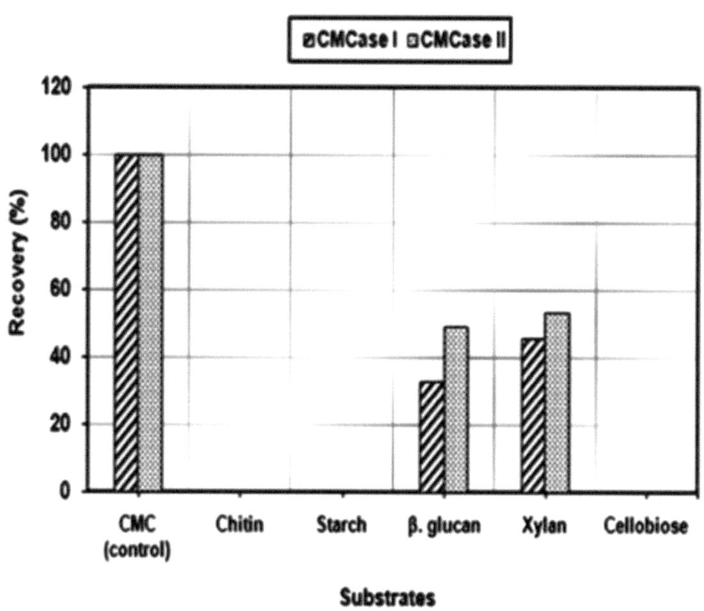


**Figure 12.** Substrate specificity of CMCase I and II

Finally, it could be concluded that fungi in generalconsidered to be potential candidates for the production of cellulases which seemed to be very important for different biotechnological aspects. The use of agro-industrial residues as the basis for cultivation media is a matter of great interest, aiming to decrease the costs of enzyme production and meeting the increase in awareness of energy conservation and recycling. In our study wood dust was selected as substrate for CMCase productionby *Trichoderma sp.* (Tri). Wood dust is a waste by-product from wood processing, as a result it is widely available, relatively inexpensive, and a highly cellulolytic substrate. Moroever, as compared to thermal degradation of agricultural, industrial wastes, biological degradation has great potential for both economic and ecological reasons.

Our result showed that using wood dust as substrate *Trichoderma sp.* (Tri) produce high levels of endoglucanases (CMCase I and II). In terms of enzyme novelty from an applications perspective, interest is focused on not only finding enzymes which could break down lignocellulose much more rapidly, but also enzymes which could withstand pH, temperature and inhibitory agents. Our result showed that both purified CMCase I and II withstand high temperature 70°C even after 90 min, making them good candidates for novel industrial applications.

## Author details

Shereen A. Soliman*
*Biology Department, Faculty of Science Jazan University, Saudi Arabia*
*Botany department, Faculty of Science, Zagazig University, Egypt*

Yahia A. El-Zawahry and Abdou A. El-Mougith
*Botany department, Faculty of Science, Zagazig University, Egypt*

## 4. References

[1] Esterbauer, H, Steiner, W, & Labudova, I. (1991). Production of Trichoderma cellulase in laboratory and pilot scale. Biores. Technol., 36, 51-65.

[2] Jørgensen, H, Erriksson, T, & Börjesson, J. (2003). Purification and characterisation of 7fivecellulases and one xylanases from Penicillium brasilianum IBT 20888. Enzyme 8 Microb. Technol., 32, 851-861.

[3] Béguin, P. (1990). Molecular biology of cellulose degradation. Annu. Rev. Microbiol, 44, 219-248.

[4] Zhang, Y, & Lynd, L. (2004). Toward an aggregated understanding of enzymatic hydrolysis of cellulose: Noncomplexed cellulase systems. Biotechnol Bioeng., 88, 797-824.

[5] Ding, S, & Himmel, M. (2006). The maize primary cell wall microfibril: A new model derived from direct visualization. J Agric Food Chem., 54, 597-606.

* Corresponding Author

[6] Lee, Y. (2005). Oxidation of Sugarcane Bagasse Using a Combination of Hypochlorite and Peroxide. A Thesis Submitted for partial fulfillment for the degree of Master of Science In Food Science.

[7] Levine, J. (1996). Biomass burning and global change. In: Levine JS (eds) (Remote sensing and inventory development and biomass burning in Africa. The MIT Press, Cambridge, Massachusetts, USA, 1, 35.

[8] Gadgil, N, Daginawala, H, Chakrabarti, T, & Khanna, P. (1995). Enhanced cellulose production by a mutant of Trichoderma reesei. Enzyme and Microbial Technology., 17, 942-946.

[9] Sun, Y, & Cheng, J. (2002). Hydrolysis of lignocellulosic materials for ethanol production: a review. Bioresource. Technol., 83, 1-11.

[10] Beauchemin, K, Colombatto, D, Morgavi, D, & Yang, W. (2003). Use of exogenous fibrolytic enzymes to improve animal feed utilisation by ruminants. J. Anim. Sci., 81, 37-47.

[11] Doi, R. (2008). Cellulases of mesophilic microorganisms: cellulosome and non-cellulosome producers. Annals of the New York Academy of Sciences., 1125, 267-279.

[12] Perez, J, Munoz-dorado, J, De La Rubia, T, & Martinez, J. (2002). Biodegradation and biological treatments of cellulose, hemicellulose and lignin: an overview Int Microbio., 5, 53-63.

[13] Robinson, T, & Nigam, S. (2001). Solid-state fermentation: A promising microbial technology for secondary metabolite production, Appl. Microbiol. Biotechnol., 55, 284-289.

[14] Milala, M, Shehu, B, Zanna, H, & Omosioda, V. (2009). Degradation of Agro- waste by cellulose from Aspergillus candidusAsian Journal of biotechnology, 1, 51-56.

[15] Liu, J, & Yang, J. (2007). Cellulase Production by T. koningii, Food Technol. Biotechnol., 45, 420-425.

[16] Karmakar, M, & Ray, R. (2010). Extra Cellular Endoglucanase Production by Rhizopus oryzae in Solid and Liquid State Fermentation of Agro Wastes.Asian Journal of Biotechnology, 2, 27-36.

[17] Walker, L, & Wilson, D. (1991). Enzymatic hydrolysis of cellulose: An Overview. Biores. Technol.,36, 3-14.

[18] Klass, L. (1998). Biomass for renewable energy, fuels, and chemicals. Academic Press, San Diego, California.

[19] Pareek, S, Azuma, J, Shimizu, Y, & Matsui, S. (2000). Hydrolysis of newspaper polysaccharides under sulfate reducing and methane producing conditions. Biodegradation., 11, 229-237.

[20] Banjo, N, & Kuboye, A. (2000). Comparion of the effectiveness of some common Agro-industrial wastes in growing three tropical edible mushrooms. Proceedings of the Internal Conference on Biotechnology: Commercialization and Food Saftly, Nigeria, 161-168.

[21] Wuyep, P, Khan, A, & Nok, A. (2003). Production and regulation of lignin degrading enzymes from Lentinus squarrosulus (mont) Singer and Psathyrella atroumbonata Pengler. Afr. J. Biotechnol., 2, 444-447.

[22] Tsao, G, Ladisch, M, Ladisch, C, Hsu, T, Dale, B, & Chou, T. (1978). Fermentation substrates from cellulosic materials: productionof fermentable sugars from cellulosic materials: Annual reports on fermentation processes. Chapter 1, Academic Press,New York, NY., 2

[23] Segal, L. (1971). Cellulose and cellulose derivatives, Wiley, New York.

[24] Bertran, M, & Dale, B. (1986). Determination of cellulose accessibility by differential scanning calorimetry. J. Appl. Polym. Sci., 32, 4241-4253.

[25] Sjöström, E. (1981). Wood chemistry: Fundamental and application. Academic Press, New York.

[26] Feller, R, Lee, S, & Bogaard, J. (1986). The kinetic of cellulose deterioration. Advances in Chemistry, 212, 329-347.

[27] Krässig, H. (1993). Cellulose: Structure, accessibility, and reactivity. Gordon and Breach Science Publishers S.A. 6-13, 187-205.

[28] Lynd, L, Weimer, P, van Zyl, W, & Pretorius, (2002): Microbial Cellulose Utilization: Fundamentals and Biotechnology. Microbiology and Molecular Biology Reviews, 66, 506-577.

[29] Wu, Z, & Lee, Y. (1997). Ammonia recycled percolation as a complementary pretreatment to the dilute-acid process. Appl. Biochem. Biotech., 65, 21-34.

[30] Jeffries, T. (1994). Biodegradation of lignin and hemicelluloses, In: Ratledge C (ed.) Biochemistry of microbial degradation. Kluwer, Dordrecht. , 233-277.

[31] Kuhad, R. (1997). Microorganisms and enzymes involved in the degradation of plant fiber cell walls Springer Verlag.

[32] Williams, B. (2001). Biotechnology; from A to Z, 2nd edition, Oxford Press, New York, 384-385.

[33] Shide, E, Wuyep, P, & Nok, A. (2004). Studies on the degradation of wood sawdust by Lentinus squarrosulus (Mont.) Singer African Journal of Biotechnology., 3, 395-398.

[34] Keith, D. (1994). Enviro-Mangement: How Smart companies turn environmental costs into profit Prentice-Hall, USA, 1-18.

[35] Liu, C, Wang, Y, Ouyang, F, Ye, H, & Li, G. (1998). Production of artemisimin by hairy root cultures of Artemisia annua L in bioreactor. Biotechnol. Lett., 20, 265-268.

[36] Chellapandi, P, & Himanshu, M. (2008). Production of endoglucanase by the native strains of Streptomyces isolates in submerged fermentation. Braz. J. Microbiol., 39, 122-127.

[37] Luo, J, Xia, L, Lin, J, & Cen, P. (1997). Kinetics of simultaneous saccharification and lactic acid fermentation processes. Biotechnol. Prog., 13, 762-767.

[38] Oksanen, T, Pere, J, Paavilainen, L, Buchert, J, & Viikari, L. (2000). Treatment of recycled kraft pulps with Trichoderma reesei hemicellulases and cellulases. J Biotechnol., 78, 39-48.

[39] Miettinen-Oinonen, J, Londesborough, V, Joutsjoki, R, & Lantto, J. Vehmaanper, (2004). Three cellulases from Melanocarpus albomyces for textile treatment at neutral pH. Enzyme Microb. Technol., 34, 332-341.

[40] Kaur, J, Chadha, B, Kumar, B, & Saini, H. (2007). Purification and characterization of two endoglucanases from Melanocarpus sp. MTCC 3922. Bioresour. Technol., 98 74-81.

[41] Zhou, J, Wang, Y, Chu, J, Zhuang, Y, Zhang, S, & Yin, P. (2008). Identification and purification of the main components of cellulases from a mutant strain of Trichoderma viride T 100-14. Bioresour. Technol., 99, 6826-6833.
[42] Omosajola, P, & Jilani, O. (2008b). Cellulase production by Trichoderma longi, Aspergillus niger, Saccharomyces cerevisae cultured on waste materials from orange. Pak. J. Biol. Sci., 11, 2382-2388.
[43] Goyal, A, Ghosh, B, & Eveleigh, D. (1991). Characterisation of fungal cellulases. Biores. Technol., 36, 37 50.
[44] Rabinovich, M, Melnik, M, & Bolobova, A. (2002a). Microbial cellulases: A review. Appl. Biochem. Microbiol., 38, 305-321.
[45] Rabinovich, M, Melnik, M, & Bolobova, A. (2002b). The structure and mechanism of action of cellulolytic enzymes. Biochemistry (Moscow), 67, 850-871.
[46] Kang, S, Eh, K, Lee, J, & Kim, S. (1999). Over production of glucosidase by Aspergillus niger mutant from lignocellulosic biomass. Biotechnol. Lett., 21, 647-650.
[47] Gao, J, Weng, H, Zhu, D, Yuan, M, Guan, F, & Xi, Y. (2008). Production and characterization of cellulolytic enzymes from the thermoacidophilic fungal Aspergillus terreus M11 under solidstate cultivation of corn stover. Bioresource Technology, 99,7623-7629.
[48] Han, L, Feng, J, Zhu, C, & Zhang, X. (2009). Optimizing cellulase production of Penicillium waksmanii F10-2 with response surface methodology African Journal of Bio-technology., 8, 3879-3886.
[49] Howard, R, Abotsi, E, & Rensburg, E. (2003). Lignocellulose biotechnology : issues of bioconversion and enzyme production African Journal of Biotechnology., 2, 602-619.
[50] Persson, I, Tjerneld, F, & Hagerdal, B. (1991). Fungal cellulolytic enzyme production. Process Biochem., 26, 65-74.
[51] Szakacs, G, & Tengerdy, R. (1997). Lignocellulolytic enzyme production on pretreated poplar wood by filamentous fungi. World J. Microbiol. Biotechnol., 13, 487-490.
[52] Wen, Z, Liao, W, & Chen, S. (2005). Production of cellulase/ β glucosidase by the mixed fungi culture Trichoderma reesei and spergillus phoenicis on dairy manure. Process Biochemistry, 40, 3087-3094.
[53] Teeri, T, Reinikainen, T, Ruohonen, L, Jones, T, & Knowles, J. (1992). Domain function in Trichoderma reesei cellobiohydrolases. J. Biotechnol., 24, 169-176.
[54] Saloheimo, M, Nakari-setälä, T, Tenkanen, M, & Penttilä, M. (1997). cDNA cloning of a Trichoderma reesei cellulase and demonstration of endoglucanase activity by expression in yeast. Eur. J. Biochem., 249, 584-591.
[55] Reinikainen, T. (1994). The cellulose-binding domain of cellobiohydrolase I from Trichoderma reesei. Interaction with cellulose and application in protein immobilization. Academic dissertation. Espoo: VTT Publications 206. 115 p. + app. 46 p.
[56] Tomme, P, Van Tilbeurgh, H, Petterson, G, Van Damme, J, Vandekerckhove, J, Knowles, J, Teeri, T, & Claeyssens, M. (1988). Studies of the cellulolytic system of Trichoderma reesei QM9414: Analysis of domain function in two cellulobiohydrolases by limited proteolysis. Eur. J. Biochem., 170, 575-581.

[57] Srisodsuk, M. (1994). Mode of Action of Trichoderma reesei Cellobiohydrolase I on Crystalline Cellulose. Finland: VTT Publications, Espoo, Durand H, Clanet M, Tiraby G. 1988 Genetic improvement of Trichoderma reesei for large scale cellulase production. Enzyme Microb Technol., 10, 341-346.

[58] Reinikainen, T, Teleman, O, & Teeri, T. (1995). Effects of pH and ionic strength on the adsorption and activity of native and mutated cellobiohydrolase I from Trichoderma reesei. Proteins, 22, 392-403.

[59] Suurnäkki, A, Tenkanen, M, Siika-aho, M, Niku-paavola, M, Viikari, L, & Buchert, J. (2000). Trichoderma reesei cellulases and their core domains in the hydrolysis and modification of chemical pulp. Cellulose., 7, 189-209.

[60] Esteghlalian, A, Srivastava, V, Gilkes, N, Kilburn, D, Warren, R, & Saddler, J. (2001). Do Cellulose Binding Domains Increase Substrate Accessibility? Applied Biochemistry and Biotechnology, 91-93, 575-592.

[61] Cowling, E. (1958). A Review Literature on the Enzymatic Degradation of Wood and Cellulose. USDA Forest Service Report, (2116)

[62] Khalid, M, & Yang, W. Kishwar, N; Rajput, Z. and Arijo A. ((2006). Study of cellulolytic soil fungi and two nova species and new medium J Zhejiang Univ SCIENCE B, 7, 459-466.

[63] Mandels, M, & Sternberg, D. (1976). Recent advances in cellulase technology. Ferment. Technol., 54, 267-286.

[64] Benoliel, B, Fabrício, B, Castelo-branco, A, De Siqueira, S, Parachin, N, & Fernando, A. (2005). Hydrolytic enzymes in Paracoccidioides brasiliensis- ecological aspects. Genetics and Molecular Research, 2, 450-561.

[65] Punnapayak, H, Kuhirun, M, & Thanonkeo, P. (1999). Cellulolytic fungi and the bioconversion of fiber from Agave sisalana Science Asia., 25, 133-136.

[66] Kirk, K, & Cullen, D. (1998). Enzymology and molecular genetics of wood degradation by white rot fungi. In: Young RA, Akhtar M (eds) Environmental friendly technologies for pulp and paper industry. Wiley, New York, 273-307.

[67] El-Zawahry, Y, & Mostafa, I. (1983). Study on the production of cellulase enzyme by non-irradiated and irradiated isolates of Trichoderma viride. Rad. Res., 15, 103-110.

[68] Singh, A, & Kumar, P. (1991). Fusarium oxysporum: status in bioethanol production. Critical Reviews in Biotechnology., 11, 129-147.

[69] Murali, H, Mohan, M, Manja, K, & Sankaran, R. (2004). Cellulolytic activity of four Fusarium spp. World Journal of Microbiology and Biotechnology., 10, 487

[70] Peij, N, Gielkens, M, Veries, R, Visser, J, & Graaff, L. (1998). The transcriptional activator XlnR Regulates both xylanolytic and endoglucanase gene expression in Aspergillus niger. Appl. Environ. Microbiol., 64, 3615-3619.

[71] Singh, A, Singh, N, & Bishnoi, N. (2009). Production of Cellulases by Aspergillus Heteromorphus from wheat straw under submerged fermentation. International Journal of Environmental Science and Engineering, 1, 23-26.

[72] Frisvad, J, & Thrane, U. (2000). Mycotoxin production by common filamentous fungi. In: Introductionto food and airborne fungi, 6th ed. (Samson et al. Eds.). Centraalbureau voor Schimmelcultures, Utrecht, Netherlands., 321-331.

[73] Saloheimo, M, Kuja-panula, J, Ylösmäki, E, Ward, M, & Penttilä, M. (2002). Enzymatic properties and intracellular localization of the novel Trichodermareesei β-glucosidase BGLII (Cel1A). Appl. Environ. Microbiol., 68, 4546-4553.
[74] Kenealy, W, & Jeffries, T. (2003). Enzyme Processes for Pulp and Paper: A Review of Recent Developments, In Wood deterioration and preservation advances in our changing world / Barry Goodell, Darrel D. Nicholas, Tor P. Schultz, Chapter 12 page , 210-239.
[75] Jacobs, C, Venditti, R, & Joyce, T. (1997). Effect of enzyme pretreatments on conventional kraft pulpingtappi pulping conference, San Francisco, 593-601.
[76] Jacobs, C, Venditti, R, & Joyce, T. (1998). Effects of enzymatic pre-treatment on the diffusion of sodium hydroxide in wood", TAPPI, 81, 260-266.
[77] Bajpai, P. (1999). Application of enzymes in the pulp and paper industry. Biotechnol. Prog., 15, 147-157.
[78] Rahkamo, L, Siika-aho, M, Vehviläinen, M, Dolk, L, Viikari, P, Nousiainen, p, & Buchert, j. (1996). Modification of hardwood dissolving pulp with purified Trichoderma reesei cellulases. Cellulose, 3, 153-163.
[79] Gubitz, G, Mansfield, S, Bohm, D, & Saddler, J. (1998). Effect of endoglucanases and hemicellulases in magnetic and flotation deinking of xerographic and laser-printed paper, J. BiotechnoL., 65, 209 215.
[80] Kim, T, Ow, S, & Eom, T. (1991). Tappi Pulping Conference Proceedings. TAPPI PRESS, Atlanta.
[81] Heikinheimo, L. (2002). Trichoderma reesei cellulases in processing of cotton. Dissertation for the degree of Doctor of Technology. Espoo. VTT Publications 483.
[82] Liu, J.; Otto, E.; Lange, N.; Husain, P.; Condon, B. and Lund, H. (2000): Selecting cellulases for Bio polishing based on enzyme selectivity and process conditions. Textile Chemist and Colorist, 32, 30–36.
[83] Asferg, L, & Videbaek, T. (1990). Softening and polishing of cotton fabrics by cellulose treatment. ITB., 2, 5-8.
[84] Tyndall, R. (1992). Application of cellulase enzymes to cotton fabrics and garments. Textile Chem. Color., 24, 23-26.
[85] Buschle-diller, G, Zeronian, S, Pan, N, & Yoon, M. (1994). Enzymatic hydrolysis of cotton, linen, ramie and viscose rayon fabrics. Textile Res. J., 64, 270-279.
[86] Paster, M. (2003). Industrial bioproduct: Today and tomorrow. U.S. Department of Energy, Office of Energy Efficiency and Renewable Energy, Office of the Biomass Program. Washington, D.C.
[87] Campbell, C, & Laherrere, J. (1998). The end of cheap oil. Sci. Am., 3, 78-83.
[88] Motavalli, J. (2012). Flex-Fuel Amendment Makes for Strange Bedfellows". The New York Times.
[89] Iyayi, E. (2004). Changes in the cellulose, sugar and crude protein contents of agro industrial by-products fermented with Aspergillus niger, Aspergillus flavus and Penicillium sp. African Journal of Biotechnology., 3, 186-188.
[90] Jalk, D, Nerud, R, & Siroka, P. (1998). The effectiveness of biological treatment of wheat straw by white rot fungi. Folia Microbiol., 43, 687-689.

[91] Gordon, G, Mcsweeney, C, & Phillips, M. (1995). An important role for ruminal anaerobic fungi in the voluntary intake of poor quality forages by ruminants. In: Wallace RJ, Lahlou-Kassi A (Eds), Rumen ecology research planning. Proc. Workshop, Addis Ababa, Ethiopia. , 91-102.
[92] Rajoka, M, & Malik, K. (1997). Cellulase production by Cellulomonas biazotea cultured in media containing different cellulosic substrates. Bioresource Technology.,59, 21-27.
[93] Miller, G. (1959). Use of dinitrosalicyclic acid reagent for determination of reducing sugar. Anal. Chem., 31, 426-428.
[94] Lowry, O, Rosebrouch, N, Farr, A, & Randall, R. (1951). Protein measurement with the Folin Phenol reagent. J. Biol. Chem., 193, 265-275.
[95] Bollag, D, & Edelstein, S. (1991). Protein Methods. New York, Chichester, Brisbane, Toronto, Singapore, Wiley-Liss.
[96] Kluepfel, D., 1988. Screening of prokaryotes for cellulose and hemicelluloses degrading enzymes. Methods Enzymol., 160: 180-186.
[97] Hurst, P, Nielsen, J, Patrick, A, & Shepherd, M. (1977). Purification and Properties of a Cellulase from Aspergillus niger. Biochem. J., 165, 33-41.
[98] Immanuel, G, Dhanusa, R, Prema, P, & Palavesam, A. (2006). Effect of different growth parameters on endoglucanase enzyme activity by bacteria isolated from coir retting effluents of estuarine environment. Int. J. Environ. Sci. Tech., 3, 25-34.
[99] Ray, A, Bairagi, A, Ghosh, K, & Sen, S. (2007). Optimization of fermentation conditions for cellulose production by Bacillus subtilis CY5 and Bacillus Circulans TP3 isolated from fish gut. Acta I Chthyologica et piscatoria., 37, 47-53.
[100] Ojumu, T, Solomon, B, Betiku, E, Layokun, S, & Amigun, B. (2003). Cellulase Production by Aspergillus flavus Linn Isolate NSPR 101 fermented in sawdust, bagasse and corncob African Journal of Biotechnology., 2, 150-152.
[101] Acharya, P, Acharya, D, & Modi, H. (2008). Optimization for cellulase production by Aspergillus nigerusing saw dust as substrate. Afr. J. Biotechnol., 7, 4147-4152.
[102] Shamala, T, & Sreekantiah, K. (1987). Successive cultivation of selected cellulolytic fungi on rice straw and wheat bran for economic production of cellulases and D-xylanase. Enzyme Microb. Technol., 9, 97-101.
[103] Nigam, P, & Singh, D . (1994). Solid-state (substrate) fermentation systems and their applications in biotechnology Journal of Basic Microbiology., 34, 405- 423.
[104] Chahal, D. (1985). Solid-State Fermentation with Trichoderma reesei for Cellulase Production. Appl Environ Microbiol., 49, 205-210.
[105] Macris, B, Kekos, D, Evangelid, X, Galiotou-panayotou, M, & Rodis, P. (1987). Solid state fermentation of straw with Neurospora Crassa for CMCase and B-glucosidase production. Biotechnology Letters, 9, 661-664.
[106] Sul, O, Kim, J, & Park, S. Jun Son, Y.; Park, B.; Chung, D.; Jeong, C. and Han, I. (2004). Characterization and molecular cloning of a novel endoglucanase from Trichoderma sp. C-4. Appl Microbiol Biotechnol., 66, 63-70.
[107] Bakare, M, Adewale, I, Ajayi, A, & Shonukan, O. (2005). Purification and characterization of cellulose from the wild-type and two improved mutants of Pseudomonas fluorescens. African Journal of Biotechnology., 4, 898-904.

[108] Tong, C, Cole, A, & Maxwell, G. (1980). Purification and properties of the cellulases from the thermophilic fungus Thermoascus aurantiacus. Biochem. J., 191, 83-94.

[109] Inglin, M, Feinberg, B, & Loewenberg, J. (1980). Partial purification and characterization of a new intracellular beta-glucosidase of Trichoderma reesei. Biochem J., 185,515-519.

[110] Sadia, M, Aslam, N, Naim, S, Saleem, H, & Jamil, A. (2005). Purification of endoglucanase from Trichoderma harzianum Pakistan Journal of Life and Social Sciences, 3, 40-44.

[111] Dutta, T, Sengupta, R, Sahoo, S, Ray, S, Bhattacharjee, A, & Ghosh, S. (2007). A novel cellulase free alkaliphilic xylanase from alkali tolerant Penicillium citrinum: production, purification and characterization. Letters in Applied Microbiology, 44, 206-211.

[112] Bhikhabhai, R, Johansson, G, & Pettersson, G. (1984). Isolation of cellulolytic enzymes from Trichoderma reesei QM 9414. J Appl Biochem., 6, 336-345.

[113] Kim, D, Jeong, Y, Jang, Y, & Lee, J. (1994). Purification and characterization of endoglucanase and exoglucanase components from Trichoderma viride Journal of fermentation and bioengineering, 77, 363-369.

[114] Petrova, S, Bakalova, N, & Kolev, D. (2009). Properties of two endoglucanases from a mutant strain Trichoderma sp. M7 with potential application in the paper. Applied Biochemistry and Microbiology, 45, 150 155.

[115] Holt, S, & Hartman, P. (1994). A zymogram method to detect endoglucanases from Bacillus subtilis, Myrothecium verrucaria and Trichoderma reesei. Journal of industrial microbiology and Biotechnology., 13, 2-4.

[116] Liu, j. and Xia, w. (2006): Purification and characterization of a bifunctional enzyme with chitosanase and cellulase activity from commercial cellulase Biochemical Engineering Journal. , 30, 82-87.

[117] Shindia, A. (1990). Studies on fungal degradation of composts in Egypt. Ph.D. Thesis, Faculty of Science, Zagazig Univ., Zagazig, Egypt.

[118] Kalra, M, Sidhu, M, & Sandhu, D. (1986). Partial purification, characterization and regulation of cellulolytic enzymes from Trichoderma longibrachiatum. Journal of Applied Microbiology., 61, 73-80.

[119] Ülker, A, & Sprey, B. (1990). Characterization of an unglycosylated low molecular weight glucan glucanohydrolase of Trichoderma reesei FEMS Microbiology Letters, 69, 215- 219., 1, 4.

[120] Gashe, B. (1992). Cellulase production and activity by Trichoderma sp. A-001 Journal of Applied Microbiology., 73, 79-82.

[121] Mandels, M, & Reese, E. (1964). Fungal cellulases and the microbial decomposition of cellulosic fabric. Dev. Ind. Microbiol., 5, 5-20.

[122] Mandels, M, & Reese, E. (1963). Inhibition of cellulases and β- g1ucosidases in Advances in Enzymic hydrolysis of cellulose and related materials. 115- 158. (Reese, E. T., Ed., Pergarnon Press, london , 299.

[123] Catriona, A, Sheila, I, & Thomas, M. (1994). Characterization of a D-glucosidase from the anaerobic rumen fungus Neocallimastix frontalis with particular reference to attack on cello-oligosaccharides. J. Biotechnol., 37, 217-227.

[124] Steiner, J, Socha, C, & Eyzaguirre, J. (1994). Culture conditions for enhanced cellulose production by a native strain of Penicillium purpurogenum. World J Microbiol and Biotech., 10, 280-284.

[125] Ali, S. and Sayed, A. (1992): Regulation of cellulase biosynthesis in Aspergillus terreus. World J. Microbiol. Biotechnol., 8, 73–75.

[126] Kotchoni, O, Gachomo, W, Omafuvbe, B, & Shonukan, O. (2006). Purification and Biochemical Characterization of Carboxymethyl Cellulase (CMCase) from a Catabolite repression insensitive mutant of Bacillus pumilus. International journal of agriculture and biology., 8, 286-292.

[127] Kim, C. (1995). Characterization and Substrate Specificity of an Endo-b-D Glucanase I (Avicelase I) from an Extracellular Multienzyme Complex of Bacillus circulans. Applied and Environmental Microbiology, 61, 959-965.

Chapter 5

# Recombinant Cellulase and Cellulosome Systems

Andrew S. Wieczorek, Damien Biot-Pelletier and Vincent J.J. Martin

Additional information is available at the end of the chapter

http://dx.doi.org/10.5772/54225

## 1. Introduction

The non-renewable fossil resources currently exploited by the oil and gas industries are the objects of growing concern owing to their finite supply and contribution to global warming. Lignocellulosic biomass is a sustainable alternative to fossil resources, and has the added advantage of not competing with human and animal nutrition. Indeed, lignocellulosic biomass, in particular its main polymer component cellulose, is a potential carbon source for the production of fuels and commodity chemicals in microbes.

Hydrolysis of cellulose polymer molecules to liberate the readily fermentable glucose they contain is a necessary step in their use as feedstock by fermenting organisms. The hydrolysis of cellulose is typically carried out by glycoside hydrolase enzymes termed cellulases, and produced by specialist microorganisms. Organisms that naturally feed on and hydrolyse cellulose are mainly found among filamentous fungi, such as the highly exploited *Trichoderma reesei*, and obligate anaerobic bacteria such as those of the *Clostridium* genera. The complete breakdown of cellulose to glucose requires the cooperation of three different types of cellulases. Endoglucanases (EGLs) cleave amorphous cellulose randomly at endo sites to release cellodextrins of various lengths. Cellobiohydrolases (CBHs), on the other hand, are required for the hydrolysis of crystalline cellulose, and release cellobiose by acting at the reducing and non-reducing ends of cellulose strands [1, 2]. Finally, β-glucosidases (BGLs) produce glucose from the hydrolysis of the cellobiose and cello-oligomers produced by EGLs and CBHs. The three types of enzymes are believed to act synergistically. EGLs cleave at random inside strands, creating termini for CBHs, which in turn contribute to loosening of cellulose crystallinity, making further material available to EGLs [2]. Some cellulases, as well as other proteins involved in cellulose degradation, carry a cellulose-binding domain (CBD) that acts to tether them to their polymeric substrate, and allows them to processively degrade cellulose by crawling along its strands [3]. Certain organisms assemble their cellulases on their cell surface as multi-enzyme complexes termed cellulosomes, notably to enhance synergy between enzymes and promote substrate channelling [4].

The bionconversion of cellulose to biofuels or commodity chemicals must proceed through several steps. Following pre-treatment of the biomass, cellulose is hydrolyzed as described in the above paragraph. The glucose liberated by cellulose hydrolysis can then be fed to microbes that produce compounds of interest, for example the yeast *Saccharomyces cerevisiae*, which ferments it to ethanol. Doing these two steps one after the other is known as sequential hydrolysis and fermentation (SHF). It requires the addition of costly cellulase cocktails separately produced by fungi, and accumulation of glucose during the hydrolysis step leads to end product inhibition. The capital cost of having multiple separate steps, and the time required for sequential conversion processes further reduce the profitability of sequential hydrolysis and fermentation [5]. Simultaneous saccharification and fermentation (SSF) reduces the number of steps and alleviates the end-product inhibition issue, however it still requires the addition of exogenous cellulases [6]. To further reduce costs, a strategy known as consolidated bioprocessing (CBP) has been proposed, which entails the *in situ* production of cellulases by the fermenting organism. This strategy consolidates enzyme production, hydrolysis and fermentation into a single step. However, CBP requires an organism efficient at both degrading cellulose and fermenting glucose to a single product at high titers. Such an organism does not exist in nature [7]. To overcome this obstacle, two solutions can be envisioned. Efficient cellulose degraders may be engineered to produce chemicals of interest, or alternatively, organisms that natively produce such compounds can be endowed with recombinant cellulase genes.

Thus, the recombinant expression of cellulases, or cellulase systems, enables CBP. It may also be used to reduce exogenous enzyme loads required by SSF, and may have benefits for the production of the cellulase cocktails used in both SHF and SSF. The recombinant and heterologous expression of cellulases in microorganisms may also benefit other industries. The textile industry, for example, uses cellulases to create stonewashing effects on cellulose-derived clothing fibres. Use of cellulase-expressing lactic acid bacteria, on the other hand, is of interest for the ensilage of hay fed to livestock. For these reasons, considerable research has been done to engineer organisms that express recombinant cellulases and cellulase systems. The aim of this chapter is to review the progress made in the engineering of such organisms. We first review the production of cellulases expressed as freely secreted or cell surface-anchored enzymes, and divide our discussion based on the types of organisms engineered (yeast, bacteria, then fungi). We then put special emphasis on the production of artificial recombinant cellulosomes and cellulosome-inspired architectures, outlining the different manners in which they can be assembled, and which microorganisms were used to do so.

## 2. Cell surface-anchored and secreted recombinant cellulase systems

The scientific literature is ripe with examples of secreted or surface-anchored recombinant cellulases and cellulase systems expressed in yeast, bacterial and fungal hosts. Most research has focused on a handful of organisms, namely *Saccharomyces cerevisiae*, the enteric bacteria *Escherichia coli* and *Klebsiella oxytoca*, the gram-negative bacterium *Zymomonas mobilis*, and the cellulolytic fungus *Trichoderma reesei*. Other species have garnered less attention, yet represent an interest to the field and should not be dismissed.

This section focuses on work aimed at producing organisms that can efficiently degrade cellulose via the expression of recombinant cellulases. Because the recombinant expression of cellulases was extensively reviewed in a number of publications in the last decade [8-10], the text is centered on the most significant outcomes, and provides an overview of the most recent work.

## 2.1. Expression of cellulases in yeast

Attempts at expressing recombinant cellulases in yeast abound owing to the traditional role of the brewer's yeast *Saccharomyces cerevisiae* in ethanol production. The use of other yeast species for recombinant expression of cellulases is also discussed in this section, namely species that display interesting metabolic capabilities or stress tolerance characteristics.

### *2.1.1. Recombinant cellulase expression in Saccharomyces cerevisiae*

A significant proportion of recombinant cellulase expression studies were performed in yeast, and almost all of that work was done in *Saccharomyces cerevisiae*. The millennia-old utilization of this organism for ethanol production, its relatively well-studied physiology, and the diversity of readily available tools for its genetic manipulation mean that it is an important candidate for the engineering of a cellulose-degrading ethanologen.

Since the 1990s, numerous cellulases from various bacterial and fungal sources were cloned and expressed in *S. cerevisiae*, and those have been reviewed elsewhere [8, 9]. Over the last thirteen years, a few studies representing significant progress towards the production of a cellulose-fermenting yeast strain were published. Cho and coworkers [11, 12] reported an early example of a recombinant yeast strain that could functionally express several cellulases. Using δ-integration, they inserted multiple copies of two cellulase genes - encoding a bifunctional endo/exo-glucanase and a BGL- into the chromosomes of *S. cerevisiae*. The recombinant organism displayed enhanced growth on cellooligosaccharides when compared to wildtype, and required reduced loads of exogenous cellulases when applied in SSF [12]. However, levels of cellulase expression were deemed low, and did not enable growth and ethanol production using cellulose as the sole carbon source. A later study similarly expressed the three types of cellulases required for cellulose degradation in *S. cerevisiae* [13]. The EGL and CBH, from *Trichoderma reesei*, and the BGL, from *Aspergillus aculeatus*, were co-displayed as α-agglutinin fusions on the surface of yeast cells, enabling the liberation of glucose from phosphoric acid swollen cellulose (PASC), and fermentation to ethanol when the cells were pre-grown in rich media. Den Haan and coworkers [14] reported similar accomplishments, co-expressing an EGL from *T. reesei* and a BGL from the yeast *Saccharomycopsis fibuligera* in *S. cerevisiae*. This study was allegedly the first report of direct conversion of cellulose to ethanol by cellulase-expressing yeast, as it was reported that the engineered strain could grow and produce modest yields of ethanol (1.0 g/L in 192 hours) from PASC in media also containing yeast extract and peptone (YP-PASC). A study published almost simultaneously by the same authors reported the low level expression of CBHs in yeast [15], but expression of these enzymes in the PASC-fermenting BGL/EGL background was not reported.

Following these milestone studies, other groups reported on the expression of cellulases in *S. cerevisiae* and their use for fermentation of cellulose to ethanol. Jeon and coworkers reported the expression of EglE from *Clostridium thermocellum* and BGL1 of *S. fibuligera* in the budding yeast. The resulting yeast strain could produce ethanol from carboxymethyl cellulose (8.56 g/L, 16 hours), β-D-glucan (9.67 g/L, 16 hours) and PASC (7.16 g/L, 36 hours) after pre-culturing in synthetic galactose medium and extensive washing in minimal media [16]. This was a progress compared to previous studies, in that it did not require yeast extract or peptone to produce ethanol from cellulosic substrates. Another study [17] compared the performance of two recombinant yeast strains in directly converting cellulose to ethanol in YP-PASC medium. A BGL from *A. aculeatus* was anchored to the cell surface, while an EGL and a CBH were either anchored or secreted. Higher ethanol yields were obtained when all three enzymes were surface-anchored. These results suggested that this configuration enhances the ability of yeast to degrade cellulose and use the resulting sugars in a manner reminiscent of cellulosome-enzyme-microbe complexes (discussed in Section 3).

Direct conversion of cellulose to ethanol poses the problem of finding the optimal ratios of the different types of cellulase. A novel strategy, termed cocktail δ-integration was recently proposed to address this issue [18]. This strategy involves the simultaneous transformation and integration in the yeast chromosomes of BGLs, EGLs and CBHs on a single DNA fragment with a single selection marker. Fragments are designed to carry varying numbers of each cellulase gene. Integrants are then compared in their ability to degrade cellulose, and those with the best ratios can be identified. The procedure can be repeated several times using different selection markers. After three rounds of cocktail δ-integration, Yamada and coworkers [18] were able to generate a strain with twice the activity on PASC, but half the number of cellulase genes than a similar strain generated using a conventional method. These results strongly argue for a successful optimization of cellulase ratio. The activity of the ratio-optimized strain was further improved by making it diploid [19]. The optimized diploid showed an ability to produce ethanol directly from PASC (7.6 g/L in 72 hours) or pretreated rice straw (7.5 g/L) in yeast peptone (YP) medium without addition of exogenous enzymes. This was the first report of direct conversion to ethanol of agricultural waste residue without exogenous enzyme addition by recombinant cellulase-expressing yeast [19]. Other strategies used to incorporate enzymes at specific ratios into artificial cellulosomes using yeast consortia are discussed later in this chapter (Section 3.4.1).

Two independent studies gave examples of improved SSF using cellulase-expressing yeast. One study [20] reported the transformation of an industrial strain with a BGL-carrying plasmid, enabling the use of cellobiose as the sole carbon source and its conversion to ethanol, producing 3.3 g/L in 48 hours. When supplementing with exogenous cellulases, the strain was shown to produce 20 g/L of ethanol from pre-treated corncobs, a yield similar to outcomes obtained with the parent strain supplemented with additional BGL. Another SSF study [21] reported the production of 7.94 g/L of ethanol in 24 hours from barley β-D-glucan using yeast co-displaying a BGL and an EGL from *Aspergillus oryzae*.

In recent years, a few thermotolerant enzymes have been expressed in *S. cerevisiae*. For example, BGL4 from *Humicola grisea* was recently cloned in the budding yeast [22]. Interestingly, the recombinant enzyme displayed resistance to glucose inhibition in addition to thermotolerance. Others have reported on the expression of thermotolerant cellulases in yeast using a mutagenesis and recombination strategies rather than a discovery approach to further improve stability and activity of the recombinant enzymes [23-25].

Inadequate secretion of cellulases by recombinant yeast is an obstacle to their successful application in an industrial context. To address this issue, a library of approximately 4800 non-essential deletion mutants was systematically transformed with a plasmid carrying an endoglucanase gene from the bacterium *C. thermocellum* [26]. Mutants were compared in their ability to degrade carboxymethyl cellulose, and 55 of them showed increased activity. The mutants covered a large spectrum of cellular functions, including transcription, translation, phospholipid synthesis, endosome/vacuole function, ER/Golgi function, nitrogen starvation response, and the cytoskeleton. The effect of a subset of these mutations was tested on the level of activity of another cellulase, a BGL from *A. aculeatus*. Interestingly, five out of the nine mutations tested increased BGL activity in addition to EGL activity, suggesting that certain mutations may increase the secretion level of several cellulases, and potentially all enzymes within a cellulase system [26].

### 2.1.2. Recombinant cellulase expression in other yeast species

While most studies expressing recombinant cellulase systems in yeast have used *Saccharomyces cerevisiae*, other species, superior to brewer's yeast in some respects, have also been used.

The yeast *Scheffersomyces stipitis* (formerly *Pichia stipitis*) is one of the organisms considered for its potential in the bioconversion of lignocellulosic biomass, owing to its native cellulase activity, but foremost to its pentose-fermenting capabilities. Indeed, hemicellulose, the second most abundant sugar polymer of plant cell walls after cellulose, is composed largely of xylose, which *S. cerevisiae* cannot ferment. *S. stipitis*, on the other hand, produces the largest yields of ethanol from xylose that have been observed to date [27]. *S. stipitis* naturally consumes lignocellulosic biomass, therefore cellulase activity, notably β-glucosidase activity, has been detected in this organism [28], while its genome was found to encode several putative cellulolytic enzymes [29]. Yet, during the development of molecular genetics tools for *S. stipitis*, recombinant cellulases were used as reporters of protein expression [30].

*Saccharomyces cerevisiae* is generally not viable in conditions of temperature optimal for cellulase activity. Indeed, cellulases from the common cellulolytic microbes *C. thermocellum* and *T. reesei* are found to lose most of their activity at temperatures below 40°C [31], while *S. cerevisiae* grows poorly above 38°C [32] and could not so far be engineered to remain productive at temperatures that exceed 42°C [33]. In addition, acids are commonly used in the pretreatment of lignocellulosic biomass, while both high temperatures and acidic conditions can be used in preventing contamination during fermentation. For these reasons, expression of recombinant cellulase systems has been attempted in a few stress tolerant species of yeast. For example, the thermotolerant *Kluyveromyces marxianus* was used in a number of SSF studies in which cellulases were added exogenously [34-37]. The strain was subsequently engineered to express three thermostable cellulases, endowing it with the ability to grow at 45°C on both cellobiose and carboxymethyl cellulose and to ferment cellobiose to ethanol [38]. The multi-stress tolerant *Issatchenka orientalis* was also successfully engineered for recombinant cellulase expression. This organism is tolerant to acid, salt and elevated temperature, in addition to being ethanol tolerant, making it a suitable candidate for cellulose bioconversion [39]. Kitagawa and coworkers [40] provided the first report of heterologous gene expression in *I. orientalis*, isolating and cloning the necessary auxotrophy markers and building a recombinant cassette for the production of *A. aculeatus* BGL. The engineered strain showed BGL activity and was able to grow and produce ethanol on cellobiose in conditions of elevated temperature, acidity and salinity. SSF trials using this strain achieved measureable ethanol outputs, albeit at lower levels than what was obtained with the parental strain supplemented with exogenous BGL. Still, to achieve similar yields, reduced BGL supplementation was required for the recombinant strain.

## 2.2. Expression of cellulases in bacteria

This section reviews recent research aimed at expressing recombinant cellulases in bacteria. Although the workhouse and longtime protein overproducing *Escherichia coli* has received significant interest, several other species with specialized functions have also been exploited. These functions include: the ability to assimilate cellulose-derived oligosaccharides, native production of biofuel molecules or organic acids, and thermophilicity.

### *2.2.1. Recombinant cellulase expression in enteric bacteria*

The enteric bacterium *E. coli* has a long history of being used for the expression of recombinant proteins, and numerous tools for the genetic engineering of this organism are readily available. Furthermore, *E. coli* has among the simplest and cheapest growth requirements. It is thus an attractive canvas for the engineering of a cellulose-utilizing industrial strain. Therefore, it comes to no surprise that studies have reported the heterologous expression of cellulase systems in this organism. Significant advances have also been reported in *Klebsiella oxytoca*, a bacterium related but superior to *E. coli* in its native ability to assimilate and use cello- and xylo-oligosaccharides.

Wildtype *E. coli* and *K. oxytoca* are not prolific ethanologens and neither have cellulolytic activity. The classical strategy to turn these organisms into ethanol producers is to endow them with an alcohol dehydrogenase and a pyruvate decarboxylase genes from the ethanologenic bacterium *Zymomonas mobilis* (Section 2.2.2) [41]. It is normally with this background that enteric bacteria have been used for recombinant cellulase expression. Several papers over the course of the last twenty years have reported the engineering of *E. coli* and *K. oxytoca* in this manner [42-47]. The most advanced examples report the expression of the endoglucanase genes *celY* and *celZ* from the phytopathogenic bacterium *Erwinia chrysanthemi* in an ethanologenic *K. oxytoca* background [46, 47]. This recombinant cellulase system, in conjunction with the native BGL activity of *K. oxytoca* enabled the direct conversion of crystalline cellulose to ethanol with addition of exogenous cellulases [46], while amorphous cellulose could be readily converted to ethanol without exogenous cellulase supplementation [47]. However, as is the case for reports of direct cellulose-to-ethanol conversion by yeast, these successes depended on the presence of yeast extract and peptone in the fermentation medium.

More recently, a proof-of-concept study by Bokinsky and coworkers [48] reported the expression of complete sets of cellulases and hemicellulases in *E. coli* for the conversion of lignocellulosic biomass to second-generation biofuels. In this study, a library of EGLs was tested for expression in *E. coli*, while collections of BGL and xylobiosidases were evaluated for their ability to enable growth of *E. coli* on cellobiose and xylobiose, respectively. The best EGL and BGL genes were introduced into *E. coli* to generate a cellulose-degrading strain. The best xylobiosidase was similarly combined with a previously identified xylanase to generate a hemicellulose-degrading strain. Growth on ionic liquid-pretreated lignocellulosic feedstock (switchgrass, eucalyptus and yard waste) was demonstrated. Combining both strains allowed enhanced growth on all substrates. The strains were further engineered to express one of three operons for the production of advanced biofuel molecules (fatty acyl ethyl esters, butanol or pinene) from ionic liquid-pretreated switchgrass, achieving modest yields. This study is the first report of a complete cellulose-to-biofuel conversion in bacteria using natural feedstock. Moreover, no exogenous cellulases were added, and all hydrolysis and fermentation experiments in this study were performed in minimal media with cellulose or hemicellulose as the sole carbon source.

### 2.2.2. Recombinant cellulase expression in *Zymomonas mobilis*

*Zymomonas mobilis* is an ethanologenic gram-negative bacterium. Unlike *S. cerevisiae*, it converts glucose to ethanol via the Entner-Doudoroff pathway, enabling ethanol yields that could more closely match theoretical yield values than the classical glycolytic pathway. It is considered superior to brewer's yeast in other respects. Indeed, it has higher tolerance to ethanol, enabling superior yields, which it produces with high productivities [7, 49-55]. Therefore, several reports of recombinant cellulase expression in *Z. mobilis* have been published [55-59]. Among early reports of recombinant cellulase expression in *Z. mobilis* [56-58], only one succeeded in exporting an EGL to the extracellular milieu using the protein's native signal [56]. In that study, approximately 10% of the EGL protein was found to be extracellular, while most of the cell-associated activity was found in the periplasm [56]. Recent studies fused recombinant cellulases to native *Z. mobilis* export signals in an attempt to direct a larger proportion of the enzymes to the extracellular milieu. In one study, a BGL from *Ruminococcus albus* was fused to the glucose–fructose oxidoreductase and gluconolactonase export signals of *Z. mobilis*, resulting in the secretion of only 4.7% and 11.2% of the protein, respectively. The resulting strain was able to use cellobiose and ferment it to ethanol [55]. A more recent study used two different secretion signals native to

*Z. mobilis*, and suggested to use distinct pathways. These endogenous signals were fused to the catalytic domain of two *Acidothermus cellulolyticus* EGLs, enabling the export of 40%-50% of the recombinant cellulases to either the periplasm or extracellular milieu [59]. This latter study did not report on the ability of the strains to grow on or convert cellulosic substrates. Interestingly, it provided a confirmation to an earlier study that suggested the presence of endogenous cellulase activity in *Z. mobilis* [60].

### *2.2.3. Recombinant cellulase expression in other bacterial hosts*

Other bacterial species with useful industrial properties have been used for the expression of recombinant cellulases. Species such as *Clostridium acetobutylicum* and *Clostridium beijerinckii* can be used in the industrial scale production of solvents and biofuels in the acetone-butanol-ethanol (ABE) process [61]. Enthusiasm for biofuels and synthetic biology in recent years has renewed interest for the high yields of solvents, in particular butanol, achieved by these organisms. The classical source of carbon for the ABE process was potato starch, however recent research has been aimed at enabling the use of cellulose, a more sustainable and industrially suitable carbon source, by solventogenic *Clostridium*. The genome of *C. acetobutylicum* encodes genes for putative cellulosome components, which will be discussed later in this chapter (Section 3.3.2). However, growth of this microbe, while successful on hemicellulose [62, 63] has so far not been observed with cellulose as the sole carbon source [64], despite observations that various substrates induce the expression of cellulases in *C. acetobutylicum* [65, 66]. Therefore, solventogenic *Clostridium* were engineered to express recombinant cellulases. Most efforts were aimed at reconstituting functional *Clostridium cellulolyticum* cellulosomes in *C. acetobutylicum*, but expression of isolated cellulases was also attempted. In an early study, an EGL from the cellulolytic bacteria *Clostridium cellulovorans* was expressed in *C. acetobutylicum* [67]. While the resulting strain could degrade carboxymethyl cellulose in Congo Red plate assays, it failed to grow on cellulose as the sole carbon source. Mingardon and coworkers expressed six *C. cellulolyticum* cellulases in *C. acetobutylicum* and found that three enzymes, those with lower molecular weights, were successfully secreted [68]. The larger enzymes failed to generate viable clones, or led to accumulation of cellulase protein in the cytoplasm. In a subsequent study, the same group reported the successful secretion of large cellulases by fusing them to sequences of scaffoldins and cellulose binding modules of *C. cellulolyticum* [69]. The related species *C. beijerinckii* was also used for the heterologous expression of recombinant cellulases. Expression of an EGL from the cellulolytic fungus *Neocallimastix patriciarum* in *C. beijerinckii* yielded results that resembled those observed with *C. acetobutylicum*. Indeed, the recombinant *C. beijerinckii* strain displayed cellulolytic activity in Congo Red plate assays, but failed to grow on cellulose. Interestingly, the fungal EGL improved growth and solvent yields of the microbe on lichenan, a polymer of glucose similar to cellulose [70].

Lactic acid bacteria (LAB) have also served as hosts for recombinant cellulase expression. The interest of LAB lies in their potential as silage inocula, probiotics, and industrial lactic acid producers. Several LAB species, including *Lactobacillus plantarum* and *Lactococcus lactis* have been engineered for the improved lactic fermentation of forage by expressing cellulose-degrading enzymes. Early studies reported the successful expression and secretion of functional EGLs from plasmids [71] or from the chromosome [72] in *L. plantarum*. Chromosome integration of a *Bacillus* sp. EGL in *L. plantarum* was later shown to elicit increased acidification of forage in micro-ensiling experiments [73]. Similarly, *L. lactis* was transformed with a cellulase gene from the rumen fungus *Neocallimastix* sp. [74]. The recombinant *L. lactis* strain enhanced the digestibility of forage when used in ensiling experiments. *Lactobacilli* species *L. gasseri* and *L. johnsonii*, natural inhabitants of the mammalian gastrointestinal tract, were also engineered to express a *C. thermocellum* endoglucanase [75]. The aim of this study was to generate probiotics that would facilitate digestion of plant cell walls by monogastric animals, thus alleviating the need for the onerous supplementation of animal feed with exogenous cellulases. The resulting strains displayed cellulase activity on carboxymethyl cellulose, and had characteristics desirable for probiotics. A lactate dehydrogenase-deficient strain of *L. plantarum* was later engineered to express a *C. thermocellum* EGL, allowing the successful hydrolysis and conversion of barley β-D-glucan to lactic acid in anaerobic conditions, achieving best yields with addition of exogenous BGL [76].

We have already mentioned the relevance and interest of thermotolerant or thermophilic enzymes for the bioconversion of cellulose. While tools for the genetic engineering of thermophilic bacteria are still in their infancy [77], one example of cellulase expression is found in the thermophile *Thermoanaerobacterium saccharolyticum* [78]. In this study, development of recombinant protein expression systems used cellulases and other glycoside hydrolases from *C. thermocellum* as test proteins, and cellulase activity was detected.

## 2.3. Expression of cellulases in fungi

Several species of fungi are superior protein secretors, and as such show high potential for the industrial-scale production of enzymes. Not surprisingly, the cellulase cocktails used in industry for the bioconversion of cellulose or for the treatment of textile fibers are typically produced by cellulolytic fungi [79]. The organism most commonly used for this purpose is the filamentous fungus *Trichoderma reesei*, because of the high titers of cellulase enzymes that it secretes [80]. Recombinant approaches have been applied to enhance the production of native cellulases or to express heterologous cellulolytic enzymes in *T. reesei* and other fungi.

To increase yields of EGL produced by *T. reesei*, Miettinen-Oinonen and Suominen reported a strategy whereby the native *cbh2* locus was disrupted to redirect the secretory capabilities of the fungus towards other proteins [81]. This CBH-deficient background was transformed with constructs of *T. reesei* EGL genes placed under the control of the strong *cbh2* promoter. These modifications, coupled to an increase in EGL copy number, were successful in augmenting the levels of secreted EGL, and in increasing the performance of the *T. reesei*-secreted enzyme in stonewashing treatment of denim fabric. A follow-up study by the same group tested the effect of promoter swapping, deletion of native enzymes, and copy number increase on the level of CBH secretion, yielding comparable results [82]. Another approach aimed at increasing the activity of the *T. reesei*-secreted cellulase cocktail was to fuse an *A.*

*cellulolyticus* EGL domain to native CBH expressed in *T. reesei*. The resulting bi-functional enzyme increased the saccharification yields of *T. reesei* [83].

Several studies have reported the heterologous expression of thermophilic fungal cellulases in mesophilic fungi, notably *T. reesei*, *Aspergillus oryzae* and *Humicola insolens* (reviewed in [84]). For example, protein variants of the Cel12A enzyme of *T. reesei* rationally designed for increased thermal stability and activity were expressed in the efficient protein secretor *Aspergillus niger* [85]. In another study, the cellobiohydrolase Cel7A from *T. reesei* was expressed in *A. niger*, and mass spectrometry was used to compare N-glycosylation between the recombinant and the native protein. The cellobiohydrolase contained six times more N-linked glycans when expressed in *A. niger*, and its activity was reduced, underlining the critical effect of post-translation modifications on recombinant cellulases [86]. Recently, a library of EGLs from various fungi were cloned and expressed in *A. niger* [87]. Both activity and level of expression were compared to that of TrCel5A, one of the major endoglucanases from *T. reesei*. This screen identified three EGLs, from species *Aureobasidium pullulans*, *Gloeophyllum trabeum* and *Sporotrichum thermophile* with expression levels and hydrolysis performances superior to those of the *Trichoderma* enzyme [87].

## 3. Recombinant cellulosomes

The degradation of recalcitrant cellulosic substrates into fermentable carbohydrates requires multiple catalytic activities [4]. Many cellulolytic fungi are capable of degrading crystalline cellulose by secreting cocktails of free hydrolytic enzymes [88]. Alternatively, the hydrolysis of cellulosic substrates can be carried out by macromolecular enzyme complexes [4]. The incorporation of enzymes in a larger multi-enzyme complex yields several benefits associated with substrate channeling as well as synergy among neighboring enzymes [89]. Substrate channeling refers to the flow of intermediate metabolites from one reaction to another, where individual catalytic activities are co-localized in a central protein scaffold. In the case of cellulose hydrolysis, longer chain polysaccharides produced by non-processive cellulases become the substrate for processive cellulases, which can produce short chain cellodextrins and cellobiose as primary products. Enzyme synergy results when the sum of individual enzyme activities is augmented by their incorporation in multi-enzyme complexes. From a biotechnological perspective, optimizing the spatial organization of enzymes through co-localization can greatly enhance the channeling of hydrolysis intermediates to enzymes that will use them as substrates. A number of cellulolytic bacteria have evolved to assemble multi-enzyme complexes such as cellulosomes. Cellulosomes have become inspiration for the engineering of recombinant complexes with defined enzyme compositions. For instance, the thermophilic bacterium *C. thermocellum*, which is documented to have one of the most efficient system for cellulose hydrolysis [89], produces one of the most thoroughly studied and well-characterized cellulosomes. The engineering of multiple cellulases into macromolecular cellulolytic complexes is a strategy that has been adopted by a number of research groups in the development of microorganisms that can degrade cellulose and produce commodity chemicals and biofuels.

### 3.1. Nature's building blocks for engineering recombinant cellulosomes

Cellulosomes are cellulose-degrading protein complexes comprised of a multitude of hydrolytic enzymes with varying catalytic activities that associate with a central scaffold protein [90]. The variability in architecture of cellulosomes from different organisms has been a significant source of inspiration for the engineering of protein scaffolds and multi-enzyme complexes [91-94]. The assembly of the cellulosome complex is mediated via non-covalent interactions between non-catalytic dockerin and cohesin domains. These domains serve as the building blocks that hold the complex together and dictate its architecture. Two characteristics of a dockerin and cohesin pair determine the specificity of the interaction: the species from which they are derived, as well as the type of interaction. Type 1 and type 2 cohesins from a single organism do not interact with dockerins of the opposite type (e.g. type 1 cohesins do not interact with type 2 dockerins, and vice-versa). In the case of *C. thermocellum*, type 1 dockerins and cohesins mediate the interaction between enzymes and scaffold proteins, while type 2 dockerins and cohesins mediate binding of scaffolds and cell surface anchor proteins. Cellulosomal enzymes carry type 1 dockerin domains which bind any of the nine type 1 cohesin domains found on the central scaffold protein CipA [95]. Cellulosomal scaffolds such as CipA typically contain a CBD that brings the complex in close proximity to the cellulose fibers, allowing the different cellulases to act in synergy on the crystalline substrate. CipA protein also carries a type 2 dockerin domain, which interacts with type 2 cohesins located on cell wall anchor proteins OlpB and SdbA [96, 97]. These anchor proteins ensure the attachment of the complex on the cell surface. In addition, cohesin and dockerin domains derived from different organisms do not bind with one another. Therefore, cohesins and dockerins from different species as well as those of different types have become the building blocks used by researchers to engineer custom-designed recombinant cellulosomes or cellulosome-inspired complexes with precise compositions. The strategies adopted by most researchers in this effort can be divided into three categories discussed in subsequent sections. These include (i) the production of recombinant enzymes and scaffolds in host strains followed by their purification and assembly *in vitro* (Figure 1A), (ii) the production of all components in a single strain resulting in the *in vivo* assembly of resulting complexes in the culture supernatant (Figure 1B), and (iii) the surface-tethering of scaffolds towards the *in vivo* assembly of artificial cellulosomes on the cell surface of the host organism (Figure 1C).

### 3.2. *In vitro* assembly of recombinant cellulosomes

The assembly of custom-designed cellulosomes initially involved the production of individual components in an organism of choice, followed by their purification and assembly *in vitro*. Desirable characteristics for a bacteria designed to overexpress individual components include ease of manipulation of the organism, and low endogenous proteolytic activity. Since multiple strains are used to generate individual components, this strategy is not limited to a single organism being used for the production of each recombinant subunit, since further purification and *in vitro* assembly of the final complex is required (Fig. 1A).

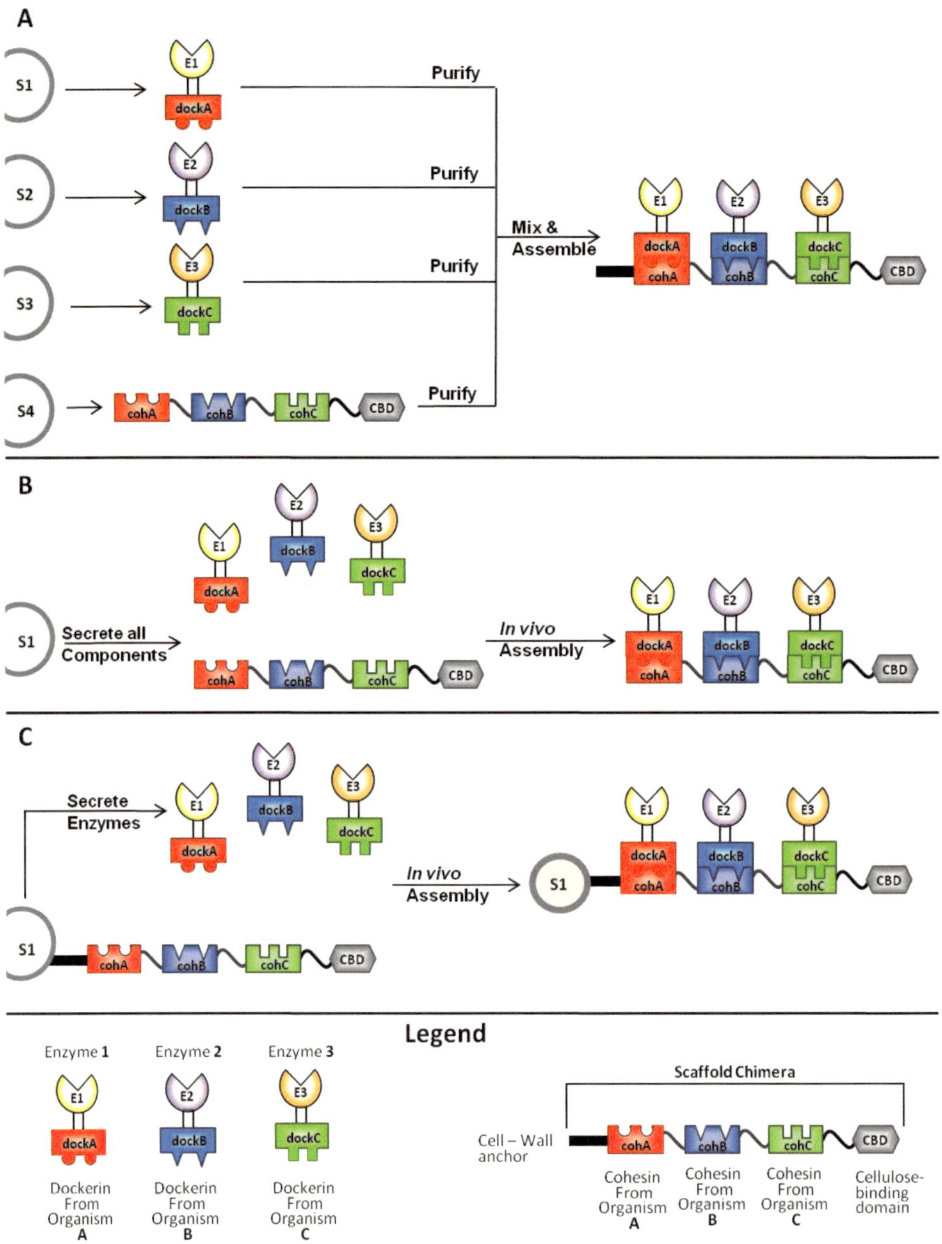


**Figure 1. Strategies for the assembly of artificial cellulosome complexes.** (A) Enzyme-dockerin fusions and scaffold chimeras are produced by different strains of a host organism (e.g. S1, S2, S3, S4), purified, and subsequently assembled *in vitro*. (B) Enzymes and scaffold subunits are secreted by a single host organism into the culture supernatant where they self-assemble into cellulosomes *in vivo*. (C) A host organism tethers a scaffold to its surface while secreting recombinant enzyme-dockerin fusions, resulting in the *in vivo* assembly of the cellulosome complex on the cell surface.

### 3.2.1. Expression of cellulosome components in E. coli

Early work on the *in vitro* assembly of cellulosomes focused mostly on demonstrating the effects of having cellulase enzymes bound to a scaffold on activity towards cellulose. In a study by Kataeva and coworkers, the EGL CelD was shown to bind stoichiometrically with fragments of the CipA scaffold protein, and CelD-CipA complexes showed increased activity on cellulose compared with free CelD enzyme. A major observation was that the activity of the complex was dependent on the presence of a cellulose binding domain (CBD), not necessarily the amount of CelD present. The authors hypothesized that the CBD located on the scaffold protein was either indirectly contributing to the hydrolysis process by optimally positioning CelD to act on the crystalline substrate, or that it was playing a more direct role, participating in the partial decomposition of the substrate and ultimately, allowing access to CelD [98]. A subsequent study by Ciruela and colleagues revealed that the binding of another EGL, CelE, with full length CipA, resulted in the assembly of artificial cellulosomes with increased activity on crystalline cellulose compared to free enzymes [99]. Interestingly, although the CBD of CipA was capable of binding both crystalline and amorphous cellulose, the increase in activity observed when CelE was complexed with CipA was only observed on the former, suggesting the pivotal role of the scaffold-enzyme complex in degrading the crystalline substrate. Both studies conducted by Kataeva and Ciruela involved the incorporation of a single enzyme into artificial cellulosomes. Murashima and coworkers used a truncated version of the *C. cellulovorans* scaffold protein CbpA (Mini-CbpA) and three enzymes, EngE, EngH, and EngS, for the *in vitro* assembly of artificial cellulosomes containing combinations of two enzymes [100]. Synergy was affected by both the type and stoichiometric ratios of enzyme used. Optimal combinations of enzymes were determined based on increased activity on crystalline cellulose. In this case, however, the effects of relative enzyme positioning within the complex could not be deduced due to the non-specific binding of each enzyme with any of the two cohesins present on the scaffold. The multiple cellulase activities required to degrade crystalline cellulose and the possibility to optimize their positioning within an artificial cellulosome prompted the construction of recombinant protein scaffolds using cohesins with different specificities.

Initial work describing the construction of chimeric scaffolds was carried out by Fierobe and coworkers, where the fusion of cohesins derived from the cellulosomes of *C. thermocellum* and *C. cellulolyticum* were used to engineer complexes with dual enzyme activities [101]. The authors engineered a total of four scaffolds that contained two divergent cohesins positioned at various locations relative to the CBD. Two *C. cellulolyticum* cellulases, CelA and CelF, were engineered to contain either native or *C. thermocellum* dockerins. All components were over-produced in *E. coli,* purified and assembled *in vitro* into three-component cellulosomes. The authors once again demonstrated the necessity of the CBD for increased hydrolysis of the cellulose substrate, and observed that the sequential or simultaneous assembly of each component yielded similar activities. Increased synergy, however, was observed when enzymes were positioned adjacent to each other, suggesting a possible mechanism of substrate channeling between catalytic domains. In a subsequent effort, Fierobe and colleagues successfully generated a library of 75 different chimeric cellulosomes and tested their activities on both crystalline and less recalcitrant substrates [102]. The enzymes incorporated into the bifunctional complexes consisted of a combination of *C. cellulolyticum* cellulases CelA, CelC, CelE, CelF, or CelG. Synergy due to enzyme assembly on the chimeric scaffolds was only observed when acting on the more recalcitrant substrates such as Avicel and bacterial microcrystalline cellulose, with less or no synergy observed when acting on the less crystalline substrates bacterial cellulose and PASC. To further augment the synergistic and overall activities of bifunctional artificial cellulosomes, Fierobe and coworkers generated trifunctional cellulosomes [91]. In order to control the relative position of the enzymes within the complexe, a third dockerin-cohesin pair derived from *Ruminococcus flavefaciens* was used in which the interaction is characterized by both high affinity and lack of cross-reactivity with other cohesin-dockerin pairs. Upon incorporation of three cellulases, the complexes demonstrated significantly higher activity than their bifunctional counterparts. The synergy among the complexed enzymes was also demonstrated.

In an effort to generate artificial cellulosome systems with novel geometries and potentially higher overall activities on cellulose, Mingardon and coworkers constructed chimeric scaffolds and cellulases designed to self-assemble in precise spatial arrangements [92]. A hybrid cellulosome consisted of enzymes targeted to a central scaffold, a covalent cellulosome was generated by covalently fusing all components together in a single polypeptide chain, and three other cellulosomes with novel architectures were engineered as well. Still, the hybrid cellulosome, which more closely resembled traditional cellulosome architectures, demonstrated significantly higher activity than all others [92]. Some other notable observations were that the least effective cellulosome contained the most CBDs and that in certain architectures, cohesin-dockerin pairs could dissociate, most probably due to conformational strain.

Cellulosic biomass is mostly composed of lignin and hemicellulose in addition to cellulose. To bestow hemicellulase activity upon engineered cellulosomes, Morais and colleagues intergraded two xylanases as well as a xylose binding domain to a scaffold containing three divergent cohesins from *Acetivibrio cellulolyticus*, *C. thermocellum*, and *R. flavefaciens* [103]. The assembled complexes demonstrated a 1.5 fold increase in activity on hatched wheat straw when compared with the free enzyme mixtures, and the authors attributed this to substrate targeting by the xylose binding domain as well as to the proximity of the enzymes within the complex [103]. This system was further improved in a subsequent study whereby another dockerin-cohesin pair derived from *Bacteriodes cellulosolvens* was incorporated resulting in a four component artificial cellulosome that could accomodate two EGLs and two xylanases [104]. An overall 2.4-fold increase in activity on hatched wheat straw was observed compared with the free enzyme mixtures.

### *3.2.2 Expression of cellulosome components in B. subtilis*

While *E. coli* remains an attractive host for the production of enzymes and scaffolds the presence of endogenous proteases can lead to the degradation of desired proteins. Another attractive host towards the production of recombinant cellulosomes is *B. subtilis*, since it can be easily genetically manipulated, is characterized by fast growth, and is an efficient protein secretor. A strain of *B. subtilis* deficient in eight major extracellular proteases, *B. subtilis* WB800, was engineered and used as a host for the production and secretion of *C. cellulovorans* EngE since this enzyme was shown to be partially degraded in *E. coli* [105]. Murashima and colleagues were successful in using this protease-deficient strain to produce EngE, and subsequent incubation with scaffold Mini-CbpA, which contains a CBD as well as two cohesins, resulted in assembly of an enzyme-scaffold complex capable of binding cellulose [105].

## 3.3. *In vivo* secretion and assembly of recombinant cellulosomes

The overexpression and purification of individual scaffolds and enzymes towards the assembly of artificial cellulosomes poses extra costs and steps towards cellulose hydrolysis. Rather, the development of a CBP-capable organism would require the production, secretion and *in vivo* assembly of artificial cellulosomes in the extracellular space (Fig. 1B).

### *3.3.1. Secretion of recombinant cellulosomes by B. subtilis*

Initial work began as an extension of Murashima and colleagues' work employing *B. subtilis* WB800 as a host for heterologous production of all components. Cho and colleagues constructed an expression cassette encoding both Mini-CbpA and EngE on a single vector which was established in *B. subtilis* WB800 [106]. The result was the secretion and subsequent assembly of both enzyme and scaffold components into an artificial cellulosome complex which was localized in the supernatant. This study was the first report of the *in vivo* assembly of artificial cellulosomes by a single organism, although the activity of this strain against cellulosic substrates was not verified. A study by Arai and colleagues used a different approach towards the *in vivo* assembly of recombinant cellulosomes. In this case, three strains of *B. subtilis* WB800 were engineered to secrete either EngB, XynB, or MiniCbpA into the culture supernatant [107]. By co-culturing enzyme and scaffold producing strains, complexes formed in the supernatant and were characterized by the appropriate enzymatic activity. This provided a novel method for assembling complexes *in vivo* based on intercellular complementation.

### *3.3.2. Secretion of recombinant cellulosomes by C. acetobutylicum*

*C. acetobutylicum* is an organism which has been employed in the production of a number of acids and solvents including acetone, butanol, and ethanol. The potential to engineer this organism to degrade cellulose as a cheap and abundant carbon source has garnered significant attention in the past decade. Interestingly, this bacterium is not cellulolytic, however investigation of its genome sequence reveals a cellulosomal gene cluster encoding a number of hydrolytic enzymes as well as a scaffold protein CipA [64, 108]. Sabathe and colleagues were successful in engineering *C. acetobutylicum* to secrete and assemble a functional minicellulosome *in vivo* [109]. Since CipA had been previously demonstrated to not be secreted in this organism, the authors replaced the original signal peptide with that of the *C. cellulolyticum* scaffold protein CipC. Overexpression and secretion of a truncated version of CipA containing two cohesin domains and a CBD resulted in its binding with endogenous cellulase Cel48A, and formation of a secreted cellulosome *in vivo* [109]. In analyzing the activity of the recombinant cellulosome on Avicel, bacterial cellulose, PASC and carboxymethyl cellulose, no detectable activity was observed when using the crystalline substrates, as is the case for native *C. acetobutylicum*. Low levels of activity were observed on carboxymethyl cellulose and PASC, however such levels did not exceed those demonstrated by the native cellulosome. A next logical step was to produce artificial scaffold chimeras in this organism, capable of binding enzymes at very precise locations via divergent cohesin domains derived from different bacterial species. Perret and colleagues first engineered this organism to produce and secrete scaffold miniCipC1 which is a truncated form of *C. cellulolyticum* scaffold CipC, and subsequently generated chimeric scaffold Scaf3 which contains cohesins from both *C. cellulolyticum* and *C. thermocellum*, as well as a CBD [93]. After visualizing the chimeric scaffold using SDS-PAGE, the protein was blotted on a nitrocellulose membrane and was subsequently shown to bind both Cel48 and Cel9 containing a dockerin from *C. cellulolyticum*, as well as Cel9 with a dockerin from *C. thermocellum*.

### 3.4. *In vivo* surface-anchoring of recombinant cellulosomes

The architecture of the cellulosome establishes proximal and synergistic effects of enzymes within the complex when associated with the substrate [95, 110, 111]. In natural and recombinant systems, these synergistic effects are further augmented by an extra level of synergy resulting from the cellulosome's association with the surface of cells, yielding cellulose-enzyme-microbe (CEM) ternary complexes [89, 112-118]. CEM ternary complexes benefit from the effects of microbe-enzyme synergy, ultimately limiting the escape of hydrolysis products and enzymes, increasing access to substrate hydrolysis products, minimizing the distance products must diffuse before cellular uptake occurs, concentrating enzymes at the substrate surface, protecting hydrolytic enzymes from proteases and thermal degradation, as well as optimizing the chemical environment at the substrate-microbe interface [89, 112-116]. In several cellulosome-producing bacteria, including *C. thermocellum*, the cellulosome is anchored to the surface of cells, resulting in one of the most efficient systems for bacterial cellulose hydrolysis [4, 116]. In an effort to mimic such a system, microbial engineers have adopted this strategy as a next logical step towards the improvement of recombinant cellulosome systems with the ultimate goal of increasing the efficiency of the bioconversion process.

#### *3.4.1. Anchoring recombinant cellulosomes on the cell surface of S. cerevisiae*

Much interest towards the development of a CBP-capable organism comes from a desire to generate biofuels such as ethanol from cheap and abundant substrates. Therefore, much attention has been directed towards engineering cellulosome systems in ethanologenic organisms such as *S. cerevisiae*. Lily and colleagues were successful in targeting hybrid scaffold Scaf3p to the cell surface of *S. cerevisiae* by fusing it with the glycosyl phosphatidylinositol (GPI) signal peptide of the Cwp2 protein for linking to the β-1,6 glucan of the yeast cell wall [119]. The scaffold contained two divergent cohesins from *C. thermocellum* and *C. cellulolyticum* as well as a CBD. Microsocopy revealed that the CBD was functional in adhering cells to filter paper, and the successful targeting of a Cel5a-dockerin fusion to the scaffold confirmed functionality of the cohesin modules. The ability to generate scaffold chimeras using non-cohesin modules was established by Ito and colleagues [120]. This research group generated artificial scaffolds by fusing the Z domain of *Staphylococcus aureus* Protein A with a cohesin from the *C. cellulovorans* cellulosome and displayed them on the cell surface [120]. The scaffold chimeras were engineered to contain two Z domains as well as two cohesins for precisely targeting different enzymes to the cell surface. The authors fused two enzymes, EGII and BGLI, to either a dockerin domain or Fc domain, which successfully targeted the enzymes to the cohesin and Z domains, respectively [120]. Hydrolysis experiments on β-glucan revealed that co-displaying EGII-FC and BGL-dock resulted in cells capable of degrading this soluble cellulosic substrate, but due to lack of a CBD on the engineered scaffold, this strain would most likely be inefficient at hydrolyzing more recalcitrant cellulosic substrates. A more direct approach to ethanol production was adopted by Tsai and coworkers, where yeast strains were engineered to display a trimeric scaffold containing three divergent cohesins from *C. thermocellum, C. cellulolyticum* and *R. flavefaciens* as well as a CBD [121]. Three enzymes, *C. thermocellum* CelA, and *C. cellulolyticum* CelE and CelG were overproduced in *E. coli* and successfully targeted to corresponding cohesin domains on the scaffold by fusion with appropriate dockerin domains, resulting in the surface-display of trifunctional cellulosomes. The anchor system used in this study consisted of displaying the Aga1 protein which interacted with the Aga2 protein fused with the scaffold. Replacing endoglucanase CelG with *C. thermocellum* β-glucosidase BglA resulted in significant increases in glucose liberation from PASC, and the resulting strain was capable of directly producing ethanol from this substrate. Incubating cells in the presence of PASC resulted in ethanol production that corresponds to 95% of the theoretically attainable ethanol yield. The authors also observed no accumulation of glucose in the medium during the fermentation assays, suggesting that the released glucose was immediately taken up by cells during the SSF process [121].

The production of both enzymes and scaffold in a single yeast strain was achieved by Wen and colleagues [94]. The scaffold contained three cohesins as well as a CBD and was successfully displayed by use of the α-agglutinin adhesion receptor. *In vivo* secretion of an EGL, CBH, and BGL resulted in the assembly of tetrameric complexes, and the resulting yeast strain was capable of directly converting PASC to ethanol at a yield of 1.8 g/L. Interestingly, the authors also observed that when Bgl1 was positioned within the complex, in close proximity to EGII and CBHII, increased activity was observed, most probably due to removal of the cellobiose at the cell surface which may have been inhibiting EGII and CBHII. In comparison with the work by Tsai and colleagues, this represented the first report of producing and assembling a trifunctional cellulosome on the cell surface by the *in vivo* production of all components. The relatively low levels of EGII and Bgl1 produced by this strain, however, suggested that burdening the secretion machinery of the organism was a potential bottleneck. To address this issue, the Chen group adopted a different approach towards the *in vivo* assembly of trifunctional complexes on the cell surface which entailed intercellular complementation by a yeast consortium [122]. In this case, one strain produced a trifunctional scaffold containing three divergent cohesins and a CBD, while each of three other strains produced an exoglucanase, EGL, or BGL which were targeted to specific sites on the artificial scaffold by fusion with corresponding dockerin domains. The authors also reported that an optimal ratio of each strain within the consortium resulted in two-fold increase in ethanol production when compared with a consortium containing equal proportions of each strain.

### *3.4.2. Anchoring recombinant cellulosomes on the cell surface of L. lactis*

While of the attention to the engineering of organisms to display artificial cellulosomes has been directed towards ethanol-producing microbes, the metabolic diversity among microorganisms suggests that such a strategy can be implemented towards the production of other commodity chemicals including organic acids. In an effort to assemble cellulosome-inspired multi-enzyme complexes on the surface of a bacterium, Wieczorek and Martin engineered a strain of *L. lactis* to anchor mini-scaffolds on the cell surface [123]. While several bacterial species non-covalently anchor cellulosomes to the cell surface by means of S-layer homologous domains, other organisms such as *R. flavefaciens* display cellulosomes by covalently anchoring them to the cell wall by sortase. Therefore, the authors in this study fused fragments of *C. thermocellum* CipA scaffold with a C-terminal LPXTG-containing anchor motif from *Streptococcus pyogenes* M6 protein, resulting in their successful surface-display. By fusing the scaffolds with the export-specific reporter, *S.aureus* nuclease NucA, the authors were able to easily detect them in the extracellular medium. Fusion of *E. coli* β-glucuronidase UidA with the dockerin from major *C. thermocellum* cellulosomal enzyme CelS, resulted in its successful targeting to the surface-displayed scaffolds. While the assembled complexes were not cellulolytic, the investigation yielded insights into parameters affecting secretion and anchoring of the recombinant scaffolds, including the observation that scaffold size was not a significant bottleneck in display efficiency. The strain used was deficient in its major extracellular housekeeping protease HtrA, which has been demonstrated to be responsible for the degradation of secreted recombinant proteins. In a subsequent study, the authors fused type 1 and type 2 cohesins to generate scaffold chimeras capable of binding UidA and *E. coli* β-galactosidase LacZ fused with type 1 and type 2 dockerins (unpublished data). This system yielded novel insights into the assembly of displayed complexes, suggesting that enzyme size and position relative to the cell surface may play a role in determining the overall net enzymatic profile of the displayed complexes.

### *3.4.3. Anchoring recombinant cellulosomes on the cell surface of B. subtilis*

The interest in *B. subtilis* as a potential candidate for the consolidated bioprocessing of cellulosic substrates into chemicals and fuels resulted in the development of recombinant cellulosome systems in this organism. The attractiveness of this host is compounded by several characteristics including its ability to metabolize $C_5$ and $C_6$ sugars as well as its natural ability to uptake long-chain cellodextrins. Anderson and colleagues used a system similar to the Martin group's by employing the sortase-mediated anchoring of proteins on the cell surface [124]. This group initially demonstrated proof of concept by displaying a single enzyme, Cel8A, and subsequently went on to display cohesin domains capable of interacting with an appropriate Cel8A-dockerin fusion. It was observed that proteolytic degradation of the displayed enzymes resulted in an 80% decrease in activity after only 6 hours, an effect hypothesized to result from the presence of the extracellular housekeeping protease WprA. Inserting this system into a WprA$^{-}$ strain resulted in a significant reduction in the observed proteolysis of the enzymes. The most complex artificial cellulosome generated by this group included a surface-anchored chimeric scaffold containing three divergent cohesins and a CBD. Incubation of cells with enzyme-dockerin fusions purified from *E. coli* resulted in the assembly of functional minicellulosomes on the cell surface. Soon afterwards, the Zhang group reported the engineering of a scaffold-displaying *B. subtilis* strain capable of binding three enzymes and the subsequent assembly of an artificial cellulosome on the cell surface [125]. These authors investigated the effect of the CEM ternary complex by comparing a cell-bound artificial cellulosome, a cell-free artificial cellulosome, and a commercial fungal cellulose mixture. Comparative enzyme assays were conducted on the recalcitrant substrate Avicel, as well as amorphous cellulose. When comparing the activity of cell-bound cellulosomes vs. cell-free cellulosomes, a larger significant increase in CEM synergy on Avicel as opposed to amorphous cellulose was observed in the cell-displayed constructs. The authors suggest this effect to be due to larger product inhibition at the boundary layer when active on crystalline cellulose. Since EGL demonstrates higher hydrolysis activity on amorphous cellulose, and CBH is more sensitive to product inhibition, the observed results suggest that the benefits of anchoring cellulosomes on the cell surface are a necessary component of a CBP-capable organism. In Table 1, successfully generated recombinant cellulosome components are listed according to host organism and assembly strategy.

| **Strategy** | **Host Microbe** | **#* Divergent Cohesins on scaffold** | **#* Enzymes** | **Scaffolds**** | **Enzymes Targeted to Scaffold** | **Ref** |
|---|---|---|---|---|---|---|
| *In vitro* **assembly** | *E. coli* | 1 | 2 | **CipA** (coh$_{th}$) | Cel D | [98] |
| | | 1 | 2 | **CipA** (coh$_{th}$) | Cel E | [99] |
| | | 1 | 2 | **Mini-CbpA** (coh$_{cv}$) | Cel E, H, S | [100] |
| | | 2 | 2 | **Scaf1-4** (coh$_{th}$ / coh$_{cl}$) | Cel A, F | [101] |
| | | 2 | 2 | **Scaf1-5** (coh$_{th}$ / coh$_{cl}$) | Cel A, C, E, F, G | [102] |
| | | 3 | 3 | **Scaf6** (coh$_{th}$ / coh$_{cl}$ / coh$_{rf}$) | Cel A, C, E, F, G | [91] |
| | | 3 | 3 | **Scaf3, Scaf6** (coh$_{th}$ / coh$_{cl}$ / coh$_{rf}$) | Cel F, G | [92] |
| | | 3 | 3 | | Cel 5, Xyn10, | [103] |

| Strategy | Host Microbe | #* Divergent Cohesins on scaffold | #* Enzymes | Scaffolds** | Enzymes Targeted to Scaffold | Ref |
|---|---|---|---|---|---|---|
| | | 4 | 4 | **ScafATF** ($coh_{ac}$ / $coh_{th}$ / $coh_{rf}$) **Scaf-BTFA** ($coh_{ac}$ / $coh_{th}$ / $coh_{rf}$ / $coh_{bc}$) | 11 Cel 5, 45, Xyn 10, 11 | [104] |
| | *B. subtilis* & *E. coli* | 1 | 1 | **Mini-CbpA** ($coh_{cv}$) | Eng B | [105] |
| *In vivo* assembly secreted | *B. subtilis* | 1 | 1 | **Mini-CbpA** ($coh_{cv}$) | Eng B | [106] |
| | | 1 | 1 | **Mini-CbpA** ($coh_{cv}$) | XynB / EngB | [107] |
| | *C. Aceto-butylicum* | 1 | 1 | **Mini-CipA** ($coh_{ca}$) | Cel 48A | [109] [93] |
| | | 1 | 1 | **Scaf3** ($coh_{cl}$ / $coh_{th}$) | Cel 48F, 9E | |
| *In vivo* assembly anchored | *S. cerevisiae* | 1 | 3 | **CipA3** ($coh_{th}$) | EGII, CBHII, BGLI | [121] [119] |
| | | 2 | 1 | **Scaf3p** ($coh_{cl}$ / $coh_{th}$) | GFP, Cel5A | [120] [121] |
| | | 2 | 4 | **ZZ-cohcoh** (Z domain / $coh_{cv}$) | EGII, BGLI | [122] |
| | | 3 | 3 | **Scaf-ctf** ($coh_{th}$ / $coh_{cl}$ / $coh_{rf}$) | Cel E, A, G | |
| | | 3 | 3 | **Scaf-ctf** ($coh_{th}$ / $coh_{cl}$ / $coh_{rf}$) | Cel E, A, CBHII, Bgl1 | |
| | *L. lactis* | 1 | 2 | **$CipA_{frags}$** ( $coh_{th}$) | UidA | [123] |
| | *B. subtilis* | 3 | 3 | **Scaf** ($coh_{th}$ / $coh_{cl}$ / $coh_{rf}$) | Cel8A, 9E, 9G | [124] [125] |
| | | 1 | 3 | **Mini-CipA** ($coh_{th}$) | Cel5, 9, 48 | |

*Corresponds to complexes containing the largest number o divergent cohesins and integrated enzymes.

**Scaffolds listed are containing the largest number of cohesin modules from that study. Names in parenthesis correspond to types of cohesins included in the most complex scaffolds. Coh: cohesin domain. Subscript indicates organism of origin: th (*C. thermocellum*), cv (*C. cellulovorans*), cl (*C. cellulolyticum*), rf (*R. flavefaciens*), ac (*A. cellulolyticum*), bc (*B. cellulosolvens*). Z domains: *S. aureus* Protein A binding domain.

**Table 1.** Strategies, organisms and successfully assembled recombinant cellulosomes

## 4. Conclusion

Recent decades have yielded significant advances in the engineering of non-cellulolytic organisms towards the degradation of cellulosic substrates into fermentable sugars. The recombinant production of cellulases is both a necessary and effective means to both characterize and utilize non-native enzymes in a host organism of choice. In addition, the recalcitrance of crystalline cellulose and complexity of hemicellulose requires multiple enzymes working together to fully achieve this bioconversion process. The potential of custom-designed recombinant cellulosomes to optimize ratio and positioning of enzymes within artificial complexes contribute to this goal. Still, significant advances are necessary in order for the cost-effective transformation of cellulose into valuable commodity chemicals such as bioethanol, non-biofuel hydrocarbons, and organic acids to become an industrial standard. For example, of significant importance is the optimizing of secretion and anchoring mechanisms in host organisms, two factors which can prove to be bottlenecks in the engineering process. Indeed, the native metabolic diversity of microbes designed to utilize cellulose as an energy source, as well as the advent of synthetic biology through which non-native and novel pathways can be introduced into these organisms, suggest that the bioconversion of cellulosic substrates into valuable chemicals is not so far from reach. Constructing more efficient recombinant cellulases, as well as the assembly of cellulosomes with complex architectures inspired by bacteria such as *R. Flavifaciens* and *A. cellulolyticus*, are possible avenues to explore in this field. With the inevitable depletion of reserves of conventional energy sources such as petroleum and other fossil fuels, it becomes more evident that cellulosic biomass is not only an attractive source for the production of alternative fuel sources, but may soon become a necessary one.

## Author details

Andrew S. Wieczorek, Damien Biot-Pelletier and Vincent J.J. Martin *
*Center for Structural and Functional Genomics, Concordia University, Montréal QC, Canada*

## 5. References

[1] Klesov A. Biochemistry and enzymology of cellulose hydrolysis. Biokhimiya. 1991;55(1) 295-318.
[2] Teeri T. Crystalline cellulose degradation: new insights into the function of cellobiohydrolases. Trends Biotechnol. 1997;15(5) 160-167.
[3] Watanabe H, Tokuda G. Cellulolytic systems in insects. Annu Rev Entomol. 2010;55 609-632.
[4] Bayer EA, Belaich JP, Shoham Y, Lamed R. The cellulosomes: multienzyme machines for degradation of plant cell wall polysaccharides. Annu Rev Microbiol. 2004;58 521-554.

* Corresponding Author

[5] Chang T, Yao S. Thermophilic, lignocellulolytic bacteria for ethanol production: current state and perspectives. Appl Microbiol Biotechnol. 2011;92(1) 13-27.
[6] Philippidis GP, Smith TK, Wyman CE. Study of the enzymatic hydrolysis of cellulose for production of fuel ethanol by the simultaneous saccharification and fermentation process. Biotechnol Bioeng. 1993;41(9) 846-853.
[7] Lynd L. Overview and evaluation of fuel ethanol from cellulosic biomass: technology, economics, the environment, and policy. Annu Rev Energy Environ. 1996;21 403-465.
[8] Lynd LR, Weimer PJ, van Zyl WH, Pretorius IS. Microbial cellulose utilization: fundamentals and biotechnology. Microbiol Mol Biol Rev. 2002;66(3) 506-577.
[9] Lynd LR, van Zyl WH, McBride JE, Laser M. Consolidated bioprocessing of cellulosic biomass: an update. Curr Opin Biotechnol. 2005;16(5) 577-583.
[10] French CE. Synthetic biology and biomass conversion: a match made in heaven? J R Soc Interface. 2009;6 Suppl 4:S547-558.
[11] Cho K, Yoo Y, Kang H. δ-integration of endo/exo-glucanase and β-glucosidase genes into the yeast chromosomes for direct conversion of cellulose to ethanol. Enzyme Microb Technol. 1999;25(1-2) 23-30.
[12] Cho KM, Yoo YJ. Novel SSF process for ethanol production from microcrystalline cellulose using deltaintegrated recombinant yeast, *Saccharomyces cerevisiae* L2612 delta GC. J Microbiol Biotechnol. 1999;9(3) 340-345.
[13] Fujita Y, Ito J, Ueda M, Fukuda H, Kondo A. Synergistic saccharification, and direct fermentation to ethanol, of amorphous cellulose by use of an engineered yeast strain codisplaying three types of cellulolytic enzyme. Appl Environ Microbiol. 2004;70(2) 1207-1212.
[14] Den Haan R, Rose SH, Lynd LR, van Zyl WH. Hydrolysis and fermentation of amorphous cellulose by recombinant *Saccharomyces cerevisiae*. Metab Eng. 2007;9(1) 87-94.
[15] Den Haan R, McBride JE, La Grange DC, Lynd LR, Van Zyl W, H. Funtional expresson of cellobiohydrolases in *Saccharomyces cerevisiae* towards one-step conversion of cellulose to ethanol. Enzyme Microb Technol. 2007;40(5) 1291-1299.
[16] Jeon E, Hyeon JE, Suh DJ, Suh YW, Kim SW, Song KH, et al. Production of cellulosic ethanol in *Saccharomyces cerevisiae* heterologous expressing *Clostridium thermocellum* endoglucanase and *Saccharomycopsis fibuligera* beta-glucosidase genes. Mol Cells. 2009;28(4) 369-673.
[17] Yanase S, Yamada R, Kaneko S, Noda H, Hasunuma T, Tanaka T, et al. Ethanol production from cellulosic materials using cellulase-expressing yeast. Biotechnol J. 2010;5(5) 449-55.
[18] Yamada R, Taniguchi N, Tanaka T, Ogino C, Fukuda H, Kondo A. Cocktail delta-integration: a novel method to construct cellulolytic enzyme expression ratio-optimized yeast strains. Microb Cell Fact. 2010;9(32) http://www.microbialcellfactories.com/content/9/1/32.
[19] Yamada R, Taniguchi N, Tanaka T, Ogino C, Fukuda H, Kondo A. Direct ethanol production from cellulosic materials using a diploid strain of *Saccharomyces cerevisiae* with optimized cellulase expression. Biotechnol Biofuels. 2011;4:8

http://www.biotechnologyforbiofuels.com/content/4/1/8

[20] Shen Y, Zhang Y, Ma T, Bao X, Du F, Zhuang G, et al. Simultaneous saccharification and fermentation of acid-pretreated corncobs with a recombinant *Saccharomyces cerevisiae* expressing beta-glucosidase. Bioresour Technol. 2008;99(11) 5099-5103.

[21] Kotaka A, Bando H, Kaya M, Kato-Murai M, Kuroda K, Sahara H, et al. Direct ethanol production from barley beta-glucan by sake yeast displaying *Aspergillus oryzae* beta-glucosidase and endoglucanase. J Biosci Bioeng. 2008;105(6) 622-627.

[22] Benoliel B, Poças-Fonseca MJ, Torres FA, de Moraes LM. Expression of a glucose-tolerant beta-glucosidase from *Humicola grisea* var. thermoidea in *Saccharomyces cerevisiae*. Appl Biochem Biotechnol. 2010;160(7) 2036-2044.

[23] Heinzelman P, Snow CD, Wu I, Nguyen C, Villalobos A, Govindarajan S, et al. A family of thermostable fungal cellulases created by structure-guided recombination. Proc Natl Acad Sci U S A. 2009;106(14) 5610-5615.

[24] Heinzelman P, Snow CD, Smith MA, Yu X, Kannan A, Boulware K, et al. SCHEMA recombination of a fungal cellulase uncovers a single mutation that contributes markedly to stability. J Biol Chem. 2009;284(39) 26229-26233.

[25] Voutilainen SP, Murray PG, Tuohy MG, Koivula A. Expression of *Talaromyces emersonii* cellobiohydrolase Cel7A in *Saccharomyces cerevisiae* and rational mutagenesis to improve its thermostability and activity. Protein Eng Des Sel. 2010;23(2) 69-79.

[26] Kitagawa T, Kohda K, Tokuhiro K, Hoshida H, Akada R, Takahashi H, et al. Identification of genes that enhance cellulase protein production in yeast. J Biotechnol. 2011;151(2) 194-203.

[27] du Preez JC, van Driessel B, Prior BA. Ethanol tolerance of *Pichia stipitis* and *Candida shehatae* strains in fed-batch cultures at controlled low dissolved-oxygen levels. Appl Microbiol Biotechnol. 1989;30:53-58.

[28] Lee JW, Rodrigues RC, Jeffries TW. Simultaneous saccharification and ethanol fermentation of oxalic acid pretreated corncob assessed with response surface methodology. Bioresour Technol. 2009;100(24) 6307-6311.

[29] Jeffries TW, Grigoriev IV, Grimwood J, Laplaza JM, Aerts A, Salamov A, et al. Genome sequence of the lignocellulose-bioconverting and xylose-fermenting yeast *Pichia stipitis*. Nat Biotechnol. 2007;25(3) 319-326.

[30] Piontek M, Hagedorn J, Hollenberg CP, Gellissen G, Strasser AW. Two novel gene expression systems based on the yeasts *Schwanniomyces occidentalis* and *Pichia stipitis*. Appl Microbiol Biotechnol. 1998;50(3) 331-338.

[31] Schwarz WH, Gräbnitz F, Staudenbauer WL. Properties of a *Clostridium thermocellum* endoglucanase produced in *Escherichia coli*. Appl Environ Microbiol. 1986;51(6) 1293-1299.

[32] Shimoda C, Itadani A, Sugino A, Furusawa M. Isolation of thermotolerant mutants by using proofreading-deficient DNA polymerase delta as an effective mutator in *Saccharomyces cerevisiae*. Genes Genet Syst. 2006;81(6) 391-397.

[33] Sridhar M, Sree NK, Rao LV. Effect of UV radiation on thermotolerance, ethanol tolerance and osmotolerance of *Saccharomyces cerevisiae* VS1 and VS3 strains. Bioresour Technol. 2002;83(3) 199-202.

[34] Ballesteros M, Oliva JM, Negro MJ, Manzanares P, Ballesteros I. Ethanol from lignocellulosic materials by a simultaneous saccharification and fermentation process (SFS) with *Kluyveromyces marxianus* CECT10875. Process Biochem. 2004;39 1843-1848.

[35] Boyle M, Barron N, McHale AP. Simultaneous saccharification and fermentation of straw to ethanol using the thermotolerant yeast strain *Kluyveromyces marxianus* imb3. Biotechnol Lett. 1997;19 49-51.

[36] Krishna S, Reddy T, Chowdary G. Simultaneous saccharification and fermentation of lignocellulosic wastes to ethanol using a thermotolerant yeast. Bioresour Technol. 2001;77 193-196.

[37] Lark N, Xia YK, Qin CG, Gong CS, Tsao GT. Production of ethanol from recycled paper sludge using cellulase and yeast, *Kluveromyces marxianus*. Biomass Bioenerg. 1997;12 135-143.

[38] Hong J, Wang Y, Kumagai H, Tamaki H. Construction of thermotolerant yeast expressing thermostable cellulase genes. J Biotechnol. 2007;130(2) 114-123.

[39] Hisamatsu M, Furubayashi T, Shuichi K, Takashi M, Naoto I. Isolation and identification of a novel yeast fermenting ethanol under acidic conditions. J Appl Glycosci 2006;53(2) 111-113.

[40] Kitagawa T, Tokuhiro K, Sugiyama H, Kohda K, Isono N, Hisamatsu M, Takahashi H, Imaeda T. Construction of a beta-glucosidase expression system using the multistress-tolerant yeast *Issatchenkia orientalis*. Appl Microbiol Biotechnol. 2010;87(5) 1841-1853.

[41] Ingram LO, Conway T, Clark DP, Sewell GW, Preston JF. Genetic engineering of ethanol production in *Escherichia coli*. Appl Environ Microbiol. 1987;53(10) 2420-2425.

[42] Wood BE, Ingram LO. Ethanol production from cellobiose, amorphous cellulose, and crystalline cellulose by recombinant *Klebsiella oxytoca* containing chromosomally integrated *Zymomonas mobilis* genes for ethanol production and plasmids expressing thermostable cellulase genes from *Clostridium thermocellum*. Appl Environ Microbiol. 1992;58(7) 2103-2110.

[43] Doran JB, Aldrich HC, Ingram LO. Saccharification and fermentation of Sugar Cane bagasse by *Klebsiella oxytoca* P2 containing chromosomally integrated genes encoding the *Zymomonas mobilis* ethanol pathway. Biotechnol Bioeng. 1994;44(2) 240-247.

[44] Wood BE, Beall DS, Ingram LO. Production of recombinant bacterial endoglucanase as a co-product with ethanol during fermentation using derivatives of *Escherichia coli* KO11. Biotechnol Bioeng. 1997;55(3) 547-555.

[45] Zhou S, Ingram L. Engineering endoglucanase-secreting strains of ethanologenic *Klebsiella oxytoca* P2. J Ind Microbiol Biotechnol. 1999;22(6) 600-607.

[46] Zhou S, Davis FC, Ingram LO. Gene integration and expression and extracellular secretion of *Erwinia chrysanthemi* endoglucanase CelY (*celY*) and CelZ (*celZ*) in ethanologenic *Klebsiella oxytoca* P2. Appl Environ Microbiol. 2001;67(1) 6-14.

[47] Zhou S, Ingram L. Simultaneous saccharification and fermentation of amorphous cellulose to ethanol by recombinant *Klebsiella oxytoca* SZ21 without supplemental cellulase. Biotechnol Lett. 2001;23 1455-1462.

[48] Bokinsky G, Peralta-Yahya PP, George A, Holmes BM, Steen EJ, Dietrich J, et al. Synthesis of three advanced biofuels from ionic liquid-pretreated switchgrass using engineered *Escherichia coli*. Proc Natl Acad Sci U S A. 2011;108(50) 19949-19954.

[49] Lee KJ, E. TD, L. RP. Ethanol production by *Zymomonas mobilis* in continuous culture at high glucose concentrations. Biotechnol Lett. 1979;1 421-426.

[50] Rogers PL, J. LK, E. TD. Kinetics of alcohol production by *Zymomonas mobilis* at high sugar concentrations. Biotechnology Letters. 1979;1 165-170.

[51] Lee KJ, Lefebvre M, Tribe DE, Rogers PL. High productivity ethanol fermentations with *Zymomonas mobilis* using continuous cell recycle. Biotechnol Lett. 1980;2 487-492.

[52] Lee KJ, Skotnicki ML, Tribe DE, Rogers PL. Kinetic studies on a highly productive strain of *Zymomonas mobilis*. Biotechnol Lett. 1980;2 339-344.

[53] Rogers PL, Lee KJ, Tribe DE. High productivity ethanol fermentations with *Zymomonas mobilis*. Process Biochem. 1980;15 7-11.

[54] Skotnicki ML, Lee KJ, Tribe DE, Rogers PL. Comparison of ethanol production by different *zymomonas strains*. Appl Environ Microbiol. 1981;41(4) 889-893.

[55] Yanase H, Nozaki K, Okamoto K. Ethanol production from cellulosic materials by genetically engineered *Zymomonas mobilis*. Biotechnol Lett. 2005;27(4) 259-263.

[56] Brestic-Goachet N, Gunasekaran P, Cami B, Baratti JC. Transfer and expression of an *Erwinia chrysanthemi* cellulase gene in *Zymomonas mobilis*. J Gen Microbiol. 1989;135 893-902.

[57] Lejeune A, Eveleigh DE, Colson C. Expression of an endoglucanase gene of *Pseudomonas fluorescens* var. cellulosa in *Zymomonas mobilis*. FEMS Microbiol Lett. 1988;49 363-366.

[58] Misawa N, Okamoto T, Nakamura K. Expression of a cellulase gene in *Zymomonas mobilis*. J Biotechnol. 1988;7 167-178.

[59] Linger JG, Adney WS, Darzins A. Heterologous expression and extracellular secretion of cellulolytic enzymes by *Zymomonas mobilis*. Appl Environ Microbiol. 2010;76(19) 6360-6369.

[60] Rajnish KN, Choudhary GM, Gunasekaran P. Functional characterization of a putative endoglucanase gene in the genome of *Zymomonas mobilis*. Biotechnol Lett. 2008;30(8) 1461-1467.

[61] Ni Y, Sun Z. Recent progress on industrial fermentative production of acetone-butanol-ethanol by *Clostridium acetobutylicum* in China. Appl Microbiol Biotechnol. 2009;83(3) 415-423.

[62] Ali MK, Rudolph FB, Bennett GN. Characterization of thermostable Xyn10A enzyme from mesophilic *Clostridium acetobutylicum* ATCC 824. J Ind Microbiol Biotechnol. 2005;32(1) 12-18.

[63] Lee SF, Forsberg CW, Gibbins LN. Xylanolytic Activity of *Clostridium acetobutylicum*. Appl Environ Microbiol. 1985;50(4) 1068-1076.

[64] Nolling J, Breton G, Omelchenko MV, Makarova KS, Zeng Q, Gibson R, Lee HM, Dubois J, Qiu D, Hitti J, Wolf YI, Tatusov RL, Sabathe F, Doucette-Stamm L, Soucaille P, Daly MJ, Bennett GM, Koonin EV, Smith DR. Genome sequence and comparative analysis of the solvent-producing bacterium *Clostridium acetobutylicum*. J Bacteriol. 2001;183(16) 4823-4838.

[65] Lopez-Contreras AM, Martens AA, Szijarto N, Mooibroek H, Claassen PA, van der Oost J, De Vos WM. Production by *Clostridium acetobutylicum* ATCC 824 of CelG, a cellulosomal glycoside hydrolase belonging to family 9. Appl Environ Microbiol. 2003;69(2) 869-877.

[66] Lopez-Contreras AM, Gabor K, Martens AA, Renckens BA, Claassen PA, Van Der Oost J, De Vos WM. Substrate-induced production and secretion of cellulases by *Clostridium acetobutylicum*. Appl Environ Microbiol. 2004;70(9) 5238-5243.

[67] Kim AY, Attwood GT, Holt SM, White BA, Blaschek HP. Heterologous expression of endo-beta-1,4-D-glucanase from *Clostridium cellulovorans* in *Clostridium acetobutylicum* ATCC 824 following transformation of the *engB* gene. Appl Environ Microbiol. 1994;60(1) 337-340.

[68] Mingardon F, Chanal A, Tardif C, Fierobe HP. The issue of secretion in heterologous expression of *Clostridium cellulolyticum* cellulase-encoding genes in *Clostridium acetobutylicum* ATCC 824. Appl Environ Microbiol. 2011;77(9) 2831-2838.

[69] Chanal A, Mingardon F, Bauzan M, Tardif C, Fierobe HP. Scaffoldin modules serving as "cargo" domains to promote the secretion of heterologous cellulosomal cellulases by *Clostridium acetobutylicum*. Appl Environ Microbiol. 2011;77(17) 6277-6280.

[70] Lopez-Contreras AM, Smidt H, van der Oost J, Claassen PA, Mooibroek H, de Vos WM. *Clostridium beijerinckii* cells expressing *Neocallimastix patriciarum* glycoside hydrolases show enhanced lichenan utilization and solvent production. Appl Environ Microbiol. 2001;67(11) 5127-5133.

[71] Bates EE, Gilbert HJ, Hazlewood GP, Huckle J, Laurie JI, Mann SP. Expression of a *Clostridium thermocellum* endoglucanase gene in *Lactobacillus plantarum*. Appl Environ Microbiol. 1989;55(8) 2095-2097.

[72] Scheirlinck T, Mahillon J, Joos H, Dhaese P, Michiels F. Integration and expression of alpha-amylase and endoglucanase genes in the *Lactobacillus plantarum* chromosome. Appl Environ Microbiol. 1989;55(9) 2130-2137.

[73] Rossi F, Rudella A, Marzotto M, Dellaglio F. Vector-free cloning of a bacterial endo-1,4-beta-glucanase in *Lactobacillus plantarum* and its effect on the acidifying activity in silage: use of recombinant cellulolytic *Lactobacillus plantarum* as silage inoculant. Antonie Van Leeuwenhoek. 2001;80(2) 139-147.

[74] Ozkose E, Akyol I, Kar B, Comlekcioglu U, Ekinci MS. Expression of fungal cellulase gene in *Lactococcus lactis* to construct novel recombinant silage inoculants. Folia Microbiol (Praha). 2009;54(4) 335-342.

[75] Cho JS, Choi YJ, Chung DK. Expression of *Clostridium thermocellum* endoglucanase gene in *Lactobacillus gasseri* and *Lactobacillus johnsonii* and characterization of the genetically modified probiotic lactobacilli. Curr Microbiol. 2000;40(4) 257-263.

[76] Okano K, Zhang Q, Yoshida S, Tanaka T, Ogino C, Fukuda H, et al. D-lactic acid production from cellooligosaccharides and beta-glucan using L-LDH gene-deficient and endoglucanase-secreting *Lactobacillus plantarum*. Appl Microbiol Biotechnol. 2010;85(3) 643-650.

[77] Pernilla Turner GM, Eva N Karlsson. Potential and utilization of thermophiles and thermostable enzymes in biorefining. Microb Cell Fact. 2012;6(1) 9.

htpp://www.microbialcellfactories.com/content/6/1/9

[78] Mai V, Wiegel J. Advances in development of a genetic system for Thermoanaerobacterium spp.: expression of genes encoding hydrolytic enzymes, development of a second shuttle vector, and integration of genes into the chromosome. Appl Environ Microbiol. 2000;66(11) 4817-4821.

[79] Adrio JL, Demain AL. Fungal biotechnology. Int Microbiol. 2003;6(3) 191-199.

[80] Schuster A, Schmoll M. Biology and biotechnology of Trichoderma. Appl Microbiol Biotechnol. 2010;87(3) 787-799.

[81] Miettinen-Oinonen A, Suominen P. Enhanced production of *Trichoderma reesei* endoglucanases and use of the new cellulase preparations in producing the stonewashed effect on denim fabric. Appl Environ Microbiol. 2002;68(8) 3956-3964.

[82] Miettinen-Oinonen A, Paloheimo M, Lantto R, Suominen P. Enhanced production of cellobiohydrolases in *Trichoderma reesei* and evaluation of the new preparations in biofinishing of cotton. J Biotechnol. 2005;116(3) 305-317.

[83] Bower B, Larenas E, Mitchenson C. Inventors; Exo-endo cellulase fusion protein 2005. US Patent WO2005093073

[84] Li D-C, Li A-N, Papageorgiou AC. Cellulases from thermophilic fungi: Recent insights and biotechnological potential. Enzyme Research. 2012;2011.

[85] Sandgren M, Stahlberg J, Mitchinson C. Structural and biochemical studies of GH family 12 cellulases: improved thermal stability, and ligand complexes. Prog Biophys Mol Biol. 2005;89(3) 246-291.

[86] Jeoh T, Michener W, Himmel ME, Decker SR, Adney WS. Implications of cellobiohydrolase glycosylation for use in biomass conversion. Biotechnol Biofuels. 2008;1(1) 10.

[87] Tambor JH, Ren H, Ushinsky S, Zheng Y, Riemens A, St-Francois C, et al. Recombinant expression, activity screening and functional characterization identifies three novel endo-1,4-beta-glucanases that efficiently hydrolyse cellulosic substrates. Appl Microbiol Biotechnol. 2012;93(1) 203-214.

[88] Zhang YH, Lynd LR. Toward an aggregated understanding of enzymatic hydrolysis of cellulose: noncomplexed cellulase systems. Biotechnol Bioeng. 2004;88(7) 797-824.

[89] Lynd LR, Weimer PJ, van Zyl WH, Pretorius IS. Microbial cellulose utilization: fundamentals and biotechnology. Microbiol Mol Biol Rev. 2002;66(3) 506-577,

[90] Gerngross UT, Romaniec MP, Kobayashi T, Huskisson NS, Demain AL. Sequencing of a *Clostridium thermocellum* gene (*cipA*) encoding the cellulosomal SL-protein reveals an unusual degree of internal homology. Mol Microbiol. 1993;8(2) 325-334.

[91] Fierobe HP, Mingardon F, Mechaly A, Belaich A, Rincon MT, Pages S, et al. Action of designer cellulosomes on homogeneous versus complex substrates: controlled incorporation of three distinct enzymes into a defined trifunctional scaffoldin. J Biol Chem. 2005;280(16) 16325-16334.

[92] Mingardon F, Chanal A, Tardif C, Bayer EA, Fierobe HP. Exploration of new geometries in cellulosome-like chimeras. Appl Environ Microbiol. 2007;73(22):7138-7149.

[93] Perret S, Casalot L, Fierobe HP, Tardif C, Sabathe F, Belaich JP, et al. Production of heterologous and chimeric scaffoldins by *Clostridium acetobutylicum* ATCC 824. J Bacteriol. 2004;186(1) 253-257.

[94] Wen F, Sun J, Zhao H. Yeast surface display of trifunctional minicellulosomes for simultaneous saccharification and fermentation of cellulose to ethanol. Appl Environ Microbiol. 2010 76(4) 1251-1260.

[95] Kruus K, Lua AC, Demain AL, Wu JH. The anchorage function of CipA (CelL), a scaffolding protein of the *Clostridium thermocellum* cellulosome. Proc Natl Acad Sci U S A. 1995;92(20) 9254-9258.

[96] Leibovitz E, Beguin P. A new type of cohesin domain that specifically binds the dockerin domain of the *Clostridium thermocellum* cellulosome-integrating protein CipA. J Bacteriol. 1996;178(11) 3077-3084.

[97] Lemaire M, Ohayon H, Gounon P, Fujino T, Beguin P. OlpB, a new outer layer protein of *Clostridium thermocellum*, and binding of its S-layer-like domains to components of the cell envelope. J Bacteriol. 1995;177(9) 2451-2459.

[98] Kataeva I, Guglielmi G, Beguin P. Interaction between *Clostridium thermocellum* endoglucanase CelD and polypeptides derived from the cellulosome-integrating protein CipA: stoichiometry and cellulolytic activity of the complexes. Biochem J. 1997;326(2):617-624.

[99] Ciruela A, Gilbert HJ, Ali BR, Hazlewood GP. Synergistic interaction of the cellulosome integrating protein (CipA) from *Clostridium thermocellum* with a cellulosomal endoglucanase. FEBS Lett. 1998;422(2) 221-224.

[100] Murashima K, Kosugi A, Doi RH. Synergistic effects on crystalline cellulose degradation between cellulosomal cellulases from *Clostridium cellulovorans*. J Bacteriol. 2002;184(18) 5088-5095.

[101] Fierobe HP, Mechaly A, Tardif C, Belaich A, Lamed R, Shoham Y, Belaich JP, Bayer EA. Design and production of active cellulosome chimeras. Selective incorporation of dockerin-containing enzymes into defined functional complexes. J Biol Chem. 2001;276(24) 21257-21261.

[102] Fierobe HP, Bayer EA, Tardif C, Czjzek M, Mechaly A, Belaich A, et al. Degradation of cellulose substrates by cellulosome chimeras. Substrate targeting versus proximity of enzyme components. J Biol Chem. 2002;277(51) 49621-49630.

[103] Morais S, Barak Y, Caspi J, Hadar Y, Lamed R, Shoham Y, et al. Contribution of a xylan-binding module to the degradation of a complex cellulosic substrate by designer cellulosomes. Appl Environ Microbiol. 2010;76(12) 3787-3796.

[104] Morais S, Barak Y, Caspi J, Hadar Y, Lamed R, Shoham Y, et al. Cellulase-xylanase synergy in designer cellulosomes for enhanced degradation of a complex cellulosic substrate. MBio. 2010;1(5) doi:10.1128.

[105] Murashima K, Chen CL, Kosugi A, Tamaru Y, Doi RH, Wong SL. Heterologous production of *Clostridium cellulovorans* engB, using protease-deficient *Bacillus subtilis*, and preparation of active recombinant cellulosomes. J Bacteriol. 2002;184(1) 76-81.

[106] Cho HY, Yukawa H, Inui M, Doi RH, Wong SL. Production of minicellulosomes from *Clostridium cellulovorans* in *Bacillus* subtilis WB800. Appl Environ Microbiol. 2004;70(9) 5704-5707.

[107] Arai T, Matsuoka S, Cho HY, Yukawa H, Inui M, Wong SL, et al. Synthesis of *Clostridium cellulovorans* minicellulosomes by intercellular complementation. Proc Natl Acad Sci U S A. 2007;104(5) 1456-1460.

[108] Sabathe F, Belaich A, Soucaille P. Characterization of the cellulolytic complex cellulosome) of *Clostridium acetobutylicum*. FEMS Microbiol Lett. 2002;217(1) 15-22.

[109] Sabathe F, Soucaille P. Characterization of the CipA scaffolding protein and in vivo production of a minicellulosome in *Clostridium acetobutylicum*. J Bacteriol. 2003;185(3) 1092-1096.

[110] Kosugi A, Amano Y, Murashima K, Doi RH. Hydrophilic domains of scaffolding protein CbpA promote glycosyl hydrolase activity and localization of cellulosomes to he cell surface of *Clostridium cellulovorans*. J Bacteriol. 2004;186(19) 6351-6359.

[111] Garcia-Campayo V, Beguin P. Synergism between the cellulosome-integrating protein CipA and endoglucanase CelD of *Clostridium thermocellum*. J Biotechnol. 1997;57(1-3) 39-47.

[112] Zverlov VV, Klupp M, Krauss J, Schwarz WH. Mutations in the scaffoldin gene, cipA, of *Clostridium thermocellum* with impaired cellulosome formation and cellulose hydrolysis: insertions of a new transposable element, IS1447, and implications for cellulase synergism on crystalline cellulose. J Bacteriol. 2008;190(12) 4321-4327.

[113] Lynd LR, van Zyl WH, McBride JE, Laser M. Consolidated bioprocessing of cellulosic biomass: an update. Curr Opin Biotechnol. 2005;16(5) 577-583.

[114] Lu Y, Zhang YH, Lynd LR. Enzyme-microbe synergy during cellulose hydrolysis by *Clostridium thermocellum*. Proc Natl Acad Sci U S A. 2006;103(44) 16165-16169.

[115] Miron J, Ben-Ghedalia D, Morrison M. Invited review: adhesion mechanisms of rumen cellulolytic bacteria. J Dairy Sci. 2001;84(6) 1294-1309.

[116] Schwarz WH. The cellulosome and cellulose degradation by anaerobic bacteria. Appl Microbiol Biotechnol. 2001;56(5-6) 634-649.

[117] Bayer EA, Kenig R, Lamed R. Adherence of *Clostridium thermocellum* to cellulose. J Bacteriol. 1983;156(2) 818-827.

[118] Ng TK, Weimer TK, Zeikus JG. Cellulolytic and physiological properties of *Clostridium hermocellum*. Arch Microbiol. 1977;114(1) 1-7.

[119] Lilly M, Fierobe HP, van Zyl WH, Volschenk H. Heterologous expression of a Clostridium minicellulosome in *Saccharomyces cerevisiae*. FEMS Yeast Res. 2009;9(8) 1236-1249.

[120] Ito J, Kosugi A, Tanaka T, Kuroda K, Shibasaki S, Ogino C, et al. Regulation of the display ratio of enzymes on the *Saccharomyces cerevisiae* cell surface by the mmunoglobulin G and cellulosomal enzyme binding domains. Appl Environ Microbiol. 2009;75(12) 4149-4154.

[121] Tsai SL, Oh J, Singh S, Chen R, Chen W. Functional assembly of minicellulosomes on he *Saccharomyces cerevisiae* cell surface for cellulose hydrolysis and ethanol production. Appl Environ Microbiol. 2009;75(19) 6087-6093.

[122] Tsai SL, Goyal G, Chen W. Surface display of a functional minicellulosome by ntracellular complementation using a synthetic yeast consortium and its application to cellulose hydrolysis and ethanol production. Appl Environ Microbiol. 2010;76(22) 7514-7520.

[123] Wieczorek AS, Martin VJ. Engineering the cell surface display of cohesins for assembly of cellulosome-inspired enzyme complexes on *Lactococcus lactis*. Microb Cell Fact. 2010 9:69. http://www.microbialcellfactories.com/content/9/1/69

[124] Anderson TD, Robson SA, Jiang XW, Malmirchegini GR, Fierobe HP, Lazazzera BA, et al. Assembly of minicellulosomes on the surface of *Bacillus subtilis*. Appl Environ Microbiol. 2011;77(14) 4849-4858.

[125] You C, Zhang XZ, Sathitsuksanoh N, Lynd LR, Zhang YH. Enhanced microbial utilization of recalcitrant cellulose by an *ex vivo* cellulosome-microbe complex. Appl Environ Microbiol. 2012;78(5):1437-1444.

Chapter 6

# Lignocellulosic Biomass Utilization Toward Biorefinery Using Meshophilic Clostridial Species

Yutaka Tamaru and Ana M. López-Contreras

Additional information is available at the end of the chapter

http://dx.doi.org/10.5772/56480

## 1. Introduction

Lignocellulosic biomass such as agricultural, industrial, and forestry residues as well as dedicated crops constitute renewable and abundant resources with great potential for a low-cost and uniquely sustainable bioconversion to value-added bioproducts. Thus, many organic fuels and chemicals that can be obtained from lignocellulosic biomass can reduce greenhouse gas emissions, enhance energy security, improve the economy, dispose of problematic solid wastes, and improve air quality. In particular, liquid biofuels are attractive candidates, since little or no change is needed to the current petroleum-based fuel technologies. However, the biorefining process remains economically unfeasible due to a lack of biocatalysts that can overcome costly hurdles such as cooling from high temperature, pumping of oxygen/stirring, and, neutralization from acidic or basic pH. Therefore, bioconversion of the lignocellulosic components into fermentable sugars is an essential step in the biorefinery.

In nature, a variety of microorganisms including bacteria and fungi have the ability to degrade lignocellulosic biomass to C-5 and/or C-6 sugars. Moreover, new concepts have been proposed to enable the overall goal of cost reduction. These include genetically modifying the cell wall composition of energy crops in order to make their conversion easier, and combining the processes of glycoside hydrolases (GHs) and polysaccharide lyases (PLs) production, saccharification, and fermentation. Several clostridial species produce an extracellular enzyme complex called the cellulosomes and free extracellular enzymes called non-cellulosomes [1,2]. The cellulosomes are particularly designed for efficient degradation of plant cell wall polysaccharides such as cellulose, hemicellulose, and pectins. The component parts of the multi-component complex are integrated by virtue of a unique family of integrating modules, the cohesins and the dockerins, whose distribution and specificity dictate the overall cellulosome architecture. On the other hand, several clostridial species are able to ferment carbohydrates to acetone, butanol, and ethanol (ABE). Industrial application of this process, also known as ABE fermentation, has a long history, but the process economics after 1960 became unfavorable compared to the petrochemical process, and its commercial exploitation was gradually abandoned. The inefficiency of the fermentation still hampers commercial reintroduction of this renewable butanol production process. However, improving the yields and productivities of the solvent products is key to its successful reintroduction.

## 2. Solvent-producing clostridia

Biological production of butanol (*n*-butanol, 1-butanol) has a long history as an industrially significant fermentation process [3]. An excellent review article by Jones and Woods on the history of acetone–butanol–ethanol (ABE) fermentation processes is available [4]. After Pasteur discovered bacterial butanol production from his landmark anaerobic cultivation in 1861, fermentative ABE production prospered during the early 20th century, and after ethanol became the second largest industrial fermentation process in the world. In 1945, two thirds of industrially used butanol was produced by fermentation in U.S. However, the ABE fermentation process lost competitiveness by the 1960s due to the increase of feedstock costs and advancement of the petrochemical industry except in Russia and in South Africa, where the substrate and labor costs were low. The ABE fermentation processes in South Africa and Russia continued to operate until the late 1980s and early 1990s [5]. It has recently been reported that the Russian fermentation industry is concentrating on the conversion of agricultural biomass into butanol[5]. The successful industrial-level butanol fermentation in these countries can provide guidelines to our current efforts to produce butanol in large-scale. Commercial solvent titres peak at about 20 g/L from 55 to 60 g/L of substrate, resulting in solvent yields of approx. 0.35 g/g sugar consumed [6]. The butanol:solvent molar ratio is typically 0.6 with an A:B:E ratio of 3:6:1[4]. *C. acetobutylicum* strain EA2018 was also developed using chemical mutagenesis and found to produce higher butanol:solvent ratios (0.7) than the parental strain (0.6) [7]. This strain has been licensed to several commercial producers in China (GBL market data). The acetone pathway has also been knocked out in this strain resulting in higher butanol:solvent ratios (0.8) but no overall increase in higher butanol titre was observed [8]. Butanol is the preferred solvent since it attracts the highest price in the chemical market. Between butanol and ethanol, butanol is a choice of fuel as compared to ethanol, mainly because of its higher energy density, lower volatility and reduced corrosiveness. In addition, butanol has relatively better compatibility for current car engines and infrastructures, offering more convenience and versatility in applications [9,10]. Thus, butanol production from lignocellulosic materials has attracted much attention from contemporary researchers in the discipline of bioenergy.

Several clostridial species such as *Clostridium acetobutylicum*, *C. beijerinckii*, *C. pasteurianum*, *C. saccharobutylicum*, and *C. saccharoperbutylacetonicum* are known to be solventogenic, producing acetone, butanol, and ethanol, but they present relatively low tolerance to butanol [5,11,12]. Among wild-type clostridial species, typical end concentrations of butanol are around 12 g/L from fermentation of glucose [12]. The fermentation efficiency was reported to be hampered due to the accumulated butanol (e.g., >7.4 g/L) [12], which could lead to cell growth inhibition and premature cessation of fermentation [13]. Such negative inhibition leads to low achievable butanol concentration and will thus increase the downstream costs associated with product purification [13]. Attempts have been made to improve the butanol concentration up to 17.8 g/L by genetically manipulating the wild-type clostridial species [12]. Nevertheless, genetically modified bacteria are usually unstable due to plasmid excision [14], leading to the deterioration of butanol-producing capability within batches of experiments. Hence, the search for novel and enhanced wild-type microbes with improved butanol tolerance is of great necessity for industrial applications [15].

## 3. Metabolic engineering of mesophilic clostridia

Synthetic biology has recently been used to introduce biosynthetic capacity for butanol into non-natural hosts. The choice between using or engineering natural function versus importing biosynthetic function has been reviewed [16]. Commonly used host strains include *Escherichia coli* and *Saccharomyces cerevisiae* that are relatively easy to genetically manipulate but do not tolerate more than 2% 1-butanol [17]. In addition, these strains do not display broad substrate ranges and cannot compete with natural or engineered clostridia for the production of 1-butanol from a broad range substrates including pentose sugars and sugars derived from cellulosic feedstocks.

For successful metabolic engineering of *C. acetobutylicum,* it is necessary to have efficient genetic engineering tools for metabolic pathway manipulation. In 2001, the complete genome sequence of *C. acetobutylicum* was published [18]. The *C. acetobutylicum* ATCC 824 genome consists of a 3.94 Mbp chromosome and a 192,000 bp megaplasmid pSOL1. A total of 3,740 and 178 ORFs were identified on the chromosome and megaplasmid, respectively. *C. acetobutylicum* has distinctive families of proteins involved in sporulation, anaerobic energy conversion, and carbohydrate degradation, which are well matched to the physiological characteristics of *C. acetobutylicum*. For butanol formation, two mechanisms have been identified in this strain; one is related to solventogenesis (ABE forming process) and the other is alcohologenesis (butanol and ethanol forming process). The key genes involved in solventogenesis are shown in **Figure 1A**. The genes involved in alcohologenesis remain unidentified. It is currently believed that the enzymes encoded by the *adhE* (aldehyde/alcohol dehydrogenase; CAP0035), *pdc* (pyruvate decarboxylase; CAP0025), and *edh* (ethanol dehydrogenase; CAP0059) genes are associated with this metabolism [12].

Several *Bacullus subtilis–C. acetobutylicum* and *E. coli–C. acetobutylicum* shuttle vectors were developed in the early 1990s [19,20]. Mermelstein et al. made a breakthrough in metabolic engineering of *C. acetobutylicum* ATCC 824 [21]. Since *C. acetobutylicum* ATCC 824 possesses a strong restriction system encoded by Cac824I (recognizing 50-GCNGC-30), which prevents efficient transformation of recombinant plasmid prepared in *E. coli*. Thus, they developed a *B. subtilis–C. acetobutylicum* shuttle vector pFNK1, which allowed higher transformation efficiency. Using this shuttle vector, the acetoacetate decarboxylase (*adc*), and the phosphotransbutyrylase (*ptb*) genes were successfully expressed at elevated levels in strain

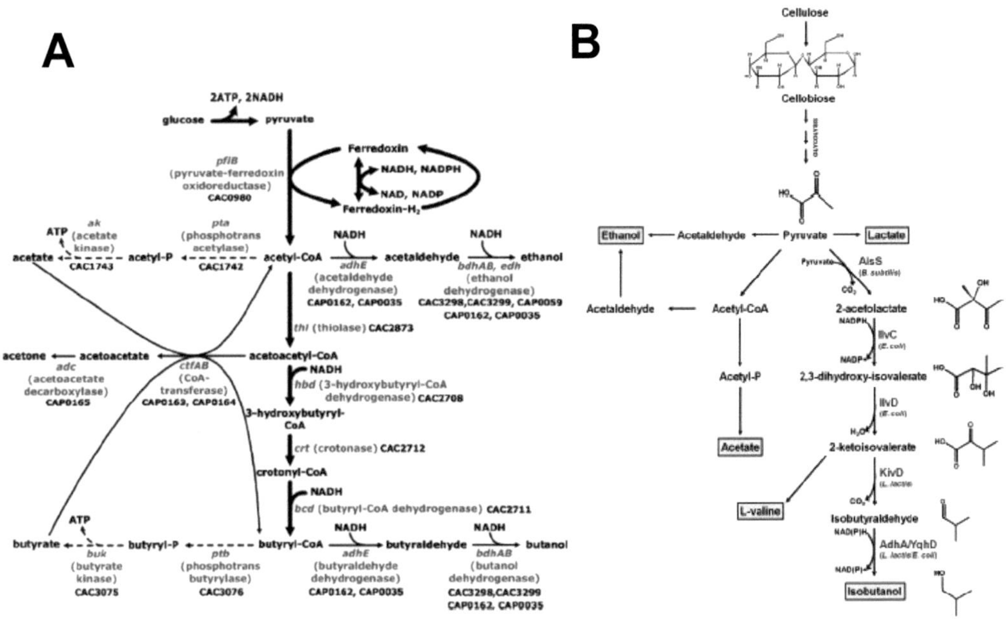


**Figure 1.** (A) Metabolic pathways in *C. acetobutylicum* [3]. Reactions which predominate during acidogenesis and solventogenesis are indicated by dotted and solid arrows, respectively. Thick arrows indicate reactions which activate the whole fermentative metabolism. Gray letters indicate genes and enzymes for the reactions. CAC and CAP numbers are the ORF numbers in genome and megaplasmid, respectively. (B) The pathway for isobutanol production in *C. cellulolyticum* [59] from cellulose. In order to achieve direct isobutanol production from pyruvate, the genes encoding *B. subtilis* α-acetolactate synthase, *E. coli* acetohydroxyacid isomeroreductase, *E. coli* dihydroxy acid dehydratase, *Lactococcus lactis* ketoacid decarboxylase, and *E. coli* and *L. lactis* alcohol dehydrogenases were cloned, respectively.

ATCC 824. The development of an *in vivo* methylation system was an important step [22]. Methylation of the shuttle vectors with w3TI methyltransferase (encoded by *B. subtilis* phage w3T) prior to transformation greatly reduces or prevents the degradation of the transforming plasmid DNA by the attack of a strong restriction system (Cac824I) present in *C. acetobutylicum* [22]. The copy number of commonly used plasmids in *C. acetobutylicum* is around 7–20 copies per cell, which seem to be suitable for metabolic engineering purposes [23]. Significant advances for *C. acetobutylicum* have been made to methods for gene integration [24]. Superior performance has also been demonstrated from genetically engineered derivatives of *C. acetobutylicum* ATCC 824 [25,26]. Methods based on a group II intron system for gene knockout have been described [27,28]. More recently an improved method, based on allele coupled exchange (ACE), has been described for stable integration of larger DNA fragments [29]. It is now possible to construct multi-step biosynthetic pathways paving the way for new synthetic clostridia.

Isobutanol is a more promising fermentation product because it is less toxic than 1-butanol. Unlike ethanol, isobutanol can also be blended at any ratio with gasoline or used directly in current engines without modification [30]. It is an attractive biofuel but cannot substitute for

1-butanol in the chemical market. One synthetic approach for isobutanol production involves the introduction of genes encoding enzymes that convert either acetyl-CoA or pyruvate to isobutanol. Alternatively, genes encoding enzymes that convert 2-keto acids intermediates (from amino acid synthesis) into isobutanol and branched-chain alcohols; 2-methyl-1-butanol and 3-methyl-1-butanol can be introduced [31,32,33]. Several companies are currently involved in scale-up and demonstration. Gevo Inc. (http://www.gevo.com) has engineered *E. coli* to produce isobutanol [34] and recently acquired a commercial-scale ethanol plant in Minnesota for retrofit to produce isobutanol. The company has also received Environmental Protection Agency certification to blend isobutanol in fossil fuels. DuPont has also engineered several biocatalysts for isobutanol [35] and assigned the technology to Butamax™ Advanced Biofuels (http://www.butamax.com), a joint venture between BP and Dupont. Butamax™ is collaborating with Kingston Research Limited, another BP–Dupont joint venture, to build a demonstration plant in the UK. Previously, the cellulosome-producing *C. cellulolyticum* has also been genetically engineered for improved ethanol production [36]. With this respect, most of the research concerning the construction of an organism for consolidated bioprocessing has focused on ethanol production. Despite this, it has been asserted that higher alcohols (i.e., alcohols with more than two carbons), such as isobutanol, are better candidates for gasoline replacement because they have energy density, octane value, and Reid vapor pressure that are more similar to those of gasoline [37].

## 4. Cellulosome-pruducing *CLostridium cellulovorans*

The anaerobic clostridia are found in the soil, on decaying plant materials, in rumens, in sewage sludge, in termite gut, in wood-chip piles, in compost piles, and at paper mills and wood processing plants (**Table 1**). Most of these bacteria occur in natural habitats such as soil and decaying plant materials, but some are enriched by human activities, such as in compost piles, in sewage plants, and at wood processing plants. Other natural habitats include the anaerobic rumen of various ruminants and the gut of termites, where they process plant materials for the host organism's nutrition. The biotechnological potential of polysaccharolytic enzymes has resulted in the isolation and characterization of a large number of anaerobic, Gram-positive, spore-forming bacteria, the majority of which have been allocated to the genus *Clostridium*. Among some clostridia, the cellulosomes produced by *Clostridium* species are particularly designed for efficient degradation of plant cell wall polysaccharides. The component parts of the multicomponent complex are integrated by virtue of a unique family of integrating modules, the cohesins and the dockerins (**Fig. 2A**), whose distribution and specificity dictate the overall cellulosome architecture [38]. The cellulosomes are characterized by the presence of two general components: (1) the nonenzymatic scaffolding protein(s) with enzyme-binding sites called cohesins and (2) a variety of cellulosomal enzymes with dockerins, which interact with the cohesins in the scaffolding protein.

Since 2002, over 100 genome sequencing projects of *Clostridium* species have been done or are being done mainly by the United States Department of Energy Joint Genome Institute

| Species | Habitat |
|---|---|
| *Clostridium acetobutylicum** | Soil |
| *Clostridium aldrichii* | Wood digester |
| *Clostridium cellobioparum* | Soil |
| *Clostridium cellulofermentans* | Soil |
| *Clostridium cellulolyticum** | Rot grass |
| *Clostridium cellulovorans** | Wood chips |
| *Clostridium herbivorans* | Pig intestine |
| *Clostridium hungate* | Soil |
| *Clostridium josui* | Compost |
| *Clostridium papyrosolvens* | Paper mill |
| *Clostridium thermocellum** | Compost, Soil |

Single asterisk (*), species whose genome sequencing is complete.

**Table 1.** Cellulolytic clostridial species from natural biomass decaying ecosystems

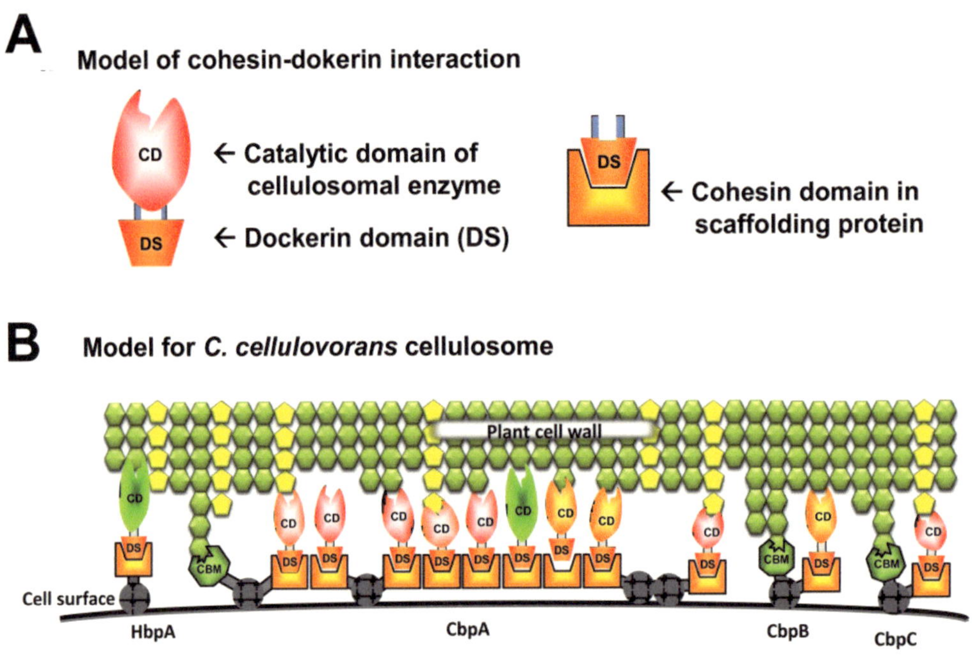


**Figure 2.** Model for *C. cellulovorans* cellulosomes. (A) Model of cohesin–dockerin interaction. (B) Recent model of cellulosomes attached to its substrate and cell surface.

(DOE-JGI). The whole genome sequences of cellulosome-producing *Clostridium* species, i.e., thermophilic *C. thermocellum* ATCC27405 and mesophilic *C. cellulolyticum* H10 were sequenced by the JGI in 2007 and 2009, respectively. In 2009 the complete genome of *C. cellulovorans* was sequenced using the next-generation DNA sequencers to compare not only cellulosomal genes but also noncellulosomal ones among cellulosome-producing clostridia [39]. *C. cellulovorans* is able to degrade native substrates in soft biomass such as corn fiber and rice straw efficiently by producing the cellulosomes. The whole genome sequence of *C.*

*cellulovorans* comprised 4,220 predicted genes in 5.10 Mbp. As a result, the genome size of *C. cellulovorans* was about 1 Mbp larger than that of other cellulosome-related clostridia, mesophilic *C. acetobutylicum* and *C. cellulolyticum*, and thermophilic *C. thermocellum*. A total of 57 cellulosomal genes were found in the *C. cellulovorans* genome (**Table 2**) and coded for not only CAZymes but also lipases, peptidases, and proteinase inhibitors [40,41]. Cellulosomal genes among clostridial genomes were identified and classified as cohesin-containing scaffolding proteins and dockerin-containing proteins. So far, the scaffolding proteins for constructing cellulosomes were found in *C. acetobutylicum* [42], *C. cellulolyticum* [43], *C. cellulovorans* [44], *C. josui* [45], and *C. thermocellum* [46].

| Organism | GenBank Accession No. | Genome size (Mb) | No. of genes | No. of cellulosomal genes | % GC |
|---|---|---|---|---|---|
| *C. cellulovorans* 743B | DF093537-DF093556 | 5.10 | 4220 | 57 | 31.1 |
| *C. acetobutylicum* ATCC 824 | AE001437 | 3.94 | 3672 | 12 | 30.9 |
| *C. cellulolyticum* H10 | CP001348 | 4.07 | 3390 | 65 | 37.4 |
| *C. thermocellum* ATCC 27405 | CP000568 | 3.84 | 3191 | 84 | 39.0 |

**Table 2.** General features of cellulosomal clostridial genomes compared with that of *C. cellulovorans*

Among a total of 57 cellulosomal genes of the *C. cellulovorans* genome, 53 dockerin-containing proteins and four cohesin-containing scaffolding proteins were found, respectively [40]. More interestingly, two scaffolding proteins, CbpB and CbpC, consisting of a carbohydrate-binding module (CBM) of family 3, a surface–layer homology domain and a cohesin domain, were recently found and tandemly localized in the *C. cellulovorans* genome, while there were no such scaffolding proteins in other cellulosomal clostridia. Thus, by examining genome sequences from multiple *Clostridium* species, comparative genomics offers new insight into genome evolution and the way natural selection molds functional DNA sequence evolution. A recent model for the *C. cellulovorans* cellulosome reveals that the enzymatic subunits are bound to the scaffolding through the interaction of the cohesins and dockerins to form the cellulosome (**Fig. 2B**).

Carbohydrate-active enzymes (CAZymes) are categorized into different classes and families in the CAZy database (for more information please visit the CAZy web page; www.cazy.org). CAZymes that cleave, build, and rearrange oligo- and polysaccharides play a central role in the biology of bacteria and fungi and are key to optimizing biomass degradation by these species. Currently, more than 2,500 GHs have been identified and classified into 115 families [47]. Interestingly, the same enzyme family may contain members from bacteria, fungi, and plants with several different activities and substrate specifications [48]. However, fungal cellulases (hydrolysis of β-1,4-glycosidic bonds) have been mostly found within a few GH families including 5, 6, 7, 8, 9, 12, 44, 45, 48, 61, and 74 [47,49]. Cellulases have a small independently folded CBM that is connected to the catalytic domain by a flexible linker [48]. The CBMs are responsible for binding the enzyme to the crystalline cellulose, and thus enhance the enzyme activity [38]. Currently, many CBMs have been identified and classified into 54 families; however, only 20 families (1, 13, 14, 18, 19, 20, 21, 24, 29, 32, 35, 38, 39, 40, 42, 43, 47, 48, 50, and 52) have been found in fungi. Among 53 cellulosomal genes encoding dockerin containing proteins in the *C. cellulovorans* genome, a total of 29 genes coded for cellulolytic, hemicellulolytic and pectin-degrading enzymes [40]. Compared with the genome-sequenced species within cellulosomal clostridia, the proteome of *C. cellulovorans* focusing on dockerin-containing proteins showed representation of many proteins with known functions. In the *C. cellulovorans* cellulosome, there are 16 cellulase genes belonging to families GH5, GH9 and GH48, six mannanase genes belonging to families GH5 and GH26, three xylanase genes belonging to families GH8, GH10 and GH11, an endo-beta-galactosidase gene belonging to family GH98, and two pectate lyase genes belonging to families PL1 and PL9.

## 5. Cellulose metabolism of *C. acetobutylicum*

Cellulosomal gene clusters were conserved only in mesophilic clostridia (**Fig. 3**) [40]. Furthermore, these cellulosomal genes were randomly distributed in the *C. cellulovorans* genome except for the cellulosomal genes related to a large cellulosomal cluster, whereas two large cellulosomal gene clusters were found in the *C. cellulolyticum* genome. Even though the organization of genes encoding cellulosome subunits differs among mesophilic cellulolytic clostridia, there is nonetheless a clear similarity, particularly when looking at the cluster of genes following the main scaffoldin gene. Such a cluster is not found in *C. thermocellum*. This would suggest that the cellulosomes of the mesophilic clostridia, including the 'ghost' cellulosome of *C. acetobutylicum*, may have arisen from a common ancestral gene cluster. However, attempts have been made to develop a *C. acetobutylicum* strain that can utilize cellulose directly. There is evidence that *C. acetobutylicum* ATCC 824 can produce an active cellulosome. The *celF* gene, encoding a unique cellulase, was found to be up-regulated in *C. acetobutylicum* ATCC 824 during growth on xylose or lichenan [50]. However, *C. acetobutylicum* ATCC 824 had no cellulolytic activity suggesting that some element of the cellulosome is missing or not expressed. In an effort to make *C. acetobutylicum* utilize cellulose more directly, the *engB* gene from *C. cellulovorans* or the gene encoding the scaffold protein from *C. cellulolyticum* and *C. thermocellum* were introduced into *C. acetobutylicum*. However, the level of expressed heterologous cellulase was rather low [51,52]. On the other hand, the *man5K* gene encoding the mannanase Man5K from *C. cellulolyticum* was cloned alone or as an operon with the gene *cipC1* encoding a truncated scaffoldin (miniCipC1) of the same origin in the solventogenic *C. acetobutylicum* [53]. The recombinant strains of the solventogenic bacterium were both found to secrete active Man5K in the range of milligrams per liter. In the case of the strain expressing only *man5K*, a large fraction of the recombinant enzyme was truncated and lost the N-terminal dockerin domain, but it remained active towards galactomannan. When *man5K* was coexpressed with *cipC1* in *C. acetobutylicum*, the recombinant strain secreted almost exclusively full-length mannanase, which bound to the scaffoldin miniCipC1, thus showing that complexation to the scaffoldin stabilized the enzyme. Moreover, the secreted heterologous complex was found to be functional: it binds to crystalline cellulose via the carbohydrate-binding module of the miniscaffoldin, and the complexed mannanase is active towards galactomannan. Taken together, these data showed that *C. acetobutylicum* is a suitable host for the production, assembly, and secretion of heterologous minicellulosomes. More studies are needed to characterize the existing cellulosomal gene cluster in *C. acetobutylicum* before further metabolic engineering.

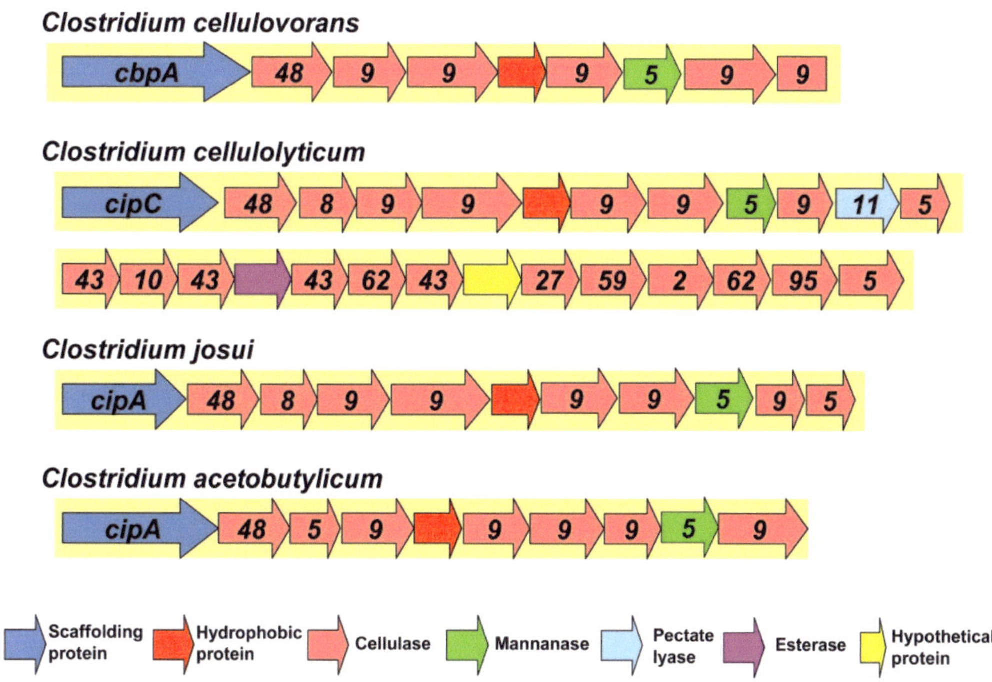


**Figure 3.** Cellulosome-related gene clusters in the genome of mesophilic clostridia.

## 6. Consolidated bioprocessing by Clostridial species

Consolidated bioprocessing, or CBP, the conversion of lignocellulose into desired products in one step without added enzymes, has been a subject of increased research effort in recent years [54]. Naturally occurring cellulolytic microorganisms are starting points for CBP organism development via the native strategy, with anaerobes being of particular interest [55]. The primary objective of such developments is to engineer product yields and titers to satisfy the requirements of an industrial process. Metabolic engineering of mixed-acid fermentations in relation to these objectives has been successful in the case of mesophilic, non-cellulolytic, enteric bacteria [56]. Far more limited work of this type has been undertaken with cellulolytic bacteria, primarily because of the absence of suitable gene-transfer techniques. Recent developments, however, appear to be removing this limitation for some organisms.

The lack of efficient genetic engineering tools including a gene knock-out system for *C. acetobutylicum* has hampered further strain improvement for a long time. As described earlier, much effort is exerted to develop genetic engineering tools for clostridia. In the mean time, Liao and collaborators recently reported metabolic engineering of *E. coli* for butanol production [57]. The mutant *E. coli* BW25113 (Δ*adhE* Δ*ldhA* Δ*frdBC* Δ*fnr* Δ*pta*) strain overexpressing the *crt*, *bcd*, *etfAB*, *hbd* and *adhE2* genes of *C. acetobutylicum*, and *atoB* gene of *E. coli* was able to produce 552 mg/L butanol using 2% (w/v) glycerol as a carbon source. In another case, *E. coli* JM109 strain overexpressing the *crt*, *bcd*, *etfAB*, *hbd*, *adhE* and *thiL* genes of *C. acetobutylicum* was developed. This engineered *E. coli* strain was able to produce 16 mM butanol using 4% (w/v) glucose as a carbon source [58]. More recently, metabolic engineering has been used for the development of *C. cellulolyticum* H10 for isobutanol synthesis directly from cellulose [59] (**Fig. 1B**). In this study, by expressing enzymes that direct the conversion of pyruvate to isobutanol using an engineered valine biosynthesis pathway, the recombinant *C. cellulolyticum* was able to produce up to 660 mg/liter of isobutanol when grown on crystalline cellulose. To our knowledge, this was the first demonstration of isobutanol production directly from cellulose.

Butanol production from crystalline cellulose by co-cultures of the thermophilic and cellulosome-producing *C. thermocellum* and the mesophilic and butanol-producing *C. saccharoperbutylacetonicum* (strain N1-4) has been reported recently [60]. Butanol was produced from Avicel cellulose after it was incubated with *C. thermocellum* for at least 24 h at 60°C before the addition of the solventogenic strain N1-4. Butanol produced by strain N1-4 on 4% Avicel cellulose peaked (7.9 g/liter) after 9 days of incubation at 30°C, and acetone was undetectable in this coculture system. Less butanol was produced by *C. acetobutylicum* and *C. beijerinckii* in co-culture with *C. thermocellum* under the same conditions than by strain N1-4, indicating that strain N1-4 was the optimal strain for producing butanol from crystalline cellulose in this system.

## 7. Conclusion

It should be noted that one of the most critical factors not only for biofuel production but also for the whole biomass biorefinery concept is securing low price substrates for the processes. To compete with the conventional fossil resource-based chemical industry, the biotechnology industry needs a reliable, cost-effective raw materials infrastructure. The cost effectiveness of biomass production and the efficient storage and transport of harvested biomass resources will be critical elements for securing raw materials. Environmental impacts and sustainability are also important issues. There is a cautious prediction that agricultural crop production may not match future industrial demand. A significant amount of research has been dedicated to engineering organisms that are capable of consolidated bioprocessing (CBP). These CBP organisms are anticipated to have the ability to efficiently degrade lignocellulose, and to convert the resulting sugars to biofuels and chemical compounds at high productivities. Towards this goal, the production of biorefinery products from lignocellulose has been shown to be feasible using mesophilic clostridia. Both the successes and problems encountered in establishing new pathways in clostridial species will aid in the adaptation of the consolidated bioprocessing strategy in related mesophilic clostridial species such as *C. acetobutylicum* and *C. cellulovorans*.

## Author details

Yutaka Tamaru*

*Department of Life Science, Graduate School of Bioresourses,*
*Department of Bioinformatics, Life Science Research Center,*
*Laboratory of Applied Biotechnology, Industrial Technology Innovation Institute, Mie University,*
*Tsu, Japan*

Ana M. López-Contreras

*Food and Biobased Research, Wageningen University and Research Centre, Wageningen,*
*The Netherlands*

## 8. References

[1] Tamaru Y, Miyake H, Kuroda K, Ueda M, Doi RH. Comparative genomics of the mesophilic cellulosome-producing *Clostridium cellulovorans* and its application to biofuel production via consolidated bioprocessing. Environ Technol 2010;31:889–903.

[2] Tamaru Y, Doi RH. Chapter 20: Bacterial strategies for plant cell degradation and their genomic information. *In* Carbohydrate Modifying Biocatalysts (Ed. Peter Grunwald). Pan Stanford Publishing Pte. Ltd. (Singapore), 2011;p.761–789.

[3] Lee SY, Park JH, Jang SH, Nielsen LK, Kim J, Jung KS. Fermentative butanol production by Clostridia. Biotechnol Bioeng 2008;101:209–228.

[4] Jones DT, Woods DR. Acetone–butanol fermentation revisited. Microbiol Rev 1986;50:484–524.

[5] Zverlov VV, Berezina O, Velikodvorskaya GA, Schwarz WH. Bacterial acetone and butanol production by industrial fermentation in the Soviet Union: Use of hydrolyzed agricultural waste for biorefinery. Appl Microbiol Biotechnol 2006;71:587–597.

[6] Jones DT, Keis S. Origins and relationships of industrial solvent producing clostridial strains. FEMS Microbiol Rev 1995;17:223–232.

[7] Zhang Y, Yang Y, Chen J, High-butanol ratio *Clostridium acetobutylicum* culturing method and its use. Chinese Patent 1997 CN 1063483C.

[8] Jiang Y, Xu C, Dong F, Yang Y, Jiang W, Yang S. Disruption of the acetoacetate decarboxylase gene in solvent-producing *Clostridium acetobutylicum* increases the butanol ratio. Metab Eng 2009;11:284–291.

[9] Dürre P. Fermentative butanol production: bulk chemical and biofuel. Ann NY Acad Sci 2008;1125:353–362.

[10] Swana J, Yang Y, Behnam M, Thompson R. An analysis of net energy production and feedstock availability for biobutanol and bioethanol. Bioresour Technol 2011;102:2112–2117.

[11] Ahn JH, Sang BI, Um Y. Butanol production from thin stillage using *Clostridium pasteurianum*. Bioresour Technol 2011;102:4934–4937.

[12] Lee SY, Park JH, Jang SH, Nielsen LK, Kim J, Jung KS. Fermentative butanol production by Clostridia. Biotechnol Bioeng 2008;101:209–228.

* Corresponding Author

[13] Ezeji TC, Qureshi N, Blaschek HP. Bioproduction of butanol from biomass: from genes to bioreactors. Curr Opin Biotechnol 2007;18:220–227.

[14] Heap JT, Pennington OJ, Cartman ST, Carter GP, Minton NP. The ClosTron: A universal gene knock-out system for the genus *Clostridium*. J Microbiol Methods 2007;70:452–464.

[15] Bramono SE, Lam YS, Ong SL, He J. A mesophilic *Clostridium* species that produces butanol from monosaccharides and hydrogen from polysaccharides. Bioresour Technol 2011;102:9558–9563.

[16] Alper H, Stephanopoulos G. Engineering for biofuels: exploiting innate microbial capacity or importing biosynthetic potential? Nat Rev Microbiol 2009;7:715–723.

[17] Knoshaug EP, Zhang M. Butanol tolerance in a selection of microorganisms. Appl Biochem Biotechnol 2009;153:13-20.

[18] Nölling J, Breton G, Omelchenko MV, Makarova KS, Zeng Q, Gibson R, Lee HM, Dubois J, Qiu D, Hitti J, Wolf YI, Tatusov RL, Sabathe F, Doucette-Stamm L, Soucaille P, Daly MJ, Bennett GN, Koonin EV, Smith DR. Genome sequence and comparative analysis of the solvent-producing bacterium *Clostridium acetobutylicum*. J Bacteriol 2001;183:4823–4838.

[19] Lee SY, Mermelstein LD, Bennett GN, Papoutsakis ET. Vector construction, transformation, and gene amplification in *Clostridium acetobutylicum* ATCC 824. Ann NY Acad Sci 1992;665:39–51.

[20] Minton, N.P., Brehm, J.K., Swinfield, T.-J., Whelan, S.M., Mauchline, M.L., Bodsworth, N., Oultram, J.D., 1993. Clostridial cloning vectors. *In:* Woods DR, editor. The clostridia and biotechnology. Stoneham: Butterworth-Heinemann. p. 119-150.

[21] Mermelstein LD, Welker NE, Bennett GN, Papoutsakis ET. Expression of cloned homologous fermentative genes in *Clostridium acetobutylicum* ATCC 824. Bio/Technology 1992;10:190–195.

[22] Mermelstein LD, Papoutsakis ET. *In vivo* methylation in *Escherichia coli* by the *Bacillus subtilis* phage w3TI methyl-transferase to protect plasmids from restriction upon transformation of *Clostridium acetobutylicum* ATCC 824. Appl Environ Microbiol 1993;59:1077–1081.

[23] Lee SY, Mermelstein LD, Papoutsakis ET. Determination of plasmid copy number and stability in *Clostridium acetobutylicum* ATCC 824. FEMS Microbiol Lett 1993;108:319–324.

[24] Green EM. Fermentative production of butanol—the industrial perspective. Curr Opin Biotechnol 2011;22:337–343.

[25] Harris LM, Blank L, Desai RP, Welker NE, Papoutsakis ET. 2001. Fermentation characterization and flux analysis of recombinant strains of *Clostridium acetobutylicum* with an inactivated *solR* gene. J Ind Microbiol Biotechnol 2001;27:322-328.

[26] Harris LM, Desai RP, Welker NE, Papoutsakis ET. Characterization of recombinant strains of the *Clostridium acetobutylicum* butyrate kinase inactivation mutant: need for new phenomenological models for solventogenesis and butanol inhibition? Biotechnol Bioeng 2000;67:1-11.

[27] Shao, L., Hu, S., Yang, Y., Gu, Y., Chen, J., Yang, Y., Jiang, W., Yang, S., 2007. Targeted gene disruption by use of a group II intron (targetron) vector in *Clostridium acetobutylicum*. Cell Res. 17, 963-965.

[28] Heap JT, Kuehne SA, Ehsaan M, Cartman ST, Cooksley CM, Scott JC, Minton NP. The ClosTron: mutagenesis in *Clostridium* refined and streamlined. J Microbiol Methods 2010;80:49-55.
[29] Heap JT, Minton NP. Methods. International Patent Application 2009. PCT/GB2009/000380.
[30] Dürre P. Biobutanol: an attractive biofuel. Biotechnol J 2007;2:1525–1534.
[31] Connor MR, Liao JC. Microbial production of advanced transportation fuels in non-natural hosts. Curr Opin Biotechnol 2009;20:307-315.
[32] Steen EJ, Chan R, Prasad N, Myers S, Petzold CJ, Redding A, Ouellet M, Keasling JD. Metabolic engineering of *Saccharomyces cerevisiae* for the production of n-butanol. Microb Cell Fact 2008;7:36.
[33] Nielsen DR, Leonard E, Yoon SH, Tseng HC, Yuan C, Prather KL. Engineering alternative butanol production platforms in heterologous bacteria. Metab Eng 2009;11:262-273.
[34] Evanko WA, Eyal AM, Glassner DA, Miao F, Aristidou A, Evans K, Gruber PR, Hawkins AC. Recovery of higher alcohols from dilute aqueous solutions. International Patent Application, 2009. PCT/US2008/088187.
[35] Donaldson GK, Eliot AC, Flint D, Maggio-Hall A, Nagarajan V. Fermentative production of four carbon alcohols. US Patent 2010, 7,851,188.
[36] Guedon E, Desvaux M, Petitdemange H. Improvement of cellulolytic properties of *Clostridium cellulolyticum* by metabolic engineering. Appl Environ Microbiol 2002;68:53–58.
[37] Cascone R. Biobutanol—a replacement for bioethanol? Chem Eng Prog 2008;104:S4–S9.
[38] Bayer EA, Lamed R, White BA, Flint HJ. From cellulosomes to cellulosomics. Chem Rec 2008;8:364–377.
[39] Tamaru Y, Miyake H, Kuroda K, Nakanishi A, Kawade Y, Yamamoto K, Uemura M, Fujita Y, Doi RH, Ueda M. Genome sequence of the cellulosome-producing mesophilic organism *Clostridium cellulovorans* 743B. J Bacteriol 2010;192:901–902.
[40] Tamaru Y, Miyake H, Kuroda K, Nakanishi A, Matsushima C, Doi RH, Ueda M. Comparison of the mesophilic cellulosome-producing *Clostridium cellulovorans* genome with other cellulosome-related clostridial genomes. Microb Biotechnol 2011;4:64–73.
[41] Meguro H, Morisaka H, Kuroda K, Miyake H, Tamaru Y, Ueda M. Putative role of cellulosomal protease inhibitors in *Clostridium cellulovorans* based on gene expression and measurement of activities. J Bacteriol 2011;193:5527–5530.
[42] Sabathe F, Bélaïch A, Soucaille P. Characterization of the cellulolytic complex (cellulosome) of *Clostridium acetobutylicum*. FEMS Microbiol Lett 2002;217:15–22.
[43] Pagès S, Bélaïch A, Fierobe HP, Tardif C, Gaudin C, Bélaïch JP. Sequence analysis of scaffolding protein CipC and ORFXp, a new cohesin-containing protein in *Clostridium cellulolyticum*: comparison of various cohesin domains and subcellular localization of ORFXp. J Bacteriol 1999;181:1801–1810.
[44] Shoseyov O, Takagi M, Goldstein M, Doi R.H. Primary sequence analysis of *Clostridium cellulovorans* cellulose binding protein A (CbpA). Proc Natl Acad Sci USA 1992;89:3483–3487.
[45] Kakiuchi M, Isui A, Suzuki K, Fujino T, Fujino E, Kimura T, Karita S, Sakka K, Ohmiya K. Cloning and DNA sequencing of the genes encoding *Clostridium josui* scaffolding

proteinCipA and cellulase CelD and identification of their gene products as major components of the cellulosome. J Bacteriol 1998;180:4303–4308.
[46] Gerngross UT, Romaniec MP, Kobayashi T, Huskisson NS, Demain AL. Sequencing of a *Clostridium thermocellum* gene (CipA) encoding the cellulosomal SL-protein reveals an usual degree of internal homology. Mol Microbiol 1993;8:325–334.
[47] Cantarel BL, Coutinho PM, Rancurel C, Bernard T, Lombard V, Henrissat B. The Carbohydrate-Active EnZymes database (CAZy): an expert resource for Glycogenomics. Nucleic Acids Res 2009;37:D233–D238.
[48] Dashtban M, Schraft H, Qin W. Fungal bioconversion of lignocellulosic residues; opportunities & perspectives. Int J Biol Sci 2009;5:578–595.
[49] Sandgren M, Stahlberg J, Mitchinson C. Structural and biochemical studies of GH family 12 cellulases: improved thermal stability, and ligand complexes. Prog Biophys Mol Biol 2005;89:246–291.
[50] López-Contreras AM, Gabor K, Martens AA, Renckens BA, Claassen PA, Van Der Oost J, De Vos WM. Substrate-induced production and secretion of cellulases by *Clostridium acetobutylicum*. Appl Environ Microbiol 2004;70:5238–5243.
[51] Kim AY, Attwood GT, Holt SC, White BA, Blaschek HP. Heterologous expression of endo-β-1,4-glucanase from *Clostridium cellulovorans* in *Clostridium acetobutylicum* ATCC 824 following transformation of the *engB* gene. Appl Environ Microbiol 1994;60:337–340.
[52] Perret S, Casalot L, Fierobe HP, Tardif C, Sabathe F, Bélaïch JP, Bélaïch A. Production of heterologous and chimeric scaffoldins by *Clostridium acetobutylicum* ATCC 824. J Bacteriol 2004;186:253–257.
[53] Mingardon F, Perret S, Bélaïch A, Tardif C, Bélaïch JP, Fierobe HP. Heterologous production, assembly, and secretion of a minicellulosome by *Clostridium acetobutylicum* ATCC 824. Appl Environ Microbiol 2005;71:1215–1222.
[54] Olson DG, McBride JE, Joe Shaw A, Lynd LR. Recent progress in consolidated bioprocessing.. Curr Opin Biotechnol 2011; Dec 14.
[55] Lynd LR, Weimer PJ, van Zyl WH, Pretorius IS. Microbial cellulose utilization: fundamentals and biotechnology. Microbiol Mol Biol Rev 2002;66:506–577.
[56] Ingram LO, Aldrich HC, Borges ACC, Causey TB, Martinez A, Morales F, Saleh A, Underwood SA, Yomano LP, York SW, Zaldivar J, Zhou S. Enteric bacterial catalysts for fuel ethanol production. Biotechnol Prog 1999;15:855–866
[57] Atsumi S, Cann AF, Connor MR, Shen CR, Smith KM, Brynildsen MP, Chou KJY, Hanai T, Liao JC. Metabolic engineering of Escherichia coli for 1-butanol production. Metab Eng 2008;10:305–311.
[58] Inui M, Suda M, Kimura S, Yasuda K, Suzuki H, Toda H, Yamamoto S, Okino S, Suzuki N, Yukawa H. Expression of *Clostridium acetobutylicum* butanol synthetic genes in *Escherichia coli*. Appl Microbiol Biotechnol 2008;77:1305–1316.
[59] Higashide W, Li Y, Yang Y, Liao JC. Metabolic engineering of *Clostridium cellulolyticum* for production of isobutanol from cellulose. Appl Environ Microbiol 2011;77:2727–2733.
[60] Nakayama S, Kiyoshi K, Kadokura T, Nakazato A. Butanol production from crystalline cellulose by cocultured *Clostridium thermocellum* and *Clostridium saccharoperbutylacetonicum* N1-4. Appl Environ Microbiol 2011;77:6470–6475.

Chapter 7

# Sulfur Trioxide Micro-Thermal Explosion for Rice Straw Pretreatment

Ri-Sheng Yao and Feng-He Li

Additional information is available at the end of the chapter

http://dx.doi.org/10.5772/50428

## 1. Introduction

With the gradual depletion of fossil energy resources in the global scope, the conversion of cellulose into biofuels attracts considerable attentions, since it is the most abundant renewable polysaccharide on the earth. The main sources of cellulose are woods, agricultural residues, hydrophytes and straws (Dieter and Thomas et al. 2003; Hon, 1996). Rice straw is a frequently abandoned crop straw in Asia and is often burned. This wastes cellulose and pollutes the air. Thus, utilizing rice straw to produce biofuel has promise. Generally, the procedure of converting rice straw into fuels is removal of the protective lignin first, followed by transformation of cellulose into glucose and at last, converting these sugars to biofuel by fermentation.

Removing lignin is critical to utilizing cellulose (Reith et al. 2003). Lignin content, crystallinity and particle size, limit the digestibility of cellulose (Hendriks and Zeeman, 2009). Therefore, various pretreatments remove or alter the hemicellulose or lignin and decrease the crystallinity of cellulose to enhance enzymatic hydrolysis efficiency (Goering, et al., 1970; Mosier et al., 2005; Zhao et al., 2009). The major methods include pretreatment by milling (Sato et al., 2009; Delgenés et al., 2002; Chang and Holtzapple, 2000; Palmowski and Muller, 1999), acid (Keikhosro K, 2006; Nguyen, Q, 2004; Ye and Jay, 2005; Taherzadeh and Karimi, 2007; Iranmahboob, F, 2002), steam explosion (Mukhopadhyay and Fangueiro, 2009; Brownell et al., 1986; Negro et al., 1992), liquid hot water (Liu and Wyman, 2005), alkali (Das and Chakraborty, 2009; Goswami et al., 2009; Fengel and Wegener, 1984; Laser et al., 2002), wet oxidation (Palonen and Thomsen, 2004; Kumar and Wyman, 2009), ammonia fiber explosion (AFEX) (Taherzadeh and Karimi, 2008; Li et al., 2009; Zheng et al., 1998; Wu et al., 1999), $SO_2$ catalyzed steam explosion (Balint and Emma, 2010) etc. Although these common pretreatments have made great successes in recycling cellulose from lignocellulosic biomass, they are not appropriate enough for biofuel production by industrialization if considering the disadvantages of them such as low efficient, huge energy consumption, high requirements for conditions of operations, environment pollution and so on.

Recently, rice straw has been pretreated with, ionic liquid (Vitz, J. and Erdmenger, 2009), low-temperature plasma (Li and Yao, 2012), microwave-assisted dilute lye (Li and Yao, 2012). In this chapter, we concentrate upon the novel pretreatment method of rice straw based on sulfur trioxide micro-thermal explosion (STEX), the moderate conditions for the operation of this pretreatment method make it a promising technology to reduce costs and increase the utilization of lignocellulosic cellulose.

## 2. Principles of sulfur trioxide micro-thermal explosion

The principles of STEX can be described by the model diagram shown in Fig.1. sulfur trioxide ($SO_3$) gas diffuses into the interior of rice straw through its capillary channels, reacts with the water in the tissue space and surface. As a result of violent exothermic process of the reaction, the gas in the limited tissue space expands and an in-situ detachment between the cellulose and the lignin will occur, moreover, lignin will be sulphonated partly during this process which may increase the solubility of the lignin in dilute lye. The operation can be performed under atmospheric pressure at about 50°C, it makes the subsequent treatment of stripping off lignin by dilute alkali solution more effectively as compared to the procedure by dilute alkali only (Yao and Hu etc. 2011).

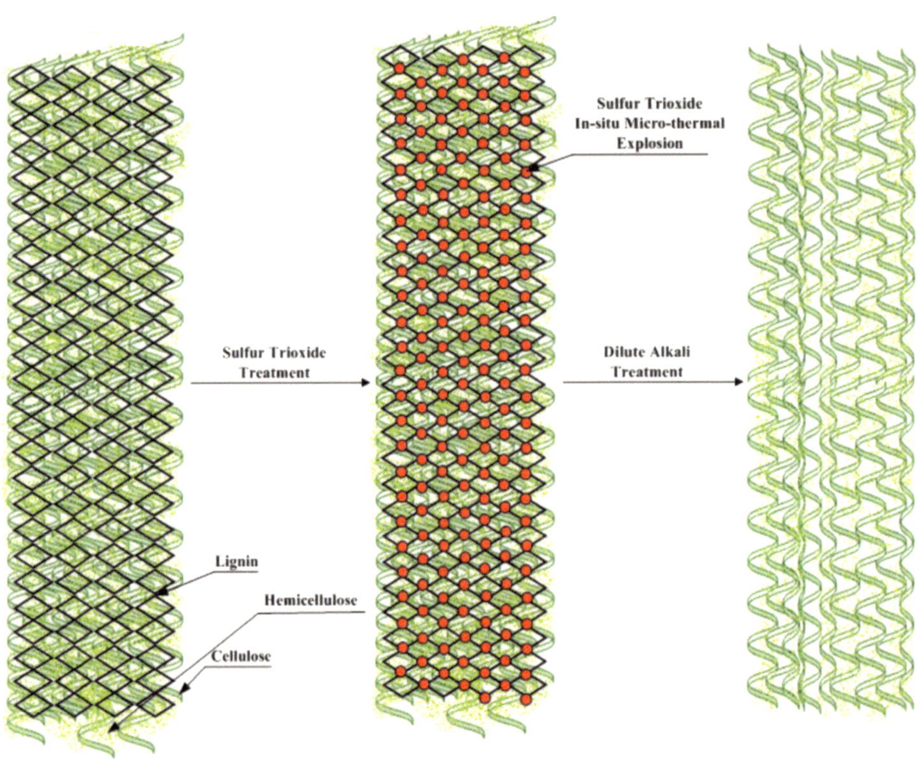


**Figure 1.** STEX collaborative dilute alkali model diagram

## 2.1. Gas diffuse and chemical exotherm in situ generating explosion

Comparing to liquid, gas diffuses more quickly and deeply into interior of rice straw through its capillary channels. When $SO_3$ gas accesses to the interior of rice straw, micro-thermo explosions take place as a result of reaction between $SO_3$ and water remaining in the pores of the rice straw, which make great changes of the interior structure of the rice straw.

The capillary channels presence in the original straw are arranged irregularly and some of its opening have been blocked (Fig.2A). The blocked openings are not conductive to the entry of cellulose hydrolysis enzyme. After straw is treated by $SO_3$ gas, the channels in the straw increase and are bundled, as if the internal structure of the straw are "hollowed out" (Fig.2B), which will promote the access of the substances during the subsequent handling. With increase of the STEX time, the channel structure becomes more obvious (Fig.2C). However, when the STEX time is more than 60 min, channels aggregate together and become not so obvious (Fig.2D and Fig.2E). The elongated STEX time may cause the internal environment of the straw more acidic, and induces acid hydrolysis, thereby major components of the rice straw degrade and the structure of the rice straw is damaged.

## 2.2. Procedure of sulfur trioxide micro-thermal explosion collaborative dilute alkali

The STEX for rice straw is convenient and can be performed under atmospheric pressure at about 50°C. A typical procedure of this method in the laboratory can be divided into two steps: First, 1g rice straw is cut into small pieces of about 2-3 cm in length, and hung over the upper portion of a test tube with 1mL oleum (Sulfur trioxide 50%) loading in the bottom, after raising the temperature of the system to 50°C SO3 gas is generated and reacts with the upper straw for 5~120min. Second, the as-prepared rice straw is soaked in NaOH solution (1%, w/v) with the solid-liquid ratio of 1:10, the mixture is stirred at 90rpm with placing the reaction flask in a water bath at 50°C for 2h, then the solids are washed with 800 mL of deionized water before the subsequent enzymatic hydrolysis procedure.

# 3. Structure of rice straw pretreated by $SO_3$ gas

STEX undermines the structure of straw at a microscopic level that makes it conducive to the subsequent alkali stripping. In this section, the results of scanning electron microscope (SEM), X-ray diffraction (XRD), and fourier transform infrared (FTIR) spectroscopy will reveal the effects of STEX on rice straw.

## 3.1. Surface morphology and crystallinity

### *3.1.1. Surface morphology observation by SEM*

The microscopic morphologies of pretreated-straws are shown in Fig.3. Regarding the untreated straw, a compact structure comprised of plant cell wall compositions such as

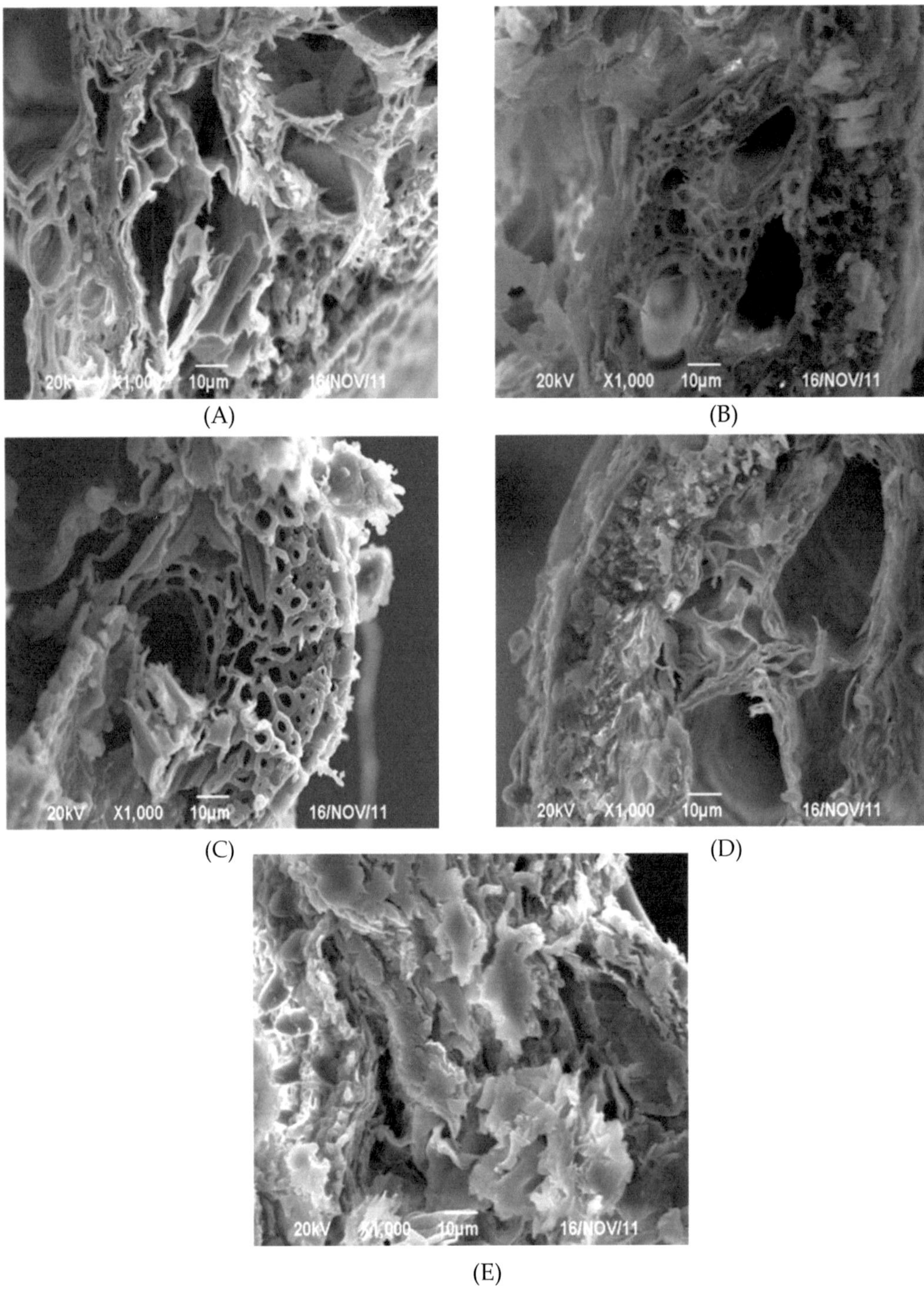


**Figure 2.** SEM of rice straw cross-sectional with the STEX, **A-**original straw, **B-**STEX 15min, **C-**STEX 30min, **D-**STEX 60min, **E-**STEX 120min.

epidermis, vascular bundles, lumens, dents and parenchyma still remains. To the rice straw treated by sulfur trioxide, the flaking traces and some holes can be observed on the straw surface (Fig.3B). As mentioned previously, gaseous sulfur trioxide can react with water in straw and release great heat in-situ at the same time. As a result, the air and water vapor in the limited interior spaces of the straw expand sharply to make a hot explosion from the interior and cause effective detachment between cellulose and lignin; it makes the subsequent removal of lignin with dilute lye more efficient. As shown in Fig.3C, vesicle shaped particles were stripped away effectively and exposures of cellulose filaments were significant after treatment of STEX treated-straw with 1% w/v NaOH for 2 hours at 50°C.

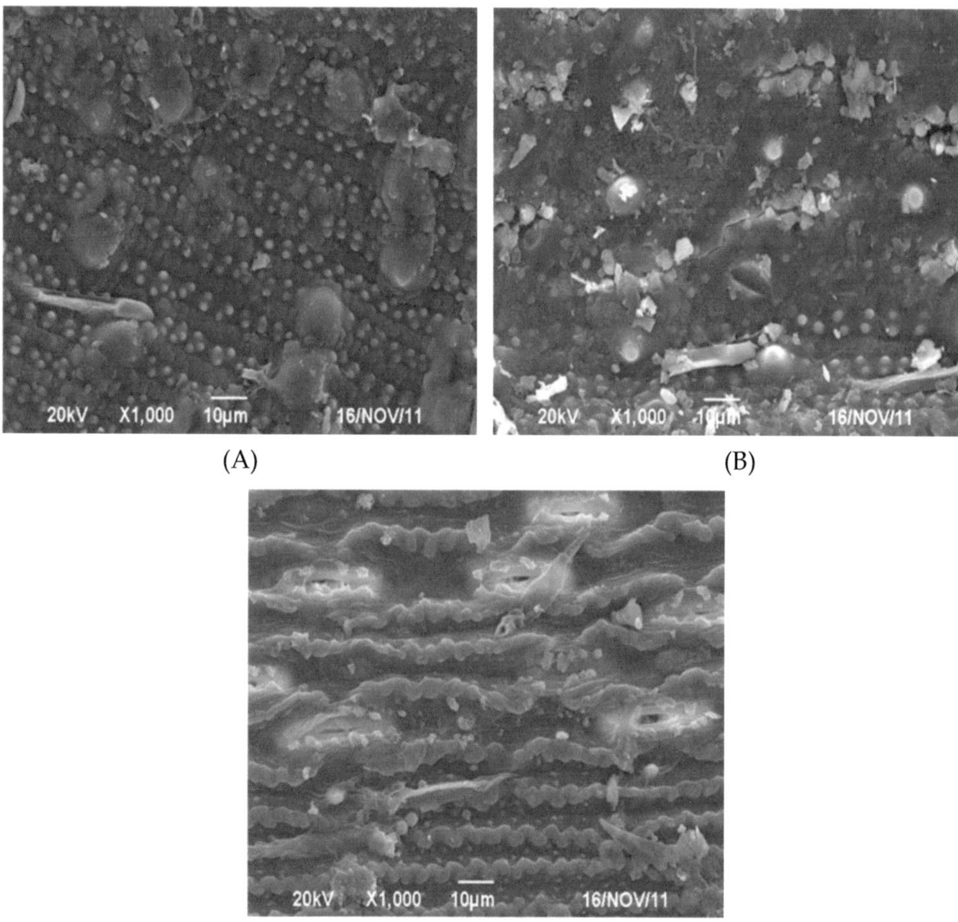


**Figure 3.** SEM imagines of straw samples with different treatment. **A**-original straw, **B**-STEX treated-straw, **C**-STEX collaborative dilute lye treated-straw

The disruption of the vesicle structure increases with reaction time between rice straw and $SO_3$ (Fig.4). In our experiments, when the reaction time was prolonged to 120 min, the structure of rice straw was almost completely ruptured.

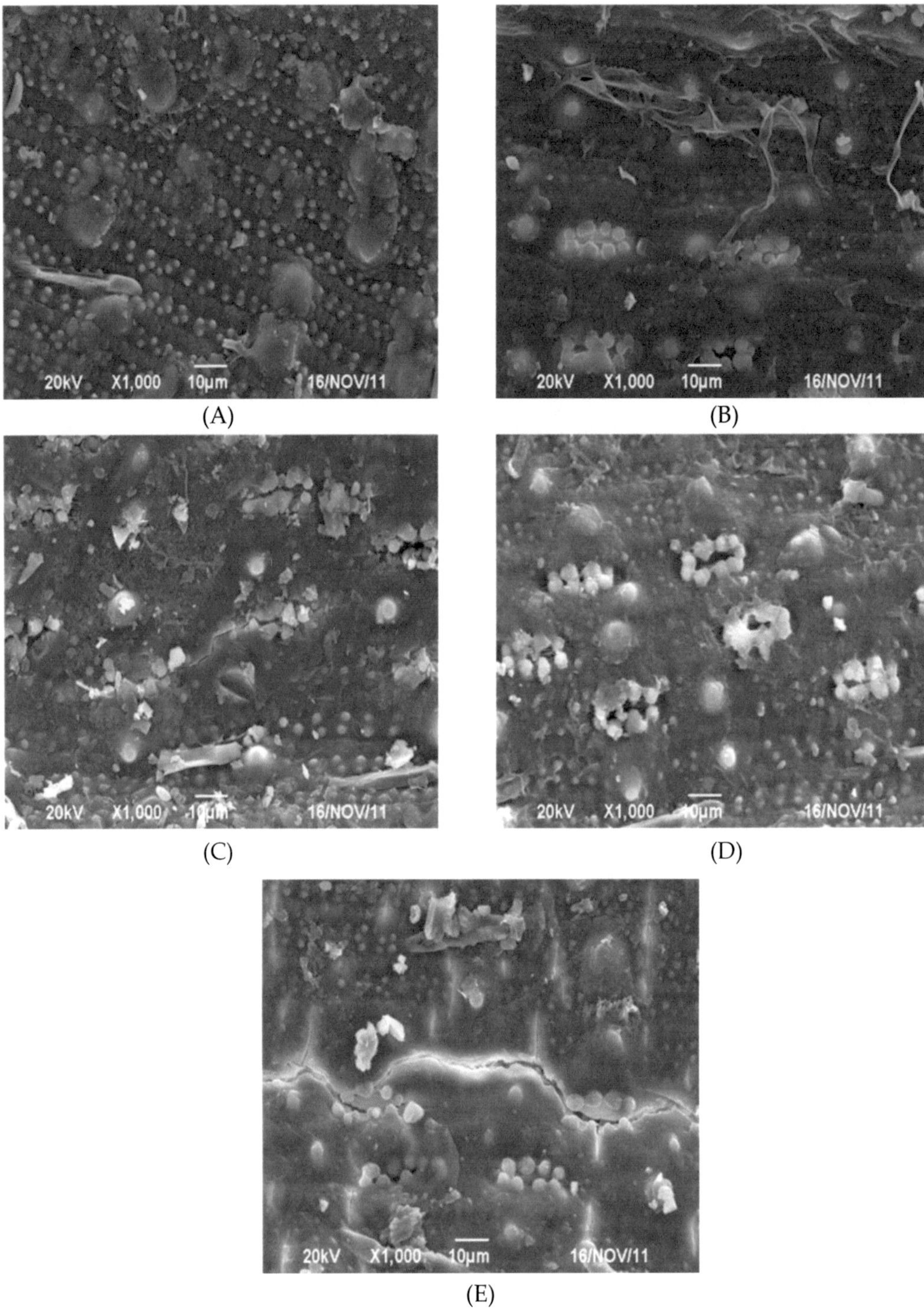


**Figure 4.** The SEM images of straw with different STEX time, **A-**untreated, **B-**STEX 15min, **C-**STEX 30min, **D-**STEX 60min, and **E-**STEX 120min

### *3.1.2. Crystalline analyzed by X-ray diffraction (XRD) and the influence of saccharification rate*

Crystallinity of cellulose is one of the main obstacles for enzymatic hydrolysis. STEX can reduce the crystallinity of rice straw effectively especially after the assistance of subsequent dilute lye. As shown in Tab.1, the effects of sulfur trioxide collaborative dilute alkali method are remarkable on the crystallinity and saccharification rate of the rice straw comparing with the ones of untreated, sulfur trioxide-treated only, alkali-treated only methods. Although sulfur trioxide-treated method decreases the crystallinity of cellulose, it doesn't significantly improve cellulose hydrolysis without the collaboration of dilute alkali. This result may imply that there are more profound changes occurring on contents and chemical structures of the straw during the process of the pretreatment.

| Pretreatment | Crystallinity(%) | Sugar conversion rate (%) |
|---|---|---|
| (A) | 64.7±0.6 | 22.2±0.8 |
| (B) | 58.5±0.3 | 41.8±0.4 |
| (C) | 55.8±0.7 | 64.3±0.6 |
| (D) | 51.3±0.4 | 91.7±0.4 |

**Table 1.** Pretreatment of rice straw had the crystallinity and the rate of saccharification

Sugar conversion rate from pretreated rice straw was calculated as follows: (EQ1)

$$SR = \frac{RG}{RS} \times 100\%$$

where RG is the dry-weight of reducing sugars in enzyme hydrolysis supernatant, RS is the dry-weight rice straw in pretreated solids.

## 3.2. Contents and chemical structures

### *3.2.1. Fourier transforms infrared (FTIR) spectroscopy*

Differences between the IR spectra of the untreated, sulfur trioxide-treated and sulfur trioxide collaborative dilute alkali-treated rice straws confirm changes of the rice straw occurring on the chemical level during STEX. As shown in Fig.6, the absorption bands at 3317cm$^{-1}$ (O-H stretch vibration), 2919 and 2850cm$^{-1}$ (C-H stretch vibration), 1630cm$^{-1}$ (C=C stretch vibration) and 1037cm$^{-1}$ (C=O stretch vibration) are characteristics of the original rice straw. After rice straw is treated by sulfur trioxide, new peaks at 1147cm$^{-1}$ (S=O stretch vibration) and 876 cm$^{-1}$(C-SO stretch vibration) appear. To the rice straw treated with sulfur trioxide collaborative dilute alkali, the peaks about 1147cm$^{-1}$(S=O stretch vibration) and 876 cm$^{-1}$(C-SO stretch vibration) disappear. These results indicate that lignin is sulphonated partly, then this part is been stripped on the processing with dilute lye. We suggest that not only the detachment between lignin and cellulose but also sulphonation of lignin during STEX may play an important role to the final lignin removal, which means the collaboration of dilute alkali is necessary.

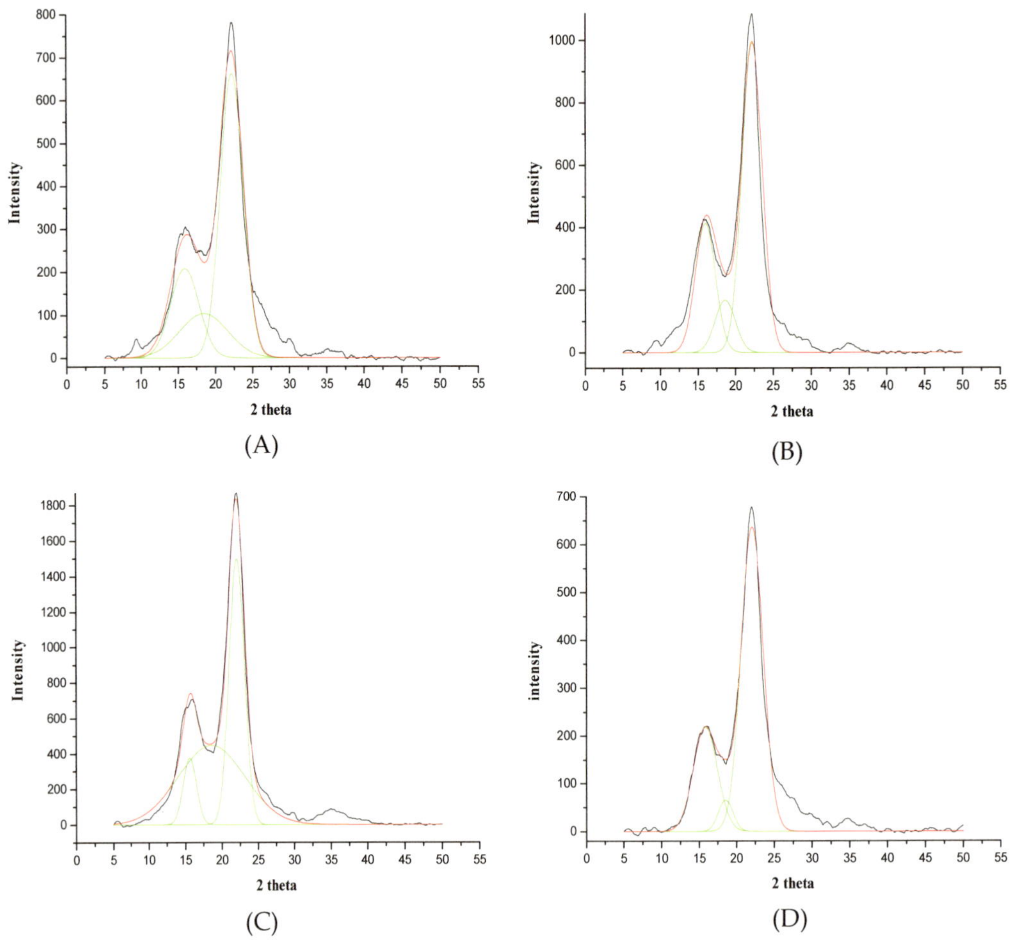


**Figure 5.** X-ray diffract diagram: (A) untreated rice straw; (B), rice straw was treated for 30min at 50°C by sulfur trioxide; (C), rice straw was treated for 7 hours at 50°C by 1% w/v NaOH. (D), straw was treated with sulfur trioxide for 30min following 1% w/v NaOH treatment for 7 hours at 50°C

### *3.2.2. Changes of main components in rice straw*

The quantities of the components of rice straw change a lot after being treated with STEX collaborative dilute alkali. Results from our work are shown in Table 2. The composition was calculated based on the dry weight of the samples and we found that water-soluble content of the STEX-treated rice straw increased. It indicates water soluble molecules such as oleoresin, lignin/hemicellulose fragments and sulphonated matter have more chances to contact water after STEX-treated. Moreover, the lignin and hemicellulose of the rice straw pretreated by STEX collaborative dilute lye decreased from 19.6% to 6.9% and 21.4% to 13.5%, respectively.

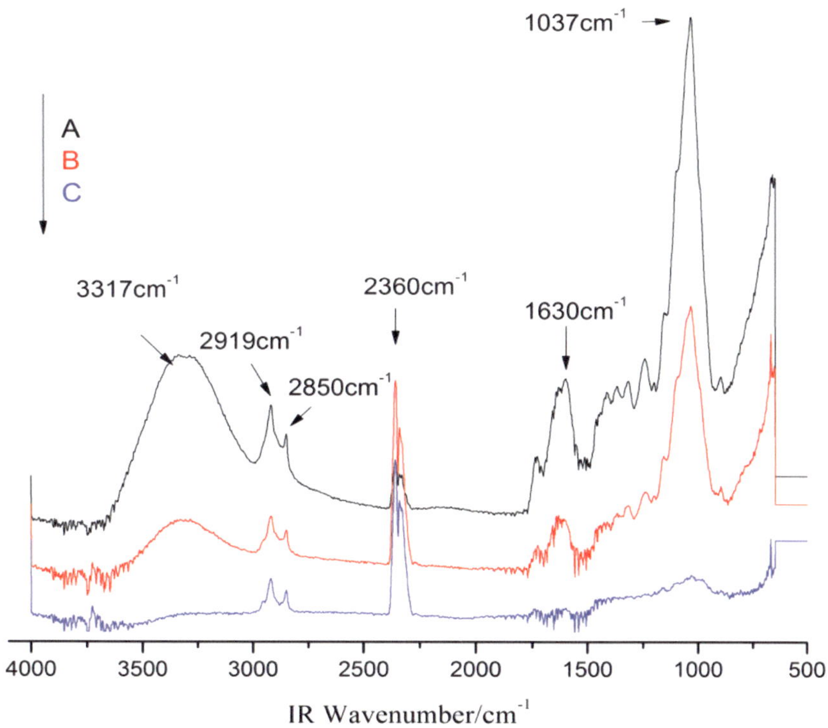


**Figure 6.** FTIR spectra: **A**-Untreated-straw, **B**-STEX treated-straw, **C**-STEX collaborative dilute alkali treated-straw.

| Composition (%) | Untreated straw (100.0 g) | STEX treated straw (100.8 g) | STEX assist lye treated straw (73.6 g) |
|---|---|---|---|
| Water-Soluble | 14.1 ± 0.2 | 15.9±0.4 | 8.6 ± 0.5 |
| Cellulose | 39.2 ± 0.7 | 39.8±0.5 | 65.8 ± 1.0 |
| Hemicellulose | 21.4 ± 0.4 | 20.4±0.6 | 13.5 ± 0.3 |
| Lignin | 19.6 ± 0.8 | 18.4±0.3 | 6.9 ± 0.6 |
| Ash | 5.7 ± 0.1 | 5.5±0.4 | 5.2 ± 0.3 |

**Table 2.** Chemical composition (percent by dry weight) of rice straw

The cellulose content increased from 39.2% to 65.8%, that is predominantly attributed to the decrease of lignin and hemicellulose. Removal of lignin can reduce the binding of lignin to hemicellulose/cellulose (Han et al. 1997; Lu et al. 2002; Ahola et al. 2008; Ma et al. 2008), and more removal of hemicellulose implies that the connections between the hemicellulose and cellulose are broken and more cellulose is exposed. In summary, STEX collaborative diluted lye pretreatment will partly break the lignocellulose structure, enhance enzymatic biocatalysis, increase the desired products yield and recycle more cellulose.

## 4. Saccharification and fermentation for bioethanol production

Producing ethanol from lignocellulosic materials need a series of chemical and biological procedures, namely opening the bundles of lignocelluloses in order to access the polymer chains of cellulose and hemicellulose, polymers hydrolysis to achieve monomer sugar solutions, fermentation of the sugars to obtain ethanol solution (mash) by microorganisms, purification of ethanol from mash by distillation and dehydration. Figure 7 is a flow figure designed for applying the technology of STEX for producing bioethanol.

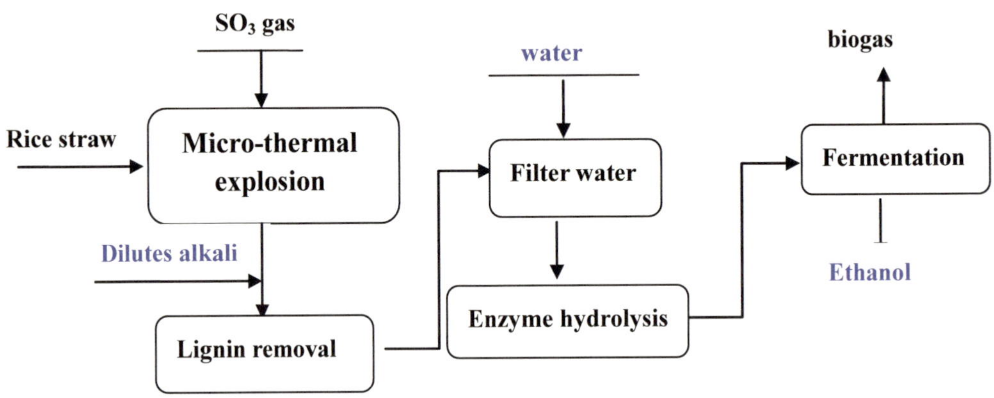


**Figure 7.** Flow diagram for bioethanol production from rice straw

### 4.1. Saccharification with enzymatic hydrolysis

#### *4.1.1. Water retention value (WRV) of celluloses*

The structure of untreated rice straw is nonporous and tight. The compact structure makes it difficult to moisturize the interior of rice straw and prevents from entering of the cellulase dissolved in water, which will decrease the efficiency of the enzyme hydrolysis. Hence, water retention value (WRV) of cellulose is often used to evaluate accessibility of cellulose for enzyme and we used it to evaluate the effects of the SO3 micro-thermal explosion. The straw water retention values were determined by centrifugation method. The straw was soaked in water at 25°C for 30 min, then centrifuged at 3000rpm for 15 min. The wet straw was weighed after centrifuging and then reweighed after being dried at 90°C for 2h, The WRV was calculated as the amount of water which retained in the rice straw after centrifuging comparing with dried weight of straw (Eq2)

$$WRV = \frac{W_w - W_d}{W_d} \times 100\%$$

#### *4.1.2. Water retention value and change of saccharification rate*

The impacts of pretreatment manners to WRV and saccharification rate are shown in Fig.8. To STEX treated rice straw, WRV and saccharification rate increased with the treating time during the initial 120min, when the treatment with sulfur trioxide duration was elongated the WRV began to decline rapidly and the saccharification rate just decreased slightly (Fig.8A), we suggest that a little of carbide might form in the rice straw when time of STEX was too long and cellulose became more hydrophobic. To dilute alkali-treated method, the WRV increased during the initial 120 min and then became stable, the saccharification rate increased within initial 180min (Fig.8B). The dilute alkali-treated method removed the lignin in lignocellulosic slowly, slightly destructed the internal structure of cellulose and the off lignin lignocellulosic structure will be opened gradually so that the WRV and the saccharification rate increased. The change of the saccharification rate and WRV of the sulfur trioxide followed by alkali-treated rice straw are shown in Fig.8C, the WRV increased within initial 60min but decreased after 60min and the saccharification rate increased and became stable after 120min.

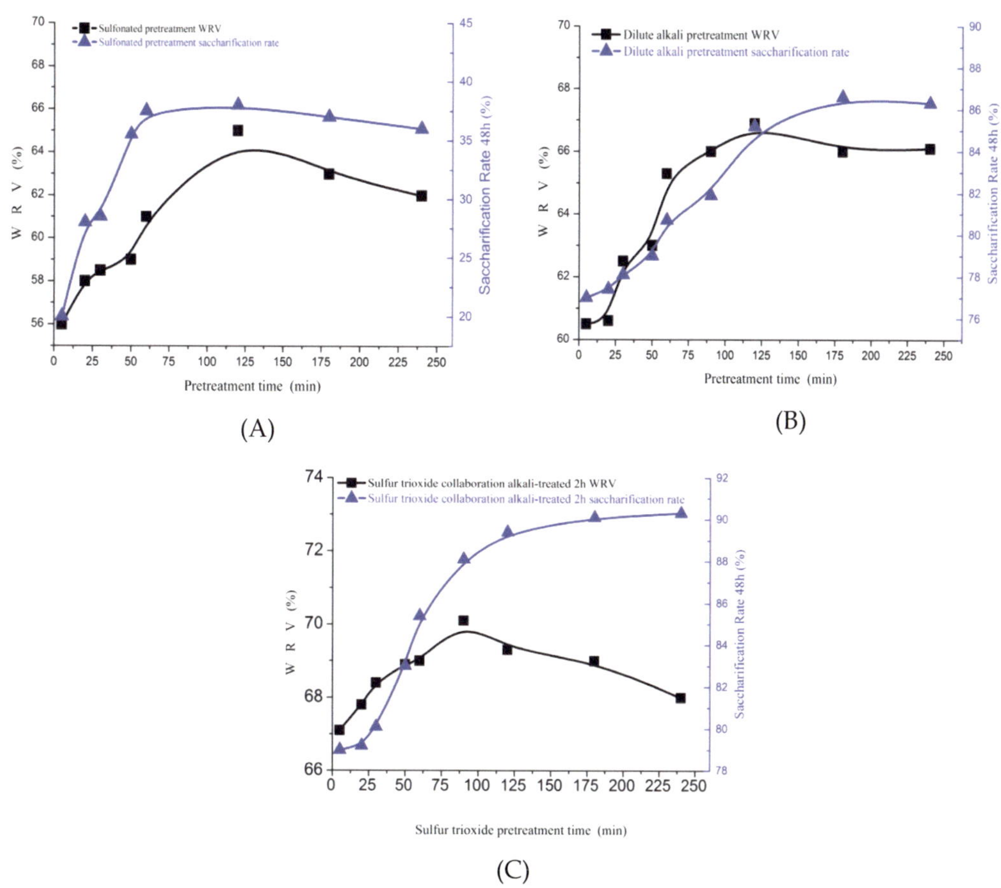


**Figure 8.** The impacts of WRV on saccharification rate by pretreatment. -▲-, saccharification rate; -●-, WRV; (A) rice straw was treated for 4 hours at 50°C by sulfur trioxide; (B) rice straw was treated for 4 hours at 50°C by 1% w/v NaOH; (C) straw was treated with sulfur trioxide for 4 hours followed by 1% w/v NaOH treatment for 2 hours at 50°C.

## 5. Future perspective

STEX for the pretreatments of cornstalk, rape straw, wheat straw had also been researched in our lab, and the STEX method exibited significant effects on the enhancement of enzyme hydrolysis of these herbs, that means STEX can act as a universal method for lignocellulosic pretreatment, and mdoderate requirements for performing STEX may improve competitiveness of the renewable energy from biomass. While there are also several aspects, which need to be researched for applying the STEX into commercial-scale application (industry), such as pretreatment equipment, clean technology in lower cost, and how to attain the utter utilization of lignocelluloses in order to manufacture high-purity cellulose. Meanwhile, the optimization of the STEX pretreatment process, clean and low-cost application of lignin black liquor and the research of high-efficiency pretreatment technology in the farmland are imminent. Commercial-scale application approach of rice straw with the STEX in the future is showed in Fig.9.

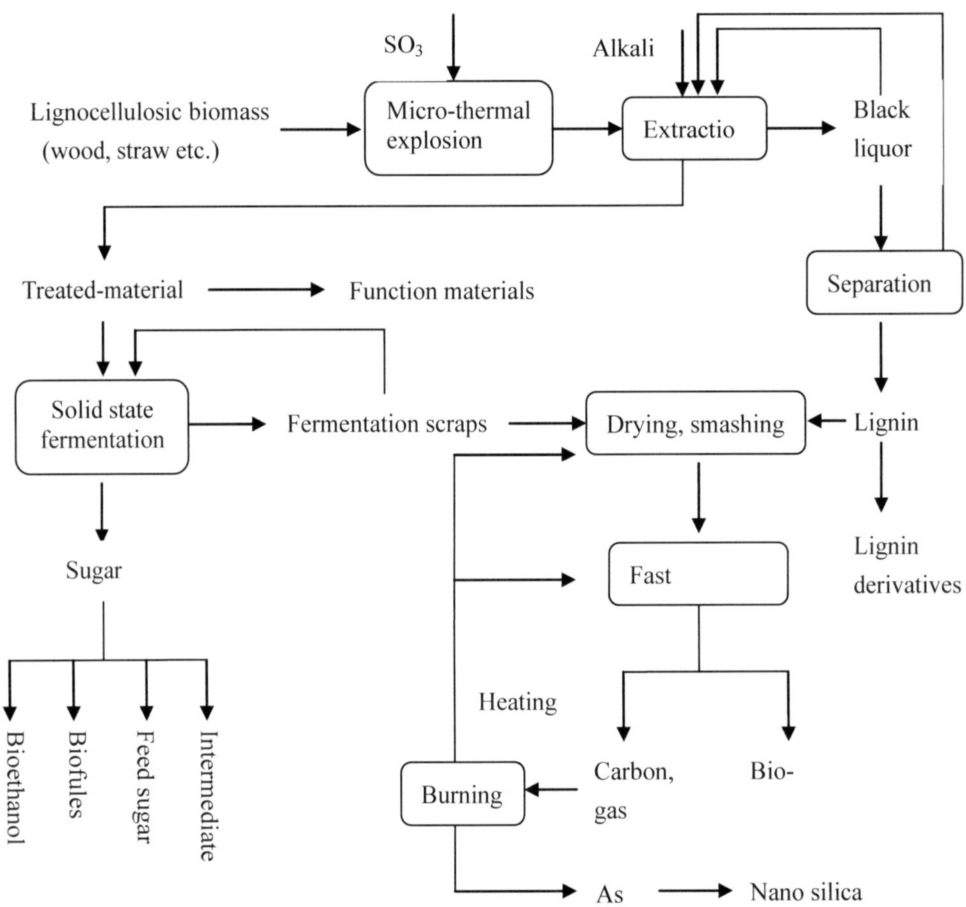


**Figure 9.** A platform of $SO_3$/alkali processing technology to achieve each component of straw and utter utilization of biomass

## Author details

Ri-Sheng Yao and Feng-He Li
*Hefei University of Technology, AnHui, China*

## 6. References

A.T.W.M. Hendriks, G. Zeeman, (2009) Pretreatments to enhance the digestibility of lignocellulosic biomass. *Bioresource Technology* 100, 10–18.

Balint Sipos, Emma Kreuger, Sven -Erik Svensson, Kati Reczey, Lovisa Bjornsson, Guido Zacchi, (2010) Steam pretreatment of dry and ensiled industrial hemp for ethanol production. *Biomass & Bioenergy* 34, 1721-1731.

Brownell, H.H., Yu, E.K.C., Saddler, J.N., (1986) Steam-explosion pretreatment of wood: effect of chip size, acid, moisture content and pressure drop. *Biotechnol. Bioeng.* 28, 792–801.

Chang, V.S., Holtzapple, M.T., (2000) Fundamental factors affecting enzymatic reactivity. *Appl. Biochem. Biotechnol.*, 5–37.

Das, M., and Chakraborty, D. (2009) Effects of alkalization and fiber loading on the mechanical properties and morphology of bamboo fiber composites. II. Resol matrix," J. *Appl. Polym. Sci.* 112, 447-453.

Delgenés, J.P., Penaud, V., Moletta, R., (2002) Pretreatments for the enhancement of anaerobic digestion of solid wastes Chapter 8. In: Biomethanization of the Organic Fraction of Municipal Solid Wastes. IWA Publishing, pp. 201–228.

Dieter K., Hans-Peter S. and Thomas H. (2003) polysaccharose II-Eukaryotes polysaccharide Chapter 10, cellulose, In: *Biopolymers,* Wiley-vch publishing, pp, 248-289.

Fengel, D., Wegener, G., (1984) Wood: Chemistry, Ultrastructure, Reactions. De Gruyter, Berlin.

Goering, H. K., and Van Soest, P. J. (1970) Forage fiber analyses (apparatus, reagent, procedure, and some application), *Agricultural Handbook* No. 379, Agricultural Research Service- United States Department of Agriculture (USDA), Washington D.C., 1-20.

Goswami, P., Blackburn, R. S., El-Dessouky, H. M., Taylor, J., and White, P. (2009) Effect of sodium hydroxide pre-treatment on the optical and structural properties of lyocell, *Eur. Polym. J.* 45, 455-465.

Hon, D.N.S. (1996) Functional polymers: a new dimensional creativity in lignocellulosic chemistry, in: *Chemical Modification of lignocellulosic materials,* pp.1-10. New York: Marcel Dekker.

Iranmahboob, F. Nadim and S. Monemi, (2002) Optimizing previous ermacid-hydrolysis: next term a critical step for production of ethanol from mixed wood chips, *Biomass and Bioenergy,* 22:401-404.

Keikhosro K., et al. (2006) Conversion of rice straw to sugars by dilute-acid hydrolysis. *Biomass and Bioenergy,* 30: 247-253.

Laser, M., Schulman, D., Allen, S.G., Lichwa, J., Antal Jr, M.J., Lynd, L.R., (2002) A comparison of liquid hot water and steam pretreatments of sugar cane bagasse for bioconversion to ethanol. *Bioresour. Technol.* 81, 33–44.

Li F.H., Hu H.J., Yao R.S., Wang H., Li M.M. (2012) Structure and saccharification of rice straw pretreated with microwave-assisted dilute lye. *Industrial & Engineering Chemistry Research.*

Li F.H., Yao R.S., (2012) Influence of low temperature plasma on the structure and saccharification of rice straw. Internal communication.

Li, R. J., Fei, J. M., Cai, Y. R., Li, Y. F., Feng, J. Q., and Yao, J. M. (2009) Cellulose whiskers extracted from mulberry: A novel biomass production, *Carbohyd. Polym.* 76, 94-99

Liu, C. G., and Wyman, C. E. (2005) Partial flow of compressed-hot water through corn stover to enhance hemicellulose sugar recovery and enzymatic digestibility of cellulose, *Bioresour. Technol.* 96, 1978-1985.

Mosier, N., Wyman, C. E., Dale, B. E., Elander, R. T., Lee, Y. Y., Holtzapple, M., and Ladisch, M. R. (2005) Features of promising technologies for pretreatment of lignocellulosic biomass, *Bioresource Technol.* 96, 673-686.

Mukhopadhyay, S., and Fangueiro, R. (2009) Physical modification of natural fibers and thermoplastic films for composites - A review. *J. Thermoplast. Compos.* 22,135-162.

Negro, M.J., Manzanares, P., Oliva, J.M., Ballesteros, I., Ballesteros, M., (2003) Changes in various physical chemical parameters of Pinus Pinaster wood after steam explosion pretreatment. *Biomass Bioenergy* 25, 301–308.

Nguyen, Q, (2004) Milestone completion report: evaluation of a two-stage dilute sulfuric previous termacid hydrolysisnext term process. Internal Report, National Renewable Energy Laboratory, Golden.

Palmowski, L., Muller, J., (1999) Influence of the size reduction of organic waste on their anaerobic digestion. In: II International Symposium on Anaerobic Digestion of Solid Waste. *Barcelona* 15–17 June, pp. 137–144.

Palonen, H., Thomsen, A.B., Tenkanen, M., Schmidt, A.S., Viikari, L., (2004) Evaluation of wet oxidation pretreatment for enzymatic hydrolysis of softwood. Appl. *Biochem. Biotechnol.* 117, 1–17.

Reith, J.H., Wijffels, R.H., Barten, H., (2003) Bio-methane & Bio-hydrogen. Status and perspectives of biological methane and hydrogen production. Production of the Dutch biological hydrogen production, The Hague.

Taherzadeh, M. J., and Karimi, K. (2008) Pretreatment of lignocellulosic wastes to improve ethanol and biogas production: A review, *Int. J. Mol. Sci.* 9, 1621-651.

Taherzadeh, M. J., and Karimi, K. (2007) Enzyme-based hydrolysis processes for ethanol from lignocellulosic materials: A review, *BioResources* 2, 707-738.

Vitz, J., Erdmenger, T., Haensch, C., Schubert, U.S. (2009) Extended dissolution studies of cellulose in imidazolium based ionic liquids. Green Chem. 11, 417–424.

Wu, M.M., Chang, K., Gregg, D.J., Boussaid, A., Beatson, R.P., Saddler, J.N. (1999) Optimization of steam explosion to enhance hemicellulose recovery and enzymatic hydrolysis of cellulose in softwoods. *Appl. Biochem. Biotechnol.*, 47-54.

Yao, R. S., Hu, H. J., Deng, S. S., Wang, H., and Zhu, H. X. (2011) Structure and saccharification of rice straw pretreated with sulfur trioxide micro-thermal explosion collaborative dilute alkali, *Bioresour. Technol.*102(10), 6340-6343.

Ye S., Jay J C. (2005) Dilute acid pretreatment of rye straw and bermudagrass for ethanol production. *Biochem. Biotechnol.*, 96(14):1599-1606.

Zhao, X. B., Cheng, K. K., and Liu, D. H. (2009) Organosolv pretreatment of lignocellulosic biomass for enzymatic hydrolysis. *App. Microbiol. Biot.* 82, 815-827.

Zheng, Y., Lin, H.M., Tsao, G.T. (1998) Pretreatment for cellulose hydrolysis by carbon dioxide explosion. *Biotechnol. Progr.*, 14, 890-896.

Chapter 8

# Raman Imaging of Lignocellulosic Feedstock

Notburga Gierlinger, Tobias Keplinger,
Michael Harrington and Manfred Schwanninger

Additional information is available at the end of the chapter

http://dx.doi.org/10.5772/50878

## 1. Introduction

The main structural plant cell wall polymers - cellulose, hemicelluloses, and lignin - rank amongst the most abundant biopolymers in Earth's carbon cycle. These three polymers form the lignocellulose complex and constitute the bulk of the cell wall with 40-50%, 10-40% and 5-30% of biomass by weight, respectively [1, 2]. Its highly ordered structure of cellulose microfibril aggregates, embedded in a matrix of hemicelluloses and lignin, provides the basis for its mechanical strength [3] and for the resistance to microbial attack [4], to which also low molecular mass extractives contribute [5]. Lignified cell walls are therefore a remarkably durable material. In nature, only higher fungi have developed biochemical systems to degrade the lignocellulose complex and perform the conversion and mineralisation of wood to carbon dioxide and water. Extensive reviews on decay pattern, chemistry and biochemistry of microbial wood degradation are available [4, 6, 7]. The natural processes occurring during fungal wood degradation may be utilised for industrial purposes and have a great potential for cellulose-producing and wood-processing industries as well as for high value-added conversion of lignocellulosic waste materials in *Biorefineries*. Particularly the molecular mechanisms of selective white-rot fungi offer a series of applications in the field of biotechnology of renewable resources [8].

Non-food biomass crops such as switchgrass (*Panicum virgntum* L.), *Miscanthus* x *giganteus,* and short-rotation coppice poplar and eucalyptus (*Populus* spp. and *Eucalyptus* spp.) [9] and willow (*Salix* spp.) offer a sustainable source of energy and platform for chemicals [10]. The absolute and relative contents of the cell wall components have a great influence on biomass quality i.e. its suitability for conversion to heat, power and chemical products. Biomass can be utilised by a number of thermochemical conversion routes (thermochemically, gasification, torrefication, flash pyrolysis) with differing feedstock demands and therefore measures of feed-stock quality are often quite specific [11, 12]. In contrast to the thermochemical processes where all plant cell wall polymers are converted to gases, the hydrolysis applications aim to convert carbohydrates into fermentable sugars. The recalcitrance to saccharification is a major limitation for conversion of lignocellulosic biomass to ethanol, which is mainly due to the lignin [13] content and composition [14]. Lignin synthesis has extensively been investigated using lignin model compounds [15-17] and lignin from plants [18-22] and has been reviewed several times [23-25] including lignin structure [26] and wood formation in general [27, 28]. The lignin content in natural *Populus* variants affects sugar release [29] and how lignin composition, structure, and cross-linking affect degradability has been reviewed for cell wall model studies [30]. High lignin contents in the feedstock necessitate harsh chemical and heat pre-treatments prior to enzymatic saccharification [11]. This increases energy inputs and often damages the polysaccharide components of the cell wall giving rise to inhibitory products [31, 32]. Therefore considerable pressure arises to optimise feedstock composition. The most feasible way to achieve this is by breeding improvement [11, 33], although agronomic practice may also influence composition [34] as well as genotypic and environmentally derived variation in the cell wall composition [35]. The characterization of different lignification genes has stimulated research programmes aimed at modifying the lignin profiles of plants through genetic engineering. The first transgenic plants with a modification of lignin composition and lignin content have been obtained in 1995 [36]. Since this time a focus on genetic engineering of lignin content and composition can be observed [37-40]. The effect of downregulation of lignin biosynthetic enzymes on wood anatomy [41] and its biomechanics [42] has been investigated [43]. In stems of transgenic alfalfa lines recalcitrance to both acid pre-treatment and enzymatic digestion was found to be directly proportional to lignin content [44]. Some transgenics yield nearly twice as much sugar from cell walls as wild-type plants. Lignin modification could bypass the need for acid pre-treatment and thereby facilitate bioprocess consolidation [45].

Both, breeding of lignocellulosic biomass and the production of transgenic plants, places huge demands on the analyst in terms of methods that can cope with the differences in polymer composition and linkages between them and large sample numbers. Wet-laboratory methods are destructive and time demanding and do not allow handling large sample numbers. Nuclear magnetic resonance (NMR) spectroscopy [46-49], analytical pyrolysis [50, 51], thioacidolysis [52], and thermogravimetry [53, 54] allow to get information about the composition and linkages between the wood polymers, and ultra-violet (UV) microscopy allows to follow the distribution of e.g. lignin within the cell wall [55, 56]. The requirements on a well suited method are: 1) fast and cheap to allow high-throughput screening 2) non-destructive to probe the native cell wall 3) to be able to analyse the content of each component (cellulose, hemicellulose, lignin) 4) to analyse their distribution within the plant tissues down to the cell wall level and 5) linkages as well as the interdependencies within and between the wood components. Vibrational spectroscopic methods such as infrared [57] and Raman [58] spectroscopy have shown potential to fulfil these requirements and can contribute to understand the actual lignocellulosic substrate and what kind of chemical and microstructural alterations take place during breeding, genetic engineering, decay or processing.

Near infrared (NIR) spectroscopy, that enables analyses of high number of samples on a day basis, was used for the prediction of the content of wood components and mechanical properties [59, 60] and the assignment of bands in the near infrared region have been reviewed recently [61]. Moreover it was shown that NIR spectroscopy can be used for the examination of the biodegradation of spruce wood by the white-rot fungi *Ceriporiopsis subvermispora* and that it is sensitive enough to differentiate between three applied strains [62]. NIR spectroscopy, which is often used in combination with multivariate data analysis, allows following the degradation of wood polymers [63-65]. Furthermore changes in interdependencies such as hydrogen bonding [66], and the accessibility of alcoholic and phenolic hydroxyl (O-H) groups to heavy water in non-degraded and brown-rot degraded spruce wood (*Piceas abies* L. Karst.) have been examined [67].

Mid infrared (MIR) spectroscopy allows similar investigations [68] as NIR spectroscopy with the advantage of better separated bands in the fingerprint region and the possibility of revealing the orientation of polymers and their interactions, which is of utmost importance in lignocellulose feedstock utilization. Dynamic Fourier transform infrared (FT-IR) spectroscopy has been shown to be appropriate for studying interactions among wood polymers and their ultrastructural organization [69-72]. In spruce wood fibres a close cooperation between cellulose and glucomannan in the fibre wall was suggested, whereas xylan showed no mechanical interaction with cellulose [69]. In primary cell walls investigations indicated a strong interaction among lignin, protein, pectin, xyloglucan, and cellulose [73]. Furthermore the orientation of cell wall polymers can be elucidated by polarised FT-IR measurements. In spruce glucomannan and xylan appear to have a parallel orientation with regard to the orientation of cellulose and, in all probability, an almost parallel orientation with regard to the fibre axis [74]. The first evidence for lignin orientation within native wood cell walls was revealed by Raman microprobe studies [75] and later confirmed in the secondary wall of tracheids fibres of thermomechanical pulp by FT-IR [70]. Very recently H-2 NMR spectroscopy was used to quantify lignocellulose matrix orientation with the ability to separately investigate oriented and unoriented amorphous domains in intact natural plant tissue [76].

Mid-infrared spectrometers can like Raman spectrometers be coupled to a microscope to reveal spatial resolution on the micron-level. Polarised FT-IR microscopy confirmed the preferential alignment of lignin in the direction of the fiber axis within the cell wall, but no orientation was found for the lignin in the middle lamella [77]. In combination with a fluid cell FT-IR microscopy was used to monitor *in-situ* the enzymatic degradation of cellulose-treated cross-sections of poplar (*Populus nigra* x *P. deltoids*) wood [78]. The accessibility of cellulose within the lignified cell wall was found to be the main limiting factor, whereas the depletion of the enzyme due to lignin adsorption could be ruled out. The fast, selective hydrolysis of the crystalline cellulose in the G-layer, even at room temperature, might be explained by the gel-like structure and the highly porous surface. Young plantation grown hardwood trees with a high proportion of G-fibres thus represent an interesting resource for bioconversion to fermentable sugars in the process to bioethanol [78]. FT-IR microscopy has been used to identify and characterise cell wall mutants [79-81] and transgenic plants altered in cell wall biosynthetic genes [82]. The localisation and characterisation of incipient brown-rot [83] and simultaneous and selective white-rot decay [84] within spruce wood cell walls was possible using FT-IR imaging microscopy. FT-IR microscopes equipped with focal plane array detectors allow very rapid chemical mapping over large areas with a spatial resolution limited by the wavelength of the infrared radiation to 10 - 5 μm. Thus single plant cells, which range generally between 10 - 50 μm in diameter, are resolved, but cellular components and cell wall layers may be substantially smaller and therefore below the limit of resolution. To overcome this limitation and get resolution on the cell wall layer level or for *in-situ* measurements of wet samples Raman microscopy imaging with a spatial resolution below 0.5 μm became the method of choice.

## 2. Basic principles, instrumentation, techniques and data analysis

### 2.1. Basic principles and Instrumentation

Raman and infrared spectroscopy monitor molecular vibrations, but are based on different principles. Raman spectroscopy involves inelastic scattering with a photon from a laser light source while IR spectroscopy involves photon absorption, with the molecule excited to a higher vibrational energy level. Thus, unlike infrared absorption, Raman scattering does not require matching of the incident radiation to the energy difference between the ground and excited states. In Raman scattering, the light interacts with the molecule and distorts (polarizes) the cloud of electrons round the nuclei to form a short-lived 'virtual state' before re-radiation. If only electron cloud distortion is involved in scattering, the photons will be scattered with very small frequency changes, as the electron mass is comparatively low. This elastic scattering process is the dominant process and called Rayleigh scattering. However, if nuclear motion is involved energy will be transferred either from the incident photon to the molecule (Stokes) or from the molecule to the scattered photon (Anti-Stokes) [85]. In the Raman scattering process the energy of the scattered photon is different from that of the incident photon (Raman-shift). Raman scattering therefore depends on changes in the polarizability due to molecular vibrations. On contrast IR absorption is based on changes in the dipole moments. Raman and IR spectroscopy thus provide "complementary" information about the molecular vibrations of a given sample. While water gives a strong absorption band in the IR (dipole), only weak Raman scattering is observed making this technique very suitable for *in-situ* studies of biological material.

The Raman scattering process is inherently a very weak process and only one of every $10^6$–$10^8$ photons is affected. It was experimentally the first time proven in 1928 and the first Raman spectra had to be recorded with very long acquisition times [86, 87]. The development of lasers in the 60's brought the method a big step forward as the Raman signal is proportional to the excitation power. Today the excitation laser power has to be adjusted well below the point where absorption leads to thermal decomposition of the sample, especially when biological materials are investigated. Furthermore, the Raman scattering intensity is proportional to $\upsilon^4$, where $\upsilon$ is the frequency of the exciting laser radiation [85]. Excitation at 400 nm ($=7.5*10^{-14}$ Hz) therefore leads to about 16 times higher Raman signal than excitation at 800 nm ($=3.75*10^{-14}$ Hz). But when measuring biological materials several components absorb the light in the lower wavelength range and therefore sample fluorescence can become problematic and swamp the Raman signal or even thermal sample decomposition may occur. Moving from the visible to the near-infrared (NIR) range, fluorescence virtually disappears as electronic absorption bands are unlikely. The use of Nd:YAG (neodymium-doped yttrium aluminum garnet) laser radiation at 1064 nm coupled with interferometers (involving Fourier transformations) led to so-called near infrared Fourier Transform (NIR-FT) Raman spectrometers [88]. Laser with wavelength in the visible range (e.g. $Ar^+$, He–Ne, $Kr^+$, doubled Nd:YAG lasers) are usually coupled with a dispersive spectrometer and a charge coupled device detector (CCD) for detection (Figure 1). These classical dispersive multichannel Raman spectrometers are nowadays often used in confocal microscope configurations with the advantage of superior rejection of fluorescence and depth resolution due to the pinhole [89].

For Raman microscopy, and especially for the imaging approach, the throughput of the radiation in the system has to be optimised in every part to acquire spectra fast and of high quality (high signal to noise (S/N) ratio). If a single spectrum is acquired, it is usually not important whether the necessary integration time is 0.1 s or 10 s. However this becomes an issue in scanning (mapping) experiments (imaging), when it becomes 15 min or 25 h. Therefore perfect coupling of the laser radiation into the microscope and out to the spectrometer is important as well as high throughput in the spectrometer and high detection efficiency of the CCD camera [90]. Optical fibres serve as ideal light pipes for connecting the elements and as a pinhole for the outgoing scattered radiation (Figure 1). Furthermore using a spectrometer optimized for the used wavelength range ("blazed" gratings) can increase the throughput as well as CCD cameras most sensitive for the chosen excitation wavelength [90, 91].

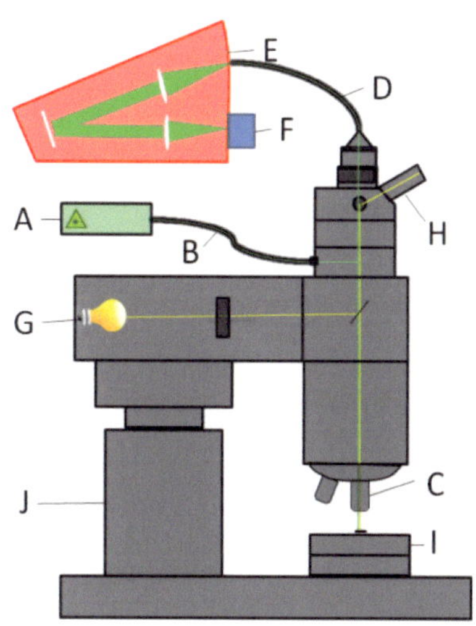


**Figure 1.** Typical set up of a confocal Raman microscope: The excitation laser (A) is focused via an optical fibre (B) and a microscope objective (C) onto the sample. The backscattered light is coupled out into a fibre (D), which acts as a pinhole. After passing the spectrometer (E) the signal is detected by a CCD camera (F). For visual inspection of the sample usually a white light source (G) and a camera (H) for picture capturing is available. For mapping/scanning the system is equipped with a piezo-driven X-Y- stage (I) and a Z-stage (J).

## 2.2. Resonance Raman spectroscopy, Surface Enhanced Raman Spectroscopy (SERS) and Coherent Anti-Stokes Scattering (CARS)

When a powerful beam of radiation is used some atoms and molecules of a sample absorb radiation at particular wavelengths and the e.g. coloured molecules become excited. Subsequently radiation of longer wavelength - termed fluorescence - is emitted. This fluorescence can be strong (intensive) and prevent the detection of the (weak) Raman signal [85]. But when the frequency of the laser beam is close to the frequency of an electronic transition, scattering enhancements of up to $10^6$ have been observed. In this resonance condition (Resonance Raman spectroscopy) the method becomes much more sensitive and since only the chromophore gives the more efficient scattering, it will also be selective for the part of the molecule involving the chromophore [85, 92, 93]. Furthermore fluorescence suppression can be achieved by using Kerr gating [93-95].

Another way of enhancing Raman intensity is to disperse the sample on metallic surfaces (either roughened wafers or colloidal solutions). The photon – plasmon interaction results in a huge signal enhancement and the technique, called surface-enhanced Raman spectroscopy (SERS), has progressed from studies of model systems on roughened electrodes to highly sophisticated studies, such as single molecule spectroscopy and molecular imaging [96-98]. The advantage is to enhance the Raman signal and besides the SERS effect leads to fluorescence quenching [99].

Another way of circumventing fluorescence is coherent anti-Stokes Raman scattering (CARS). This technique allows vibrational imaging with high sensitivity, high spectral resolution and three-dimensional sectioning capabilities. It is a nonlinear diagnostic technique that relies on inducing Raman coherence in the target molecule using two lasers, probed by a third laser which generates a coherent signal in the phase-matching direction at a blue-shifted frequency. Because of this nonlinear intensity dependence the photo-damage of the sample is reduced and the efficient background rejection improves the quality of the spectra [100]. CARS microscopy has already been used for imaging a number of delicate biological samples and processes, e.g. imaging of C–H stretching vibration present in the lipid bilayer of the cell membranes [101-103]. Two other Raman imaging techniques with great potential have evolved recently: Stimulated Raman scattering spectroscopy and Ultrafast Raman loss spectroscopy [104-107].

## 2.3. Spatial resolution and Tip-Enhanced Raman Spectroscopy (TERS)

The spatial resolution in Confocal Raman microscopy is limited by the diffraction of radiation and defined by the distance between the central maximum and the first minimum of the diffraction pattern, which is given by $r = 0.61 \lambda / NA$ ($\lambda$ = wavelength of the radiation, NA = numerical aperture of the objective) [108]. If high spatial resolution is sought-after, a laser in the visible range (e.g. 532 nm versus 1085 nm) and a microscope objective with a high numerical aperture (NA>1) have to be chosen. NA is defined by the refractive index of the medium (n) in which the optics are immersed (e.g. 1.0 for air and up to 1.56 for oils) and the half-angle of the maximum cone of radiation that enters or exits the condenser or objective (θ) (NA = n.sinθ). Two objects are completely resolved if they are separated by 2r and barely if they are separated by r (Rayleigh criterion of resolution) [108]. Therefore, the highest spatial resolution can be achieved with oil immersion objectives with high NA. Also if depth resolution is important, immersion objectives (oil, water) give better results [109]. Generally the axial resolution is around twice the lateral resolution [110].

Tip-enhanced Raman spectroscopy (TERS), which is based on the surface plasmonic enhancement and confinement of light near a metallic nanostructure, can overcome the so-called diffraction limit and produce optical images far beyond. It has been demonstrated that a spatial resolution as high as 4 nm could be achieved [111]. Consequently, nucleobases, proteins, lipids, and carbohydrates can be identified and localized in a single measurement. This has been shown in the last few years for different biological samples ranging from single DNA strand investigations to cell membrane studies [111-113].

## 2.4. Raman approaches for imaging

The main methods for Raman imaging are scanning (mapping) methods (Point-by-Point and Line scanning) and Wide-field source illumination approaches [114-117].

In Point-by-Point scanning the sample is scanned with a laser beam using X, Y, Z scanning stages. At each position of the raster a Raman spectrum is acquired and out of these spectra an image generated. The laser and the scattered light are often focused through so-called pinholes in order to know the exact position of the excitation and the collection volumes from the samples. The limiting factor of the Point-by-Point scanning approach is the fact that quite long measuring times are necessary because the duration is proportional to the number of pixels. Nevertheless the main advantage is that the whole Raman spectrum is acquired at each point and available for detailed analysis [114].

In the Line scanning approach the laser is elongated (1 dimension) to form a line with the help of a moving mirror or cylindrical optic devices. As a result the sample is illuminated with a laser line, which is parallel oriented to an entrance slit of a dispersive spectrograph. Scanning of the sample is still required, but only in the direction perpendicular to the laser line. This leads to a reduced experiment time [114]. It is the most efficient method if the spectral information from areas with perimeters of typically a few millimetre is required [116].

In Wide-field Raman imaging the whole sample field is illuminated with laser light. The experimental time depend primarily on the number of spectral channels or wavenumber positions at which an entire image is recorded [116]. There are numerous Wide-field Raman imaging approaches, such as liquid crystal tuneable filters (LCTFS) or the Fibre Array Spectral Translator (FAST). In FAST the received Raman light from a globally illuminated sample field is focused on a 2-dimensional array of optical fibres, which is followed/reduced to a one dimensional array on the distal end. This end is imaged through a dispersive spectrometer with a CCD detector. This method makes it possible to reduce two spatial dimensions data to a single dimension, which is afterwards dispersed fibre by fibre onto the CCD camera [114]. To characterize a sample´s chemical heterogeneity often only a few global Raman images need to be recorded at well-defined wavenumber positions, which are known either a priori or from spectral analysis of data obtained in point or line scanning [116].

As a non-destructive technique in general minimal or no sample preparation is necessary. Nevertheless to refer intensity changes in imaging approaches directly to changes in content or composition the same Raman scattering volume has to be probed at any position and this requires a flat surface. Otherwise a reference band for normalisation or the use of band ratios becomes necessary. Depending on the biological material to be probed microcutting or polishing might be the method of choice to achieve such a flat surface, with or without embedding [118, 119].

## 2.5. Processing of Raman spectra and image generation

To take the advantage of the scanning (mapping) method to have a molecular fingerprint (whole spectrum) at every pixel sophisticated data analysis has to be applied. Typically in each scanning experiment thousands of spectra are acquired and extracting the relevant information needs usually pre-processing of the spectra (e.g. cosmic ray removal, background subtraction, smoothing…) followed by univariate or multivariate data analysis methods to generate images.

### *2.5.1. Spectra pre-processing*

Raman instruments utilizing CCD detectors suffer from occasional spikes caused by cosmic rays. Cosmic rays are high energy particles from outer space which interact with atoms and molecules in the earth´s atmosphere and may generate a false signal in the shape of a very sharp peak in the spectrum. Various mathematical methods can be used to filter the cosmic rays from the spectra [120-122]. As the spikes are usually quite high and may overlay with bands of interest they have to be removed to avoid influences on the final results.

Smoothing algorithms are used to reduce noise in the recorded Raman spectra. They rely on the fact that spectral data are assumed to vary somewhat gradually when going from one spectral data point to the next, whereas associated noise typically changes very quickly. Different algorithms can be chosen (e.g. Savitzky-Golay [123], wavelet transformation [124], maximum entropy filter [122]) and especially before multivariate data analysis smoothing might become necessary.

Baseline correction and background subtraction can be performed based on linear models or on more complex mathematical functions. For removing background coming from the measured material (fluorescence) or signal from the substrate different methods have been developed that are capable of handling irregularly shaped baselines [125-128]. Baseline correction of Raman spectra is especially important prior to multivariate methods and different solutions to improve baseline correction methods have been developed [125, 129, 130].

Additional pre-treatments can be carried out to enhance certain properties of the image data set. The choice depends on the spectral structure and the goal of the data analysis. Derivatives can be carried out to stress subtle differences in spectral features among spectra. For pixel classification purposes, when the focus is on comparing the shapes of the pixel spectra independently from their global intensity, spectra normalization represents a useful option [126].

### 2.5.2. *Univariate and multivariate image generation*

In univariate data analysis each spectrum determines one value of the corresponding pixel in the image. The value of each pixel is determined by simple filters or by fitting procedures [122]. The most important of the simple filters is the integrated intensity (sum) filter evaluating the integrated intensity of various specific peaks found in the spectra of the image scan. The amount and scattering strength of a certain band attributed to a specific component is displayed and gives information on its spatial distribution. Filters can also plot changes in peak width, which can give a measure of crystallinity and structural orientation or changes in peak position (i.e., centre of mass position) as a measure for the strain within the material [131].

Many different multivariate methods exist and are described in detail elsewhere [126, 132-136]. Here only the very basics of the most commonly used ones, principal component analysis (PCA) and cluster analysis, are introduced.

PCA is the underlying method for many other multivariate methods since it is very effective for data reduction. It may be used to reduce the data set to 5–15 principal components (PC) and the residual error. Principal components are new, uncorrelated, and approximately normally distributed variables that provide faithful representations of the image, which can be used later as input information for exploration, segmentation, classification and other purposes. Compression by using principal components keeps all the relevant (image) information and, at the same time, allows understanding the relationship among the variables used to build the model by analysing the internal correlation structures provided by the loadings [132].

Cluster analysis applied to Raman images is essentially the sorting of the tens of thousands of spectra in a data set according to their similarities [122]. There are various ways of clustering, e.g. distance calculation (Euclidean, Manhattan), hierarchical cluster analysis, K-means Cluster Analysis, Fuzzy Clustering and each has its advantages and disadvantages [136].

In section 4 and 5 exemplary results for univariate analysis (band integration) and cluster analysis are shown.

## 3. Raman spectra of plant cell walls: What information can we gain?

Plant cell walls are nanocomposites based on cellulose microfibrils embedded in different matrix polymers (hemicelluloses, pectin and lignin) [137]. Besides water plays an essential role in the native plant cell walls and as water has relatively low polarizabilities weak Raman intensities are observed. Consequently, water saturated samples can be measured without problems. Raman investigations of cellulosic feedstock started in the 80s with the acquisition of spectra of cellulose fibres and wood [138-140].

### 3.1. Cellulose microfibrils: The structural elements of the cell wall

The cellulose microfibrils give a Raman signature comprising about 15 different significant bands (Figure 2, ramie fibre: almost pure cellulose). If these microfibrils are aligned with a preferred orientation, the Raman intensity of the cellulose bands depends on the angle between the orientation of the cellulose microfibrils and the laser polarisation direction [138]. The investigated Ramie fibres have almost perfect alignment of the cellulose microfibril parallel to the fibre axis and high crystallinity (X-ray results, not shown). Changing the laser polarization from parallel with respect to the fibre axis (0°) to perpendicular to the fibre (90°) results in severe changes of the Raman intensity of almost all characteristic bands (Figure 2) except the two bands at 1377 and 437 $cm^{-1}$. The bands at 1457 $cm^{-1}$ (HCH and HOC bending), 517 $cm^{-1}$, 499 $cm^{-1}$ and 378 $cm^{-1}$ (heavy atom stretching) show higher intensity in 90° arrangement and thus a more perpendicular alignment of these groups (Table 1). The β–(1→4)-glycosidic linkages in cellulose molecules, the methine groups of the glucopyranose rings and the methylene groups of the glucopyranose side are heavily orientation-dependent and reflect the cellulose molecule orientation along the fibre axis.

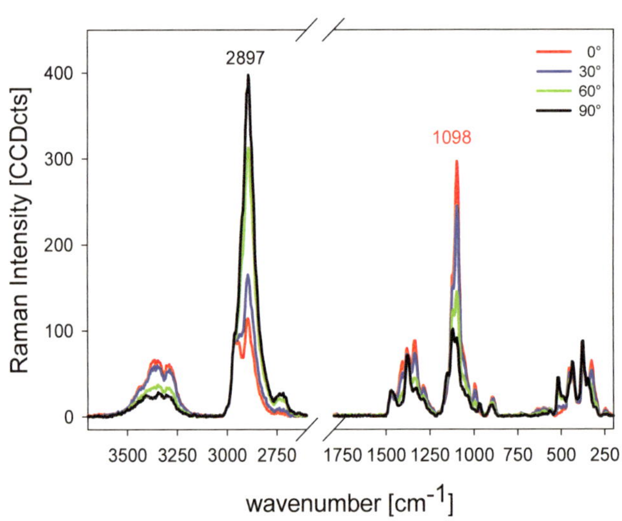


**Figure 2.** Raman spectra (532 nm excitation, 0.25 s integration time, 10 accumulations, baseline corrected) acquired from a Ramie fibre (>95% cellulose, high crystallinity, microfibrils aligned parallel to the fibre axis) with changing the polarization direction of the incident laser from 0° (parallel to the fibre axis and cellulose fibrils) to 90° (perpendicular to the fibre axis).

| Band [$cm^{-1}$] | Assignment | Molecular orientation * |
|---|---|---|
| 330 | δ (CCC) ring | ‖ s |
| 380 | δ (CCC) ring | ⟂ m |
| 436 | Γ (COC) def | ⟂ m |
| 497 | ν (COC) glycosidic | ⟂ m |
| 520 | δ (COC) glycosidic | ⟂ s |
| 902 | υ (COC) in plane, sym | ‖ m |
| 970 | ϱ ($CH_2$) | ⟂ m |
| 998 | ϱ ($CH_2$) | ‖ s |
| 1098 | υ (COC) glycosidic, asym | ‖ s |
| 1121 | υ (COC) glycosidic, sym | ⟂ m |
| 1340 | ω ($CH_2$) | ‖ s |
| 1380 | δ ($CH_2$) | ∘ |
| 1472 | δ ($CH_2$); δ (COH); | ⟂ m |
| 2897 | υ (CH) | ⟂ s |
| 3200-3500 | υ (OH) | ‖ s |

**Table 1.** Position and assignment [138, 141] of the cellulose bands (Ramie fibre, Figure 2) and their preferred molecular orientation parallel (II) or perpendicular (⟂) to the fibre axis direction (=laser polarisation direction). The sensitivity of the bands to intensity changes due to orientation is referred to as no changes (∘) and strong (s) or medium (m). (def: deformation, sym: symmetric, asym: antisymmetric)

Because of the orientation-dependence of the cellulose band intensities, the fibre direction (plant axis) and the laser polarization have to be adjusted in a known and defined way in every plant cell wall Raman experiment. As the intensity of multiple bands change in a characteristic way (up and down, Figure 2), it is possible to distinguish between intensity changes due to alterations in fibre orientation from those resulting from different cellulose content (all bands increase or decrease). To eliminate intensity changes due to different focal plane during rotating the polarizer or drift of the scan stage band height ratios or band area ratios can be calculated for a more detailed analysis (Figure 3A). These ratios also allow the comparison of Raman measurements with different integration times and thus intensity. The ratios (2897/1095, 378/1095 and 1377/1095) reveal a clear dependency of the cellulose band intensities and the angle of the incident laser polarization. The strong relationship can be described by a cosine function and a quadratic regression ($R^2>0.99$). Based on the band height ratios (Figure 3B) or partial least square regression models the angle of the cellulose molecule with respect to the laser polarisation direction and consequently the microfibril angle can be calculated [142]. Noteworthy, changes in fibre orientation often correspond to alterations in cellulose crystallinity. The effect of changes in crystallinity on the shape of the cellulose Raman bands was also investigated in detail: amorphous cellulose results in a significant decline in band heights accompanied by band broadening [143].

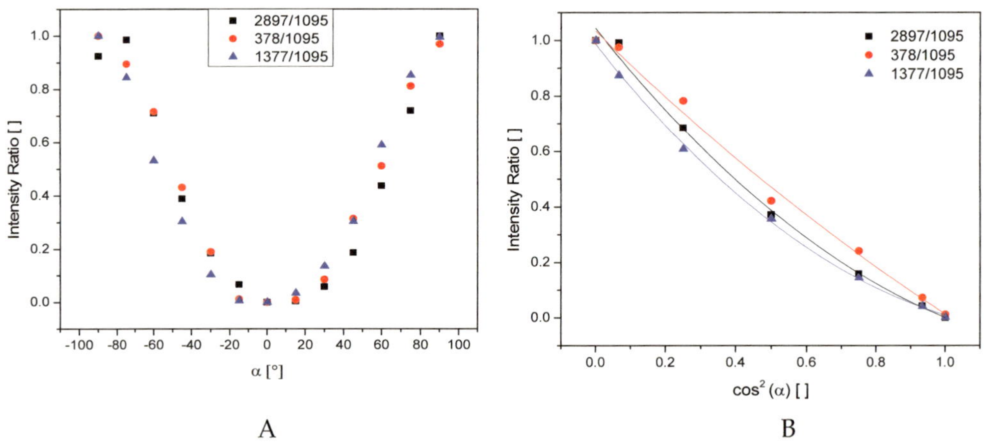


**Figure 3. A-B.** Relationship between the intensity ratios of cellulose bands and the angle of laser polarization $\alpha$ (A, B: accounting for a cosine function with a quadratic regression)

## 3.2. Carbohydrate matrix polymers: Hemicelluloses and pectin

Hemicelluloses and cellulose have similar functional groups and chemical bonds and therefore the Raman contributions are overlapping. Due to the more amorphous nature of hemicelluloses the Raman signal intensity is less and the bands are usually broader [144]. According to Himmelsbach et al. [145] the weak bands between 870–800 and 515–475 $cm^{-1}$ offer the potential to distinguish between cellulose and xylan in flax fibres. In *Miscanthus* a characteristic band at 478 $cm^{-1}$, assigned to HCC and HCO bending at C6 of hemicelluloses was used to present the hemicellulose distribution [146]. Nevertheless within wood and pulp samples the bands typical for glucomannan and xylan were hardly detected [144]. Distinguishing and probing the hemicelluloses in cell walls with the Raman imaging approach is therefore not that straightforward and needs sometimes more sophisticated data analysis tools to resolve the overlapping and less intense characteristic bands.

While cellulose and hemicelluloses have β-glycosidic bonds, pectins are composed of α-glycosidic linkages. In the Raman spectrum the region between 860-825 $cm^{-1}$ corresponds to equatorial anomeric H (α-anomers and α-glycosides), whereas the band at about 900–880 $cm^{-1}$ corresponds to axial anomeric H (β-anomers and β-glycosides) [147]. The sharp Raman band between 860 and 854 $cm^{-1}$ is characteristic for pectin and shows no overlap with the other plant cell wall polymers and can therefore be used as a marker band in the imaging approach [148, 149]. Furthermore the exact position of this band is sensitive to the state of uronic carboxyls and to *O*-acetylation thus providing insight into pectin structure; decreasing with methylation (min. 850 $cm^{-1}$) and increasing with acetylation (max. 862 $cm^{-1}$) [150].

## 3.3. The aromatic lignin polymer: Fluorescence and diversity

The structure of lignin is comprised of a variety of different types of covalent bonds derived from oxidative coupling of three different types of phenolic precursor units, p-coumaryl, coniferyl, and sinapyl alcohols [151, 152]. The structural organisation of lignin is a subject of much debate, both in terms of chemical structure (H (p-hydroxyphenyl), G (guaiacyl) and S (syringyl) units/monomers and the bondings) and in terms of the degree to which lignin is ordered within its cell wall environment. Beside NMR and IR spectroscopy also Raman microscopy has shown high potential for non-invasive investigation of *in-situ* cell wall lignin structure during the last years e.g. [153, 154].

Laser-induced autofluorescence from lignin can be the major hindrance to acquire reasonably good Raman spectra because the fluorescence intensity can be several orders of magnitude larger than the Raman scattering intensity. Traditionally, two sampling procedures were used to effectively reduce the autofluorescence: water immersion technique (usable for woody tissues) [140] and oxygen flushing technique [155]. Fluorescence problems can be reduced by choosing the near-IR Fourier transform (NIR-FT) Raman technique, using a NIR laser source with the photon energy well below troublesome low energy electronic transitions of lignin. Good quality spectra, relatively free of fluorescence interference, have been acquired from various lignin-containing materials [156-161]. Today, also more sophisticated spectroscopic methods can overcome this problem. UV resonance Raman spectroscopy exploits the combined benefit of the resonantly enhanced Raman signal and the usually relatively much longer wavelengths of fluorescence emission compared to Raman photons [153, 162-164]. By Kerr-gated Raman spectroscopy the different time-domain characteristics of fluorescence and Raman emission allow the detector only to see a narrow time-domain window centred on the excitation laser pulse [93, 164]. Also CARS gives spectra free of background from one-photon-excited fluorescence and has been used to study lignin modification in alfalfa [165]. All the significant instrumental developments opened up new fields for investigating lignified samples.

Improvements in Confocal Raman mapping/imaging approaches have provided insights into lignin distribution on the microscale. Due to the high spatial resolution it is possible to acquire spectra comprising only the chemistry of the cell corner, which is in lignocellulosics dominated by lignin contribution (Figure 4). The imaging approach requires short integration times (e.g. 0.1-0.4 s) and therefore not all lignin bands are resolved and the spectra are dominated by the strong band around 1600 $cm^{-1}$ (Figure 4A), which is assigned to aryl stretching vibrations [166]. As this band has no overlap with the carbohydrate bands it can be used as a marker to image lignin distribution on the micron level [167, 168]. Depending on laser excitation and lignin structure, more or less background or resonance enhancement is observed. Due to the different chemical structure of softwood and hardwood lignin particular laser excitation (e.g. 532 nm) results in a higher fluorescence background and stronger 1607 $cm^{-1}$ band intensity in spruce than in poplar. In softwood species, the most abundant precursor is coniferyl alcohol, which leads to an aromatic substitution by one methoxyl group, known as a guaiacyl structure (G lignin). In hardwood additionally sinapyl alcohol leads to syringyl structures (S lignin) with two methoxyl groups attached to the aromatic ring. Additionally, during the formation of the middle lamella p-coumaryl alcohol precursors are present and p-hydroxphenyl lignin without methoxyl groups is formed. The differences in lignin structure in the cell corner of spruce and poplar are reflected by the different intensity and band shape in the region at about 1600 $cm^{-1}$ as well as in the other bands (Figure 4A). Contribution from coniferaldehyde units is expected at 1623 and 1660 $cm^{-1}$, whereas coniferyl alcohol contributes at 1654 $cm^{-1}$ as well as other chromophores [169]. Using these bands the amount of coniferyl alcohol and aldehyde groups compared to the total amount of lignin was imaged in pine and spruce wood samples [170]. For the S units the intense band at 1328 $cm^{-1}$ is characteristic, while in spruce the band is found at 1334 $cm^{-1}$ [144] and accompanied by bands (shoulders) below and above (Figure 4A, Table 2). The relatively intensive band at about 1150 $cm^{-1}$ in poplar wood was tentatively assigned (Table 2). On the contrary, in spruce the band at 1139 $cm^{-1}$ is stronger (Figure 4A).

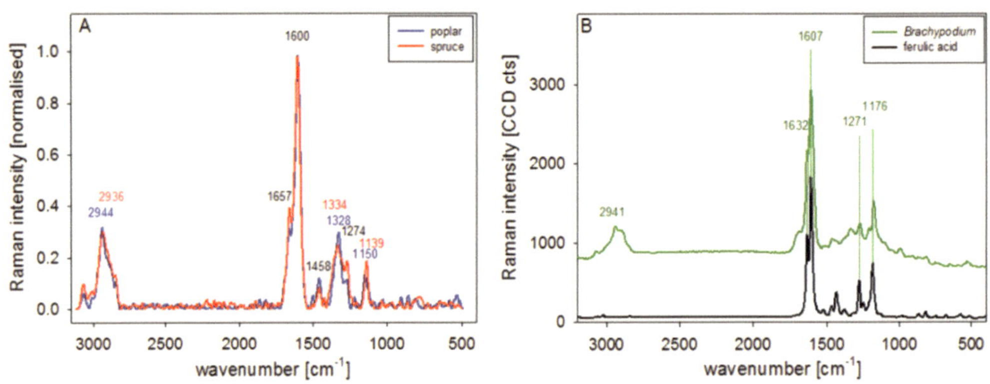


A B

**Figure 4. A-B**. Raman spectra (532 nm excitation, 0.1-0.4 s integration time) acquired from the cell corner of wood (poplar and spruce, A) and *Brachypodium* (B), reflecting the different lignin structures in softwood, hardwood and grasses. Additionally a reference spectrum of ferulic acid is plotted (B, black line)

Grasses have Type II cell walls, which in addition to other cell wall polymers, typically contain arabinoxylans and phenolics [171-173]. Noteworthy, grass xylans play an important role in the cell wall by helping to facilitate the assembly of cellulose microfibrils or/and the cross-linking of lignin to polysaccharides with the aid of hydroxycinnamic acids [174]. When compared to dicots, a high amount of hydroxycinnamic acids (ferulic and p-coumaric acid) is characteristic for grass cell walls. Therefore the cell corner spectrum of *Brachypodium* (Figure 4B) differs remarkably from the wood spectra (Figure 4A). The very strong aryl stretching vibration at 1607 $cm^{-1}$ is a clear doublet with a second band at 1632 $cm^{-1}$, accompanied by a weaker band at 1701 $cm^{-1}$. This doublet peak as well as the bands at 1176 $cm^{-1}$ and 1276 $cm^{-1}$ are typical for ferulic acid [175, 176] (Figure 4B). So the cell corner spectra of *Brachypodium* clearly reflect contributions from ferulic acid. Similar bands have been found in corn stover, although not recognized to be due to ferulic acid [177].

| Spruce | Poplar | *Brachypodium* | Assignment [Reference] |
|---|---|---|---|
| Wavenumber [$cm^{-1}$] | | | |
| 2936 | 2944 | 2941 | antisymmetric C-H str. in $OCH_3$ (SW) [178] and (HW) [179]; symmetric C-H str. in $CH_3$ in FA [180] |
| 1657 | 1657 | | ring-conjugated C=C str. of coniferyl alcohol plus C=O str. of coniferaldehyde [178, 179] |
| | | 1632 | str. of C=C from propenoic acid side chain of FA [180] |
| 1599 | 1600 | 1607 | symmetric aryl ring str. [144, 166-168] |
| 1503 | 1498 | 1505 | antisymmetric aryl ring str. [178], in FA |
| 1458 | 1458 | | C-$H_3$ def. in O-$CH_3$ [179]; C-$H_2$ scissoring; guaiacyl ring vibration (SW) [178, 179] and to C-$H_3$ def. in O-$CH_3$ (HW) [166, 179] |
| 1334 | 1328 | | aliphatic O-H bend. (SW) [144, 178], and S-lignin (HW) [177], possibly contribution from cellulose |
| 1271 | 1274 | 1271 | aryl-O str. of aryl-OH and aryl-O-$CH_3$; guaiacyl ring (with C=O group) mode (SW) [144, 178, 179], HW [177, 179] |
| | | 1176 | aryl-H def. [180] |
| | 1150 | | eventually O-$CH_3$ def. [166]; possibly contribution from carbohydrate [177] |
| 1139 | | | a mode of coniferaldehyde unit (SW) [178]; aromatic C-H in plane def. (guaiacyl type) [181] |

**Table 2.** Position and assignment of lignin bands in spectra acquired from the cell corners of spruce, poplar, and *Brachypodium*; FA: ferulic acid; SW: softwood; HW: hardwood; str: stretching vibration; bend: bending vibration; def: deformation vibration

## 4. Raman imaging of wood: Revealing lignification on the micron level

In the future, wood will play a crucial role in carbon capture and is a fundamental feedstock for bio-based fuels, chemicals, materials, and power. Currently, the greatest processing challenge is to develop efficient deconstruction and separation technologies that enable the release of sugar and aromatic compounds 'locked in' the intricacy of wood cell wall macromolecular structures [182]. To tackle this challenge detailed knowledge on the molecular composition of the cell walls within the different cell wall tissues and layers is of importance and Raman microscopy may contribute to make progress. As a non-destructive method, characterisation of the native cell walls is possible as well as the *in-situ* monitoring of the deconstruction during different treatments.

By calculating the integral of the bands in the Raman spectra of plant cell walls the distribution of different molecular structures can be imaged on the micron-level [148, 149,

167, 183]. Figure 5A shows an example of imaging water uptake of cell walls in young poplar wood (*Populus nigra x Populus deltoids*) by plotting the integral of the OH stretching vibration. As the sample was water saturated, highest intensity (blue colour) is observed in the water filled lumen of the cells. The border to the cell wall is slightly visible as less blue, followed by remarkable high intensity of the inner cell wall layer. Clear distinct layers separating the cells are visualised black and thus less hydrated areas. Integrating the strong aromatic aryl stretching vibration gives the opposite picture (Figure 5B): high intensity (red colour) in the cell corner (CC) and compound middle lamella (CML) and the rest of the cell walls are displayed black as intensity is less than to be seen with this scaling. In contrast the inner layer (gelatinous (G-) layer) of the poplar tension wood can be visualised selectively by integrating the 1380 $cm^{-1}$ cellulose band. This band is not sensitive to changes in cellulose orientation (Figure 2A). In lignified samples the band becomes a shoulder the more the adjacent lignin band rises and therefore only in cellulosic cell wall regions without (or very minor) lignin the band is clearly resolved and shows high intensity by integration. So the black regions of the images not necessarily represent non-cellulosic regions, but regions where cellulose is accompanied by higher lignin content. Based on the lignin and cellulose images three cell wall regions have been selected by using an intensity threshold (Figure 5D): 1) lignin intensity higher than 3000 cts (red), 2) lignin intensity between 1000 and 3000 cts (pink) and 3) cellulose intensity higher than 150 cts (green). The highest lignin content comprises the CC regions as well as part of the vessel walls (v) (Figure 5D, red). Medium lignin content displays the CML (and probably part of the S2 layer) as well as the ray cells (Figure 5D, pink) and the non-(or minor) lignified region is restricted to the inner cell wall (Figure 5D, green). For a detailed analysis of the cell wall regions, average spectra can be calculated from the defined cell wall regions of the sample (Figure 5E). For better comparison spectra have been baseline corrected and normalised to the aromatic lignin stretching vibrations. The spectra show that lignin is present in the green coloured cell wall region (G-layer) in minor amounts and the carbohydrate bands (C) are in relation much higher as well as the OH and CH stretching vibrations. Numerous OH-groups are present in the carbohydrates and show a contribution in the cellulose Raman spectra (Figure 2). But the height and form of the OH-bands in the G-layer are more comparable to spectra from the free water in the lumen and thus point to access of water uptake in this inner layer. In the medium lignified region (pink), similar cellulose bands are detected, but only one third of intensity compared to the inner layer. In the highly lignified region the lignin signature dominates (Figure 5E, red line). Yet carbohydrate bands are abundant, mainly because also the highly lignified vessel walls (v) are partly included in the average spectra calculation. Vessel spectra show higher amount of carbohydrates than spectra from CML (not shown). A zoom into the "lignin region" shows that not only the amount changes on the micron level, but also the lignin structure: a slight shift in band position of the aromatic stretching vibration is observed and the height of the shoulder at about 1657 $cm^{-1}$ changes. As the sampled region is near the cambium and the higher shoulder points to the alcohol and aldehyde lignin precursors in the G-layer and S2, lignification process in these regions might be in progress. Interestingly also spectra derived from substances within the rays point to aromatic compounds. These deposits could be visualised selectively by integrating from

785-486 cm-1. Detailed analysis of the spectra and band assignment will reveal the composition of these ray components. Several studies have shown that molecular structures, e.g. lignin structure are reflected in the Raman signature [154, 161, 170, 184, 185].

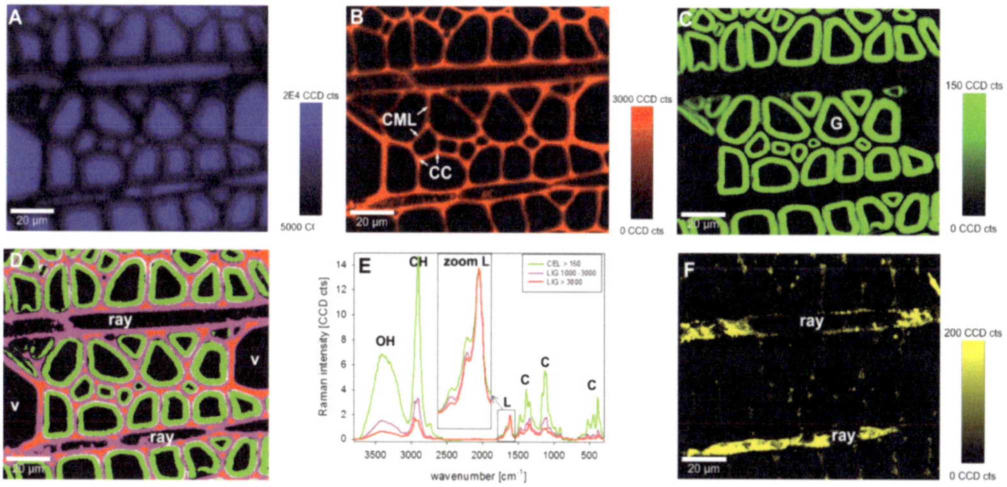


**Figure 5.** **A-F**: Raman images of poplar tension wood based on integrating the OH stretching region (A), the lignin marker band at about 1600 cm-1 (B) and the cellulose band at about 1388 cm-1 (C). In a combined image (D), regions with lignin intensity higher 3000 cts (red) and between 1000 and 3000 cts (pink) and with cellulose intensity >150 cts (green) are displayed. From these defined regions average spectra (baseline corrected and normalised to the 1600 cm-1 band) have been calculated (E). The content of the rays is visualized by integrating from 785-486 cm-1. (Experimental: 532 nm excitation, 100x oil immersion objective, scan area: 120 µm x 100 µm, Pixels: 360x300, integration time: 0.4 s) (CC: cell corner, CML: compound middle lamella, G: gelatinous layer, v: vessel)

The example of poplar tension wood showed the potential of Raman imaging to get a detailed view on the molecular structure on the micron level. Distinguishing cell wall types based on their chemistry gives an overview of the tissue composition. Furthermore extracting the underlying Raman spectra for detailed analysis can elucidate specific insights into the molecular structure and composition. The position resolved micro-resolution opens up new ways for understanding biosynthesis (especially lignification) as gradients in developing tissues can be followed cell by cell and cellular components investigated together with the cell wall itself. Different performance and reactions upon treatment can be resolved on the cell wall level and help to understand recalcitrance of wood. Different species have different chemical composition and lignin structures and recently a clear distinction between pine and spruce in terms of the distribution of coniferyl alcohol and coniferyl aldehyde was recognized using the Raman imaging approach [170]. Furthermore changes due to environmental growth conditions or genetic engineering can be evaluated. By comparing lignification in wild-type and lignin-reduced 4CL transgenic *Populus trichocarpa* stem wood, it was shown that transgenic reduction of lignin is particularly pronounced in the S2 wall layer of fibres, suggesting that such transgenic approach may help overcome cell wall recalcitrance to wood saccharification [186]. A higher volume of water was found in the cell wall of transgenic aspen compared with wild-type aspen, indicating an increase in the hydrophilicity of the cell wall [187].

## 5. Heterogeneity of lignin in the internode of the model grass *Brachypodium distachyon*

Grasses (or *Poaceae,* monocotyledon) are important plants on earth as they promote life and provide nutrients to both humans and animals. It is estimated that cereal grasses, i.e., corn, wheat, rice, oats, and barley, cover roughly 20% of the world's land surface and that the demand for these plants will remain high as the human population continues to increase [188]. Furthermore cereal grasses, including their grain and straw, are today not only used as sources of food but are also considered viable material for bioenergy. The grasses dedicated to biomass production, i.e., Switchgrass (*Panicum virgatum*) and Miscanthus (*Miscanthus sacchariflorus, M. sinensis* or *M. giganteus*) are mainly C4 plants that when grown in warm-environmental conditions, are more efficient in photosynthesis, nutrient-use and water-use [33]. The major limitations to the direct study of C4 grasses include the large size of the plants (requires land space), long generation times, and demanding growth requirements. Therefore several model plants, e.g., rice and more recently *Brachypdium distachyon,* were domesticated for laboratory studies (Figure 6A). *Brachypodium* now serves as an ideal experimental model for studying cereal grasses as it has many of the desirable traits for plant model systems including a small genome size, short generation time, the ability to self-pollinate, minimal input requirements and more importantly an amenability to forward and reverse genetic techniques [189]. Since 2001, *Brachypodium* has been established as a model however only recently have researchers begun to take advantage of this plant to study the monocot cell wall.

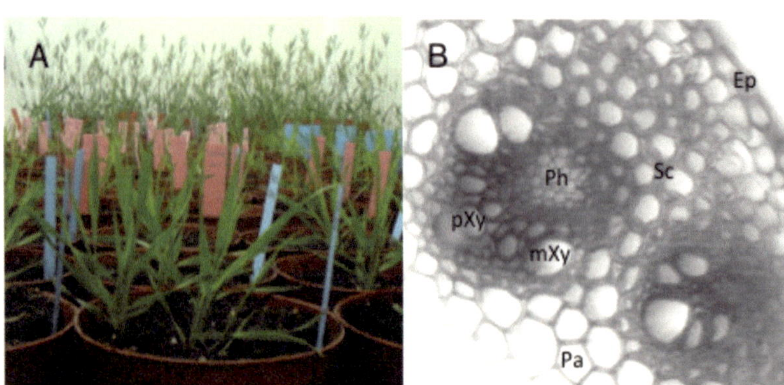


**Figure 6. A-B**: *Brachypodium distachyon*: Lateral view of *Brachypodium* development in the growth chamber (A). Cross section of a seven week old basal internode labelled with phloroglucinol-HCl (B). Dark labelling shows lignified cell walls. (Cell type abbreviations: Ep: epidermis; Sc: sclerenchyma; Ph: phloem; mXy: metaxylum; pXy: protoxylum; Pa: parenchyma)

The study of lignification in the monocot cell wall is of particular interest as several studies have demonstrated that lignin and phenolics bound to cell walls counter productively to saccharification yield and ruminant digestibility by reducing the accessibility of degrading/digestive enzymes to polysaccharides in the cell wall [45, 190]. Furthermore, a secondary, unintended effect of pre-treatments commonly used to reduce lignin content prior to saccharification for bioethanol production results in residual byproducts that inhibits growth of microorganisms used during fermentation. Therefore the natural resistance of lignocellulosic plant material to degradation serves as a major obstacle to efficient conversion of cellulose into fermentable sugars used for bioenergy [31, 191]. Within the monocot stems lignin is found in many tissues and cell types; the highest amount in the xylem tissue (Figure 6B). Phlorglucinol HCl staining gives insight into thevariability of lignification, but it is unspecific to different lignin and phenolic acid structures. As Raman images are based on underlying spectra, which represent a molecular fingerprint at every point within the acquired images, more detailed information can be gained.

Raman images of young (3 week old) basal internodes show point-wise accumulations of aromatic substances within the xylem tissues, while in the surrounding sclerenchyma fibres cell walls are visualized to be more homogenous (Figure 7A). In the lumen of the xylem cells remarkably high aromatic intensity is observed from deposits, which have not yet been further analysed. By integrating the marker band of ferulic acid at 1176 $cm^{-1}$ (Figure 4B) again the point-wise accumulation within the xylem becomes visible, but less intensity is observed in the sclerenchyma fibres (Figure 7B). As stated previously grass cell walls contain high amounts of p-hydroxycinnamic acids, particularly ferulic (FA) and p-coumaric (pCA). Previous studies have demonstrated that both pCA and FA play an important functional role in the incorporation of lignin into the cell wall by aiding to establish ester or/and ether-linkages to cell wall polymers [192]. It was shown that ferulate esters act as initiation or nucleation sites of lignin deposition in grasses [193]. Ferulate molecules connect lignin to arabinoxylans primarily through ester-ether bonds and form dimeric structures cross-linking arabinoxylan chains to polysaccharides [194].

By integrating the carbohydrate band at about 903 $cm^{-1}$, slightly higher intensity in the sclerenchyma cells was observed (Figure 7C). On the contrary to the poplar wood cells (Figure 5A-F) no clear cell corners are seen within the young xylem and sclerenchyma fibres and differences in the distribution of aromatic and carbohydrate substances within the scanned cell wall area are less pronounced when applying the band integration approach for image calculation.

Nevertheless, a cluster analysis performed with derivatives of baseline corrected spectra reveals high heterogeneity of the spectra in the lumen and on the border of the cell walls (Figure 8A) as well as clear separation of the xylem cell wall and the sclerenchyma cell wall (Figure 8B). Based on the found clusters, average spectra can be calculated corresponding to the separated regions. By this (Figure 8C-D) spectral (molecular) characterization of each cell wall region is possible on the micron level. Characteristic bands are observed e.g. for the lumen deposits (yellow line, Figure 8C), which can give insights into the chemical nature of these deposits. Furthermore, the gradual change recognized by the cluster analysis from the lumen towards the xylem cell wall can be analysed in detail. Comparing the xylem and sclerenchyma spectra (Figure 8D) it becomes clear that ferulic acid is much more accumulated in the xylem cells than in the surrounding sclerenchyma cells at this developmental stage of *Brachypodium*. The aryl stretching vibration as well as the marker bands at 1271 and 1176 $cm^{-1}$ are clearly reduced in the spectrum corresponding to the sclerenchyma cells, whereas the carbohydrate distribution appears comparable (Figure 8D). The spot-wise pattern in the xylem was observed clearly by integrating the area of the bands (Figure 7A-B) and not in the cluster analysis (Figure 8B). Therefore it can be concluded that the spots reflect more changes in intensity (amount) than compositional (structural) changes. By studying the different cell wall types as well as cell lumen ingredients during different developmental stages, insights into lignification will be gained.

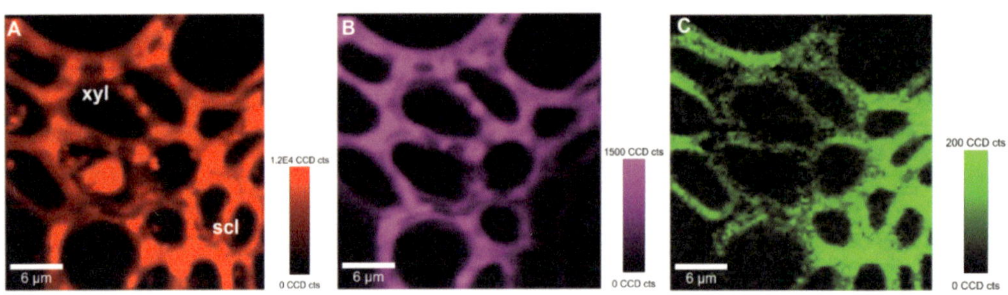


**Figure 7. A-C**: Raman images of a cross section of the basal internode of *Brachypodium distachyon* (three weeks old). High aromatic contribution is visualised by integrating from 1535 to 1674 $cm^{-1}$ in the xylem (xyl) and surrounding sclerenchyma (scl) cells. When the typical band for ferulic acid at 1176 $cm^{-1}$ is integrated (B) no lumen deposits are seen and less intensity in the sclerenchyma cells. The carbohydrate band at 903 $cm^{-1}$ (925-887 $cm^{-1}$) shows higher intensity in the sclerenchyma cells (C). (Experimental: 532 nm excitation, 100x oil immersion objective, scan area: 30 µm x 30 µm, Pixels: 90x90, integration time: 0.3 s)

Recent Raman study on corn stover revealed that lignin and cellulose abundance varies significantly among different cell types: 5-times higher in sclerenchymea cells, 3-times higher in epidermal cells than bundle sheaths and parenchyma cells [177]. They also noted characteristic bands at 1428, 1271, and 1175 $cm^{-1}$ in corn stover and although not assigned to ferulic acid, it seems that also in corn stover spectral contributions of ferulic acid have been reflected.

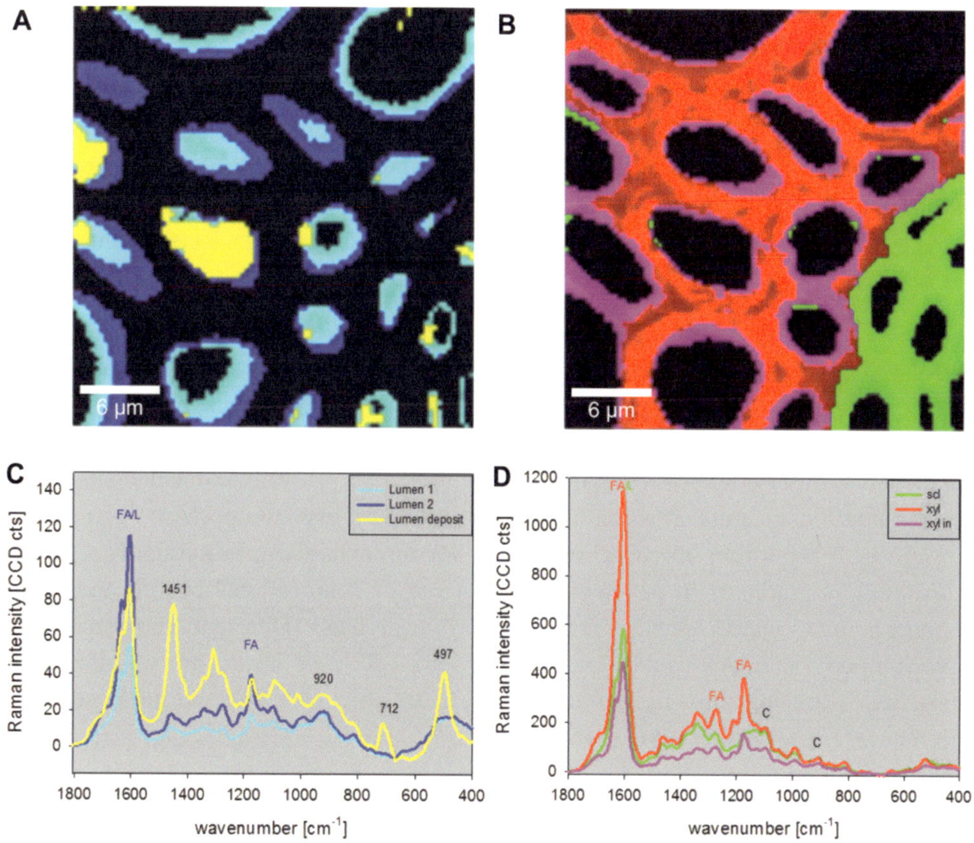


**Figure 8.** **A-D**: Raman images of *Brachypodium distachyon* based on a Cluster analysis using derivatives of baseline corrected spectra of the same measurement as shown in Figure 7. For clarity reasons the 6 calculated clusters are not shown within one image, but in two: A: representing the clusters of the cell lumen and borders towards the cell wall, B: the inner border of the xylem wall (pink) and the xylem wall itself (red) clearly separated from the sclerenchyma cells (green) High variability of the spectra within the cell lumen and its borders is seen (C) as well as clear differences between the sclerenchyma (scl) and xylem (xyl, xyl in) spectra (D).

## 6. Conclusion

The demand for plant biomass feedstock will increase as renewable resources get more and more attractive and further fields of utilizations open up. The mechanical performance as well as the recalcitrance of plant biomass to degradation is a function of which cell wall polymers are abundant and how they are cross-linked and aggregated within the walls. For understanding of biomass resources and an optimized utilization these higher order structures have to be probed in their native state on the micro- and nano level. The amount of cellulose as well as its crystallinity, structural arrangement (orientation) and interaction with other wood polymers play a key role in any utilization aspect. The recalcitrance to saccharification is a major limitation for conversion of lignocellulosic biomass to ethanol, which is mainly due to the lignin content and composition. Therefore improving feedstocks for both animal consumption and for starting material for bioethanol production is proposed through breeding and genetic engineering of lignin. High throughput methods to characterize plant cell walls have become more and more important in order to follow the native variability as well as engineered changes. Both, FT-IR and Raman spectroscopy have given important insights into plant cell wall polymers during the last years. While Raman has the advantage of higher spatial resolution (<0.5 μm) to reveal changes on the cell wall layer level and the possibility of investigating the samples in the wet state, FT-IR is more sensitive to the functional group of hemicelluloses.

The examples of poplar tension wood and *Brachypodium* have shown the potential of Raman imaging to get a detailed view on the molecular structure on the micron level. Distinguishing cell wall types based on their chemistry gives an overview of the tissue composition. Furthermore extracting the underlying Raman spectra for detailed analysis can elucidate specific insights into the molecular structure and composition. The position resolved micro-resolution opens up new ways for understanding biosynthesis (especially lignification) as gradients in developing tissues can be followed cell by cell and cellular components investigated together with the cell wall itself. Different performance and reactions upon treatment can be resolved on the cell wall level and so-called *in-situ* approaches, watching directly the effect of treatment, will help to understand the performance (e.g. recalcitrance) of plant cell walls.

## Author details

Notburga Gierlinger
*BOKU - University of Natural Resources and Life Sciences,*
*Department of Material Sciences and Process Engineering, Vienna, Austria*

Notburga Gierlinger and Tobias Keplinger
*Johannes Kepler University Linz, Institute of Polymer Science, Linz, Austria*

Michael Harrington
*INRA, UMR 1318, Institut Jean Pierre Bourgin, France*
*AgroParisTech, Institut Jean Pierre Bourgin, France*

Manfred Schwanninger
*BOKU – University of Natural Resources and Life Sciences,*
*Department of Chemistry, Vienna, Austria*

## Acknowledgement

N.G. thanks Pierre Conchon and Catherine Coutand (INRA, Clermont Ferrand, France) for providing the poplar wood cross-section. MH is supported by the National Science Foundation (IRFP # 1002683).

## 7. References


[1] McKendry P (2002) Energy production from biomass (part 1): overview of biomass. Bioresour. Technol. 83(1): 37-46.

[2] Fengel D, Wegener G (1989) Wood: chemistry, ultrastructure, reactions. Berlin: Walter de Gruyter & Co., Berlin. 613 p.

[3] Salmén L, Burgert I (2009) Cell wall features with regard to mechanical performance. A review. Holzforschung 63: 121-129.

[4] Daniel G (2003) Microview of Wood under Degradation by Bacteria and Fungi. In: Goodell B, Nicholas DD, Schultz TP, editors. Wood Deterioration and Degradation. Advances in Our Changing World: ACS Symposium Series; p. 34 - 72.

[5] Zabel RA, Morrell JJ (1992) Wood Microbiology - Decay and its Prevention. Academic Press I, editor. San Diego: Academic Press, Inc. 476 p.

[6] Eriksson K-EL, Blanchette RA, Ander P (1990) Microbial and Enzymatic Degradation of Wood and Wood Components. Timell TE, editor. Berlin: Springer. 313 p.

[7] Goodell B (2003) Brown-Rot Fungal Degradation of Wood: Our Evolving View. In: Goodell B, Nicholas DD, Schultz TP, editors. Wood Deterioration and Degradation. Advances in Our Changing World: ACS Symposium Series; p. 97 - 118.

[8] Bajpai P (2012) Biotechnology for Pulp and Paper Processing. New York: Springer. 414 p.

[9] Mizrachi E, Mansfield SD, Myburg AA (2012) Cellulose factories: advancing bioenergy production from forest trees. New Phytol. 194(1): 54-62.

[10] Sims REH, Hastings A, Schlamadinger B, Taylor G, Smith P (2006) Energy crops: current status and future prospects. Global Change Biol 12(11): 2054-2076.

[11] Gordon GA (2011) Application of Fourier transform mid-infrared spectroscopy (FTIR) for research into biomass feed-stocks. In: Nikolic GS, editor. Fourier Transforms - New Analytical Approaches and FTIR Strategies. Rijeka, Croatia: lnTech; p. 71-88.

[12] Allison GG, Robbins MP, Carli J, Clifton-Brown J, Donnison I (2010) Designing biomass crops with improved calorific content and attributes for burning: a UK perspective. In: P. Mascia, Schefrran J, Widhalm JM, editors. Plant Biotechnology for Sustainable Production of Energy and CoProducts. Heidelberg, Germany: Springer; p. 25-56.

[13] McCarthy JL, Islam A (2000) Lignin chemistry, technology, and utilization: A brief history. In: Glasser WG, Northey RA, Schultz TP, editors. Lignin : Historical, Biological, and Materials Perspectives: American Chemical Society; p. 2-99.

[14] Ko JH, Kim HT, Han KH (2011) Biotechnological improvement of lignocellulosic feedstock for enhanced biofuel productivity and processing. Plant Biotechnol. Rep. 5(1): 1-7.

[15] Kishimoto T (2009) Synthesis of lignin model compounds and their application to wood research. Mokuzai Gakkaishi 55(4): 187-197.

[16] Holmgren A, Norgren M, Zhang L, Henriksson G (2009) On the role of the monolignol gamma-carbon functionality in lignin biopolymerization. Phytochemistry 70(1): 147-155.

[17] Hafren J, Westermark U, Lennholm H, Terashima N (2002) Formation of C-13-enriched cell-wall DHP using isolated soft xylem from *Picea abies*. Holzforschung 56(6): 585-591.
[18] Monties B (2005) Biological variability of lignin. Cell. Chem. Technol. 39(5-6): 341-367.
[19] Zhong RQ, Morrison WH, Himmelsbach DS, Poole FL, Ye Z-H (2000) Essential role of caffeoyl coenzyme A *O*-methyltransferase in lignin biosynthesis in woody poplar plants. Plant Physiol. 124(2): 563-577.
[20] Liu CJ (2012) Deciphering the enigma of lignification: Precursor transport, oxidation, and the topochemistry of lignin assembly. Mol Plant 5(2): 304-317.
[21] Marjamaa K, Kukkola EM, Fagerstedt KV (2009) The role of xylem class III peroxidases in lignification. J. Exp. Bot. 60(2): 367-376.
[22] Chen YR, Sarkanen S (2010) Macromolecular replication during lignin biosynthesis. Phytochemistry 71(4): 453-462.
[23] Baucher M, Monties B, Van Montagu M, Boerjan W (1998) Biosynthesis and genetic engineering of lignin. Crit. Rev. Plant Sci. 17(2): 125-197.
[24] Whetten R, Sederoff R (1995) Lignin biosynthesis. Plant Cell 7: 1001-1013.
[25] Umezawa T (2010) The cinnamate/monolignol pathway. Phytochem. Rev. 9(1): 1-17.
[26] Vanholme R, Demedts B, Morreel K, Ralph J, Boerjan W (2010) Lignin biosynthesis and structure. Plant Physiol. 153(3): 895-905.
[27] Wang C, Wang YC, Diao GP, Jiang J, Yang CP (2010) Isolation and characterization of expressed sequence tags (ESTs) from cambium tissue of birch (*Betula platyphylla* Suk). Plant Mol. Biol. Rep. 28(3): 438-449.
[28] Samuels AL, Rensing KH, Douglas CJ, Mansfield SD, Dharmawardhana DP, Ellis BE (2002) Cellular machinery of wood production: differentiation of secondary xylem in *Pinus contorta* var. *latifolia*. Planta 216(1): 72-82.
[29] Studer MH, DeMartini JD, Davis MF, Sykes RW, Davison B, Keller M, Tuskan GA, Wyman CE (2011) Lignin content in natural Populus variants affects sugar release. P. Natl. Acad. Sci. USA 108(15): 6300-6305.
[30] Grabber JH (2005) How do lignin composition, structure, and cross-linking affect degradability? A review of cell wall model studies. Crop Sci. 45(3): 820-831.
[31] Carroll A, Somerville C (2009) Cellulosic biofuels. Annu. Rev. Plant Biol. 60: 165-182.
[32] Chang MCY (2007) Harnessing energy from plant biomass. Curr. Opin. Chem. Biol. 11(6): 677-684.
[33] Clifton-Brown J, Robson P, Allison G, Lister S, Sanderson R, Hodgson E, Farrar K, Hawkins S, Jensen E, Jones S, Huang L, Roberts P, Youell S, Jones B, Wright A, Valantine J, Donnison I (2008) Miscanthus: breeding our way to a better future. In: E. Booth, M. Green, A. Karp, I. Shield, D. Stock, Turley D, editors. Biomass and Energy Crops III. Warwick, UK: Association of Applied Biologists; p. 199-206.
[34] Hodgson EM, Fahmi R, Yates N, Barraclough T, Shield I, Allison G, Bridgwater AV, Donnison IS (2010) *Miscanthus* as a feedstock for fast-pyrolysis: Does agronomic treatment affect quality? Bioresour. Technol. 101(15): 6185-6191.
[35] Hodgson EM, Lister SJ, Bridgwater AV, Clifton-Brown J, Donnison IS (2010) Genotypic and environmentally derived variation in the cell wall composition of *Miscanthus* in relation to its use as a biomass feedstock. Biomass. Bioenerg. 34(5): 652-660.

[36] Boudet AM, Grima-Pettenati J (1996) Lignin genetic engineering. Mol. Breeding 2(1): 25-39.
[37] Boudet AM, Lapierre C, Grima-Pettenati J (1995) Tansley review No-80 - Biochemistry and molecular-biology of lignification. New Phytol. 129(2): 203-236.
[38] Chiang VL (2006) Monolignol biosynthesis and genetic engineering of lignin in trees, a review. Environ. Chem. Lett. 4(3): 143-146.
[39] Harris D, DeBolt S (2010) Synthesis, regulation and utilization of lignocellulosic biomass. Plant Biotechnol. J. 8(3): 244-262.
[40] Higuchi T (2000) The present state and problems in lignin biosynthesis. Cell. Chem. Technol. 34(1-2): 79-100.
[41] Horvath B, Peszlen I, Peralta P, Kasal B, Li LG (2010) Effect of lignin genetic modification on wood anatomy of Aspen trees. Iawa J. 31(1): 29-38.
[42] Koehler L, Telewski FW (2006) Biomechanics and transgenic wood. Am. J. Bot. 93(10): 1433-1438.
[43] Leple JC, Dauwe R, Morreel K, Storme V, Lapierre C, Pollet B, Naumann A, Kang KY, Kim H, Ruel K, Lefebvre A, Joseleau JP, Grima-Pettenati J, De Rycke R, Andersson-Gunneras S, Erban A, Fehrle I, Petit-Conil M, Kopka J, Polle A, Messens E, Sundberg B, Mansfield SD, Ralph J, Pilate G, Boerjan W (2007) Downregulation of cinnamoyl-coenzyme a reductase in poplar: Multiple-level phenotyping reveals effects on cell wall polymer metabolism and structure. Plant Cell 19(11): 3669-3691.
[44] Pu YQ, Chen F, Ziebell A, Davison BH, Ragauskas AJ (2009) NMR characterization of C3H and HCT down-regulated alfalfa lignin. Bioenerg. Res. 2(4): 198-208.
[45] Chen F, Dixon RA (2007) Lignin modification improves fermentable sugar yields for biofuel production. Nat Biotechnol. 25(7): 759-761.
[46] Maunu SL (2002) NMR studies of wood and wood products. Prog. Nucl. Magn. Reson. Spectrosc. 40: 151-174.
[47] Lu F, Ralph J (2003) Non-degradative dissolution and acetylation of ball-milled plant cell walls: high-resolution solution-state NMR. Plant J. 35: 535-544.
[48] Yelle DJ, Ralph J, Li F, Hammel KE (2008) Evidence for cleavage of lignin by a brown rot basidiomycete. Environ. Microbiol. 10(7): 1844-1849.
[49] Yelle DJ, Wei D, Ralph J, Hammel KE (2011) Multidimensional NMR analysis reveals truncated lignin structures in wood decayed by the brown rot basidiomycete *Postia placenta*. Environ. Microbiol. 13(4): 1091-1100.
[50] Alves A, Schwanninger M, Pereira H, Rodrigues J (2006) Analytical pyrolysis as a direct method to determine the lignin content in wood - Part 1: Comparison of pyrolysis lignin with Klason lignin. J. Anal. Appl. Pyrol. 76(1-2): 209-213.
[51] Alves A, Gierlinger N, Schwanninger M, Rodrigues J (2009) Analytical pyrolysis as a direct method to determine the lignin content in wood Part 3. Evaluation of species-specific and tissue-specific differences in softwood lignin composition using principal component analysis. J. Anal. Appl. Pyrol. 85(1-2): 30-37.
[52] Anterola AM, Lewis NG (2002) Review: Trends in lignin modification: a comprehensive analysis of the effects of genetic manipulations/mutations on lignification and vascular integrity. Phytochemistry 61: 221–294.

[53] Korosec RC, Lavric B, Rep G, Pohleven F, Bukovec P (2009) Thermogravimetry as a possible tool for determining modification degree of thermally treated Norway spruce wood. J. Therm. Anal. Calorim. 98(1): 189-195.
[54] Taneda K, Nishiyama Y, Uparivong S (1995) An evaluation of kinetic-parameters by derivative thermogravimetry and Its application to oood and other bioresources. Mokuzai Gakkaishi 41(4): 414-424.
[55] Gindl W, Grabner M (2000) Characteristics of spruce [*Picea abies* (L.) Karst] latewood formed under abnormally low temperatures. Holzforschung 54(1): 9-11.
[56] Gindl W, Grabner M, Wimmer R (2000) The influence of temperature on latewood lignin content in treeline Norway spruce compared with maximum density and ring width. Trees-Struct. Funct. 14(7): 409-414.
[57] Griffiths PR, Haseth JAD (2007) Fourier Transform Infrared Spectrometry. 2nd ed. New York: Wiley. 529 p.
[58] Schrader B (1995) Infrared and Raman Spectroscopy: Methods and Applications. Weinheim: Wiley-VCH Verlag GmbH 788 p.
[59] Tsuchikawa S (2007) A review of recent near infrared research for wood and paper. Appl. Spectrosc. Rev. 42: 43-71.
[60] Tsuchikawa S, Schwanninger M (2011) A review of recent near infrared research for wood and paper (Part 2). Appl. Spectrosc. Rev. (in print): DOI: 10.1080/05704928.05702011.05621079.
[61] Schwanninger M, Rodrigues J, Fackler K (2011) A review of band assignments in near infrared spectra of wood and wood components. J. Near Infrared Spectrosc. 19(287-308).
[62] Schwanninger M, Hinterstoisser B, Gradinger C, Messner K, Fackler K (2004) Examination of spruce wood biodegraded by *Ceriporiopsis subvermispora* using near and mid infrared spectroscopy. J. Near Infrared Spectrosc. 12(6): 397-409.
[63] Fackler K, Schmutzer M, Manoch L, Schwanninger M, Hinterstoisser B, Ters T, Messner K, Gradinger C (2007) Evaluation of the selectivity of white rot isolates using near infrared spectroscopic techniques. Enzyme Microb. Technol. 41: 881-887.
[64] Fackler K, Schwanninger M, Gradinger C, Srebotnik E, Hinterstoisser B, Messner K (2007) Fungal decay of spruce and beech wood assessed by near infrared spectroscopy in combination with uni- and multivariate data analysis. Holzforschung 62: 223-230.
[65] Fackler K, Gradinger C, Hinterstoisser B, Messner K, Schwanninger M (2006) Lignin degradation by white rot fungi on spruce wood shavings during short-time solid-state fermentations monitored by near infrared spectroscopy. Enzyme Microb. Technol. 39(7): 1476-1483.
[66] Fackler K, Schwanninger M (2010) Polysaccharide degradation and lignin modification during brown-rot of spruce wood: a polarised Fourier transfrom near infrared study. J. Near Infrared Spectrosc. 18: 403-416.
[67] Fackler K, Schwanninger M (2011) Accessibility of hydroxyl groups of brown-rot degraded spruce wood to heavy water. J. Near Infrared Spectrosc. 19: 359-368.
[68] Fackler K, Schwanninger M, Gradinger C, Hinterstoisser B, Messner K (2007) Qualitative and quantitative changes of beech wood degraded by wood rotting

basidiomycetes monitored by Fourier transform infrared spectroscopic methods and multivariate data analysis. FEMS Microbiol. Lett. 271: 162-169.
[69] Åkerholm M, Salmén L (2001) Interactions between wood polymers studied by dynamic FT-IR spectroscopy. Polymer 42: 963-969.
[70] Åkerholm M, Salmén L (2003) The oriented structure of lignin and its viscoelastic properties studied by static and dynamic FT-IR spectroscopy. Holzforschung 57(5): 459-465.
[71] Hinterstoisser B, Åkerholm M, Salmén L (2001) Effect of fiber orientation in dynamic FTIR study on native cellulose. Carbohydr. Res. 334: 27-37.
[72] Hinterstoisser B, Salmén L (2000) Application of dynamic 2D FTIR to cellulose. Vib. Spectrosc. 22(1-2): 111-118.
[73] Stevanic JS, Salmén L (2008) Characterizing wood polymers in the primary cell wall of Norway spruce (*Picea abies* (L.) Karst.) using dynamic FT-IR spectroscopy. Cellulose 15(2): 285-295.
[74] Stevanic JS, Salmén L (2009) Orientation of the wood polymers in the cell wall of spruce wood fibres. Holzforschung 63(5): 497-503.
[75] Atalla RH, Agarwal UP (1985) Raman microprobe evidence for lignin orientation in the cell walls of native woody tissue. Science 227: 636-638.
[76] Chowdhury S, Madsen LA, Frazier CE (2012) Probing alignment and phase behavior in intact wood cell walls using H-2 NMR spectroscopy. Biomacromolecules 13(4): 1043-1050.
[77] Salmén L, Olsson A-M, Stevanic JS, Simonovic J, Radotic K (2012) Structural organization of the wood polymers in the wood fibre structure Bioresources 7(1): 521-532.
[78] Gierlinger N, Goswami L, Schmidt M, Burgert I, Coutand C, Rogge T, Schwanninger M (2008) *In situ* FT-IR microscopic study on enzymatic treatment of poplar wood cross-sections. Biomacromolecules 9: 2194-2201.
[79] Chen LM, Carpita NC, Reiter WD, Wilson RH, Jeffries C, McCann MC (1998) A rapid method to screen for cell-wall mutants using discriminant analysis of Fourier transform infrared spectra. Plant J. 16(3): 385-392.
[80] Mouille G, Robin S, Lecomte M, Pagant S, Hofte H (2003) Classification and identification of *Arabidopsis* cell wall mutants using Fourier-Transform InfraRed (FT-IR) microspectroscopy. Plant J. 35(3): 393-404.
[81] Robin S, Lecomte M, Hofte H, Mouille G (2003) A procedure for the clustering of cell wall mutants in the model plant *Arabidopsis* based on Fourier-transform infrared (FT-IR) spectroscopy. J. Appl. Stat. 30(6): 669-681.
[82] McCann MC, Defernez M, Urbanowicz BR, Tewari JC, Langewisch T, Olek A, Wells B, Wilson RH, Carpita NC (2007) Neural network analyses of infrared spectra for classifying cell wall architectures. Plant Physiol. 143(3): 1314-1326.
[83] Fackler K, Stevanic JS, Ters T, Hinterstoisser B, Schwanninger M, Salmén L (2010) Localisation and characterisation of incipient brown-rot decay within spruce wood cell walls using FT-IR imaging microscopy. Enzyme Microb. Technol. 47: 257-267.

[84] Fackler K, Stevanic JS, Ters T, Hinterstoisser B, Schwanninger M, Salmén L (2011) FT-IR imaging microscopy to localise and characterise simultaneous and selective white-rot decay within spruce wood cells. Holzforschung 65: 411-420.

[85] Smith E, Dent G (2005) Modern Raman Spectroscopy - A practical approach. Manchester: John Wiley & Sons Ltd. 210 p.

[86] Landsberg G, Mandelstam L (1928) Light scattering in crystals. Zeitschrift Für Physik 50(11-12): 769-780.

[87] Raman CV, Krishnan KS (1928) A new type of secondary radiation. Nature 121: 501-502.

[88] Hirschfeld T, Chase B (1986) FT-Raman spectroscopy - development and justification. Appl. Spectrosc. 40(2): 133-137.

[89] Das RS, Agrawal YK (2011) Raman spectroscopy: Recent advancements, techniques and applications. Vib. Spectrosc. 57(2): 163-176.

[90] Hollricher O (2010) Raman Instrumentation for Confocal Raman Microscopy. In: Diening T, Hollricher O, Toporski J, editors. Confocal Raman microscopy. Berlin Heidelberg: Springer-Verlag; p. 43-60.

[91] Dieing T, Hollricher O (2008) High-resolution, high-speed confocal Raman imaging. Vib. Spectrosc. 48(1): 22-27.

[92] Withnall R, Chowdhry BZ, Silver J, Edwards HGM, de Oliveira LFC (2003) Raman spectra of carotenoids in natural products. Spectroc. Acta Pt. A-Molec. Biomolec. Spectr. 59: 2207-2212.

[93] Saariaho AM, Jääskeläinen AS, Matousek P, Towrie M, Parker AW, Vuorinen T (2004) Resonance Raman spectroscopy of highly fluorescing lignin containing chemical pulps: Suppression of fluorescence with an optical Kerr gate. Holzforschung 58(1): 82-90.

[94] Matousek P, Towrie M, Stanley A, Parker AW (1999) Efficient rejection of fluorescence from Raman spectra using picosecond Kerr gating. Appl. Spectrosc. 53(12): 1485-1489.

[95] Matousek P, Towrie M, Parker AW (2002) Fluorescence background suppression in Raman spectroscopy using combined Kerr gated and shifted excitation Raman difference techniques. J. Raman Spectrosc. 33: 238-242.

[96] Sharma B, Frontiera RR, Henry AI, Ringe E, Van Duyne RP (2012) SERS: Materials, applications, and the future. Mater. Today 15(1-2): 16-25.

[97] Zhang Y, Hong H, Myklejord DV, Cai WB (2011) Molecular imaging with SERS-active nanoparticles. Small 7(23): 3261-3269.

[98] McNay G, Eustace D, Smith WE, Faulds K, Graham D (2011) Surface-Enhanced Raman Scattering (SERS) and Surface-Enhanced Resonance Raman Scattering (SERRS): A review of applications. Appl. Spectrosc. 65(8): 825-837.

[99] Rösch P, Popp J, Kiefer W (1999) Raman and surface enhanced Raman spectroscopic investigation on Lamiaceae plants. J. Mol. Struct. 481: 121-124.

[100] Cheng JX, Xie XS (2004) Coherent anti-Stokes Raman scattering microscopy: Instrumentation, theory, and applications. J. Phys. Chem. B 108(3): 827-840.

[101] Chen JX, Volkmer A, Book LD, Xie XS (2002) Multiplex coherent anti-stokes Raman scattering microspectroscopy and study of lipid vesicles. J. Phys. Chem. B 106(34): 8493-8498.

[102] Le TT, Yue SH, Cheng JX (2010) Shedding new light on lipid biology with coherent anti-Stokes Raman scattering microscopy. J. Lipid Res. 51(11): 3091-3102.
[103] Pezacki JP, Blake JA, Danielson DC, Kennedy DC, Lyn RK, Singaravelu R (2011) Chemical contrast for imaging living systems: molecular vibrations drive CARS microscopy. Nat. Chem. Biol. 7(3): 137-145.
[104] Fu D, Lu FK, Zhang X, Freudiger C, Pernik DR, Holtom G, Xie XS (2012) Quantitative chemical imaging with multiplex stimulated Raman scattering microscopy. J. Am. Chem. Soc. 134(8): 3623-3626.
[105] Mallick B, Lakhsmanna A, Umapathy S (2011) Ultrafast Raman loss spectroscopy (URLS): instrumentation and principle. J. Raman Spectrosc. 42(10): 1883-1890.
[106] Freudiger CW, Min W, Holtom GR, Xu BW, Dantus M, Xie XS (2011) Highly specific label-free molecular imaging with spectrally tailored excitation-stimulated Raman scattering (STE-SRS) microscopy. Nat. Photonics 5(2): 103-109.
[107] [107] Saar BG, Zeng YN, Freudiger CW, Liu YS, Himmel ME, Xie XS, Ding SY (2010) Label-free, real-time monitoring of biomass processing with stimulated Raman scattering microscopy. Angew. Chem.-Int. Edit. 49(32): 5476-5479.
[108] [108] Griffith PR (2009) Infrared and Raman Instrumentation for Mapping and Imaging. In: Salzer R, Siesler HW, editors. Infrared and Raman Spectroscopic Imaging. Weinheim: Wiley-VCH Verlag GmbH & Co. KGaA; p. 3-64.
[109] Everall N, Lapham J, Adar F, Whitley A, Lee E, Mamedov S (2007) Optimizing depth resolution in Confocal Raman microscopy: A comparison of metallurgical, dry corrected, and oil immersion objectives. Appl. Spectrosc. 61(3): 251-259.
[110] Bruneel JL, Lassegues JC, Sourisseau C (2002) In-depth analyses by confocal Raman microspectrometry: experimental features and modeling of the refraction effects. J. Raman Spectrosc. 33(10): 815-828.
[111] Verma P, Ichimura T, Yano T, Saito Y, Kawata S (2010) Nano-imaging through tip-enhanced Raman spectroscopy: Stepping beyond the classical limits. Laser Photon. Rev. 4(4): 548-561.
[112] Deckert-Gaudig T, Deckert V (2011) Nanoscale structural analysis using tip-enhanced Raman spectroscopy. Curr. Opin. Chem. Biol. 15(5): 719-724.
[113] Elfick APD, Downes AR, Mouras R (2010) Development of tip-enhanced optical spectroscopy for biological applications: a review. Anal. Bioanal. Chem. 396(1): 45-52.
[114] Nelson MP, Treado PJ (2010) Raman imaging instrumentation. In: Sasic S, Ozaki Y, editors. Raman, Infrared, and Near-Infrared Chemical Imaging. Hoboken, New Jersey: John Wiley & Sons, Inc.; p. 23-55.
[115] Toytman I, Simanovskii D, Palanker D (2009) On illumination schemes for wide-field CARS microscopy. Opt. Express 17(9): 7339-7347.
[116] Schlucker S, Schaeberle MD, Huffman SW, Levin IW (2003) Raman microspectroscopy: A comparison of point, line, and wide-field imaging methodologies. Anal. Chem. 75(16): 4312-4318.
[117] McCreery RL (2000) Raman microscopy and imaging. In: Winefordner JD, editor. Raman Spectroscopy for Chemical Analysis. New York: John Wiley & Sons, Inc.; p. 293-332.

[118] Gamsjaeger S, Kazanci M, Paschalis EP, Fratzl P (2009) Raman application in bone imaging. In: Amer MS, editor. Raman Spectroscopy for soft matter applications. New Jersey: Wiley VCH; p. 227–267.

[119] Gierlinger N, Keplinger T, Harrington M (2012) Imaging of plant cell walls by Confocal Raman microscopy. Nat. Protoc.: in review.

[120] Zhang DM, Jallad KN, Ben-Amotz D (2001) Stripping of cosmic spike spectral artifacts using a new upper-bound spectrum algorithm. Appl. Spectrosc. 55(11): 1523-1531.

[121] Katsumoto Y, Ozaki Y (2003) Practical algorithm for reducing convex spike noises on a spectrum. Appl. Spectrosc. 57(3): 317-322.

[122] Diening T, Ibach W (2010) Software Requirements and Data Analysis in Confocal Raman Microscopy. In: Diening T, Hollricher O, Toporski J, editors. Confocal Raman microscopy. Berlin Heidelberg: Springer-Verlag; p. 61-89.

[123] Savitzky A, Golay MJE (1964) Smoothing and differentiation of data by simplified least squares procedures. Anal. Chem. 36: 1627 - 1639.

[124] Ramos PM, Ruisanchez I (2005) Noise and background removal in Raman spectra of ancient pigments using wavelet transform. J. Raman Spectrosc. 36(9): 848-856.

[125] Liland KH, Rukke EO, Olsen EF, Isaksson T (2011) Customized baseline correction. Chemometrics Intell. Lab. Syst. 109(1): 51-56.

[126] de Juan A, Maeder M, Hancewicz T, Duponchel L, Tauler R (2009) Chemometric Tools for Image Analysis. In: Salzer R, W. SH, editors. Infrared and Raman Spectroscopic Imaging. Weinheim: WILEY-VCH Verlag GmbH & Co. KGaA; p. 65-108.

[127] Schulze G, Jirasek A, Yu MML, Lim A, Turner RFB, Blades MW (2005) Investigation of selected baseline removal techniques as candidates for automated implementation. Appl. Spectrosc. 59(5): 545-574.

[128] Prakash BD, Wei YC (2011) A fully automated iterative moving averaging (AIMA) technique for baseline correction. Analyst 136(15): 3130-3135.

[129] Schulze HG, Foist RB, Okuda K, Ivanov A, Turner RFB (2011) A model-free, fully automated baseline-removal method for Raman spectra. Appl. Spectrosc. 65(1): 75-84.

[130] Zhang ZM, Chen S, Liang YZ (2010) Baseline correction using adaptive iteratively reweighted penalized least squares. Analyst 135(5): 1138-1146.

[131] Schmidt U, Ibach W, Muller J, Weishaupt K, Hollricher O (2006) Raman spectral imaging - A nondestructive, high resolution analysis technique for local stress measurements in silicon. Vib. Spectrosc. 42(1): 93-97.

[132] Geladi P, Grahn H, Manley M (2010) Data analysis and chemometrics for hyperspectral Imaging. In: Sasic S, Ozaki Y, editors. Raman, Infrared, and Near-Infrared Chemical Imaging. Hoboken, New Jersey: John Wiley & Sons, Inc.; p. 93-109.

[133] Shinzawa H, Awa K, Kanematsu W, Ozaki Y (2009) Multivariate data analysis for Raman spectroscopic imaging. J. Raman Spectrosc. 40(12): 1720-1725.

[134] Næs T, Isaksson T, Fearn T, Davies T (2002) A User-Friendly Guide to Multivariate Calibration and Classification. first ed. Chichester: NIR Publications. 344 p.

[135] Geladi P (2003) Chemometrics in spectroscopy. Part 1. Classical chemometrics. Spectroc. Acta Pt. B-Atom. Spectr. 58: 767-782.

[136] Hastie T, Tibshirani R, Friedman J (2009) The Elements of Statistical Learning. New York: Springer. 739 p.

[137] Cosgrove DJ (2005) Growth of the plant cell wall. Nat. Rev. Mol. Cell Biol. 6(11): 850-861.

[138] Wiley JH, Atalla RH (1987) Band assignments in the Raman spectra of celluloses. Carbohydr. Res. 160: 113-129.

[139] Agarwal UP, Atalla RH (1986) In-situ Raman microprobe studies of plant cell walls - Macromolecular organization and compositional variability in the secondary wall of *Picea mariana* (Mill) Bsp. Planta 169(3): 325-332.

[140] Atalla RH, Agarwal UP (1986) Recording Raman-spectra from plant cell walls. J. Raman Spectrosc. 17(2): 229-231.

[141] Schenzel K, Fischer S (2001) NIR FT Raman spectroscopy - a rapid analytical tool for detecting the transformation of cellulose polymorphs. Cellulose 8(1): 49-57.

[142] Gierlinger N, Luss S, Konig C, Konnerth J, Eder M, Fratzl P (2010) Cellulose microfibril orientation of *Picea abies* and its variability at the micron-level determined by Raman imaging. J. Exp. Bot. 61(2): 587-595.

[143] Agarwal UP, Reiner RS, Ralph SA (2010) Cellulose I crystallinity determination using FT-Raman spectroscopy: univariate and multivariate methods. Cellulose 17(4): 721-733.

[144] Agarwal UP, Ralph SA (1997) FT-Raman spectroscopy of wood: Identifying contributions of lignin and carbohydrate polymers in the spectrum of black spruce (*Picea mariana*). Appl. Spectrosc. 51(11): 1648-1655.

[145] Himmelsbach DS, Khahili S, Akin DE (1999) Near-infrared–Fourier-transform–Raman microspectroscopic imaging of flax stems. Vib. Spectrosc. 19: 361-367.

[146] Chu LQ, Masyuko R, Sweedler JV, Bohn PW (2009) Base-induced delignification of Miscanthus x giganteus studied by three-dimensional confocal Raman imaging. Bioresour. Technol. 101(13): 4919-4925.

[147] Mathlouthi M, Koenig JL (1986) Vibrational Spectra of Carbohydrates. Adv. Carbohydr. Chem. Biochem. 44: 7-89.

[148] Richter S, Mussig J, Gierlinger N (2011) Functional plant cell wall design revealed by the Raman imaging approach. Planta 233(4): 763-772.

[149] Gierlinger N, Sapei L, Paris O (2008) Insights into the chemical composition of *Equisetum hyemale* by high resolution Raman imaging. Planta 227(5): 969-980.

[150] Synytsya A, Copikova J, Matejka P, Machovic V (2003) Fourier transform Raman and infrared spectroscopy of pectins. Carbohydr. Polym. 54(1): 97-106.

[151] Arjyal BP, Katerelos DG, Filiou C, Galiotis C (2000) Measurement and Modeling of Stress Concentration around a Circular Notch. Exp. Mech. 40(3): 248-255.

[152] Lewis NG, Yamamoto E (1990) Lignin: occurrence, biogenesis and biodegradation. Annu. Rev. Plant Physiol. Plant Mol. Biol. 41: 455-496.

[153] Barsberg S, Matousek P, Towrie M (2005) Structural analysis of lignin by resonance Raman spectroscopy. Macromol. Biosci. 5(8): 743-752.

[154] Perera PN, Schmidt M, Chiang VL, Schuck PJ, Adams PD (2012) Raman-spectroscopy-based noninvasive microanalysis of native lignin structure. Anal. Bioanal. Chem. 402(2): 983-987.

[155] Agarwal UP, Atalla RH (1994) Raman spectral features associated with chromophores in high-yield pulps. J. Wood Chem. Technol. 14(2): 227-241.

[156] Agarwal UP (1999) An Overview of Raman Spectroscopy as Applied to Lignocellulosic Materials. In: Argyropoulos DS, editor. Advances in Lignocellulosics Characterization. Atlanta, GA: TAPPI Press; p. 209-225.

[157] Agarwal UP, Landucci LL (2004) FT-Raman investigation of bleaching of spruce thermomechanical pulp. J. Pulp Pap. Sci. 30(10): 269-274.

[158] Stewart D, Yahiaoui N, McDougall GJ, Myton K, Marque C, Boudet AM, Haigh J (1997) Fourier-transform infrared and Raman spectroscopic evidence for the incorporation of cinnamaldehydes into the lignin of transgenic tobacco (*Nicotiana tabacum* L) plants with reduced expression of cinnamyl alcohol dehydrogenase. Planta 201(3): 311-318.

[159] Himmelsbach DS, Akin DE (1998) Near-infrared Fourier-transform Raman spectroscopy of flax (Linum usitatissimum L.) stems. J. Agric. Food Chem. 46(3): 991-998.

[160] Ona T, Sonoda T, Ito K, Shibata M, Katayama T, Kato T, Ootake Y (1998) Non-destructive determination of lignin syringyl/guaiacyl monomeric composition in native wood by Fourier transform Raman spectroscopy. J. Wood Chem. Technol. 18(1): 43-51.

[161] Sun L, Varanasi P, Yang F, Loque D, Simmons BA, Singh S (2012) Rapid determination of syringyl:guaiacyl ratios using FT-Raman spectroscopy. Biotechnol. Bioeng. 109(3): 647-656.

[162] Barsberg S, Matousek P, Towrie M, Jorgensen H, Felby C (2006) Lignin radicals in the plant cell wall probed by Kerr-gated resonance Raman spectroscopy. Biophys. J. 90(8): 2978-2986.

[163] Saariaho AM, Jääskeläinen AS, Nuopponen M, Vuorinen T (2003) Ultra violet resonance Raman spectroscopy in lignin analysis: determination of characteristic vibrations of p-hydroxyphenyl, guaiacyl, and syringyl lignin structures. Appl. Spectrosc. 57(1): 58-66.

[164] Jääskeläinen AS, Saariaho AM, Vyorykka J, Vuorinen T, Matousek P, Parker AW (2006) Application of UV-Vis and resonance Raman spectroscopy to study bleaching and photoyellowing of thermomechanical pulps. Holzforschung 60(3): 231-238.

[165] Zeng Y, Saar BG, Friedrich MG, Chen F, Liu Y-S, Dixon RA, Himmel ME, Xie XS, Ding S-Y (2010) Imaging Lignin-Downregulated Alfalfa Using Coherent Anti-Stokes Raman Scattering Microscopy. Bioenerg. Res. 3(3): 272-277.

[166] Larsen KL, Barsberg S (2010) Theoretical and Raman spectroscopic studies of phenolic lignin model monomers. J. Phys. Chem. B 114(23): 8009-8021.

[167] Agarwal UP (2006) Raman imaging to investigate ultrastructure and composition of plant cell walls: distribution of lignin and cellulose in black spruce wood (*Picea mariana*). Planta 224(5): 1141-1153.

[168] Gierlinger N, Schwanninger M (2007) The potential of Raman microscopy and Raman imaging in plant research - review. Spectroscopy 21: 69-89.

[169] Agarwal UP, Ralph SA (2008) Determination of ethylenic residues in wood and TMP of spruce by FT-Raman spectroscopy. Holzforschung 62(6): 667-675.

[170] Hanninen T, Kontturi E, Vuorinen T (2011) Distribution of lignin and its coniferyl alcohol and coniferyl aldehyde groups in *Picea abies* and *Pinus sylvestris* as observed by Raman imaging. Phytochemistry 72(14-15): 1889-1895.

[171] Carpita NC, Gibeaut DM (1993) Structural models of primary cell walls in flowering plants: consistency of molecular structure with the physical properties of the walls during growth. Plant J. 3(1): 1-30.

[172] Carpita NC, McCann MC (2000) The cell wall. In: Buchanan BB GW, Jones RL, editor. American Society of Plant Biologists; Rockville, MD: Biochemistry and Molecular Biology of Plants.

[173] Carpita NC (1996) Structure and biogenesis of the cell walls of grasses. Annu. Rev. Plant Physiol. Plant Mol. Biol. 47: 445-476.

[174] Scalbert A, Monties B, Lallemand J-Y, Guittet E, Rolando C (1985) Ether linkage between phenolic acids and lignin fractions from wheat straw. Phytochemistry 24(6): 1359-1362.

[175] Piot O, Autran J-C, Manfait M (2001) Investigation by confocal Raman microspectroscopy of the molecular factors responsible for grain cohesion in the *Triticum aestivum* bread wheat. Role of the cell walls in the starchy endosperm. J. Cereal Sci. 34(2): 191-205.

[176] Ram MS, Dowell FE, Seitz LM (2003) FT-Raman spectra of unsoaked and NaOH-soaked wheat kernels, bran, and ferulic acid. Cereal Chem. 80(2): 188-192.

[177] Sun L, Simmons BA, Singh S (2011) Understanding tissue specific compositions of bioenergy feedstocks through hyperspectral Raman imaging. Biotechnol. Bioeng. 108(2): 286-295.

[178] Agarwal UP (1999) Chapter 9: An Overview of Raman Spectroscopy as Applied to Lignocellulosic Materials. In: Argyropoulos DS, editor. Advances in Lignocellulosics Characterization. Atlanta: TAPPI PRESS; p. 201-225.

[179] Agarwal UP, McSweeny JD, Ralph SA (2011) FT-Raman investigation of milled-wood lignins: Softwood, hardwood, and chemically modified black spruce lignins. J. Wood Chem. Technol. 31(4): 324-344.

[180] Sebastian S, Sundaraganesan N, Manoharan S (2009) Molecular structure, spectroscopic studies and first-order molecular hyperpolarizabilities of ferulic acid by density functional study. Spectroc. Acta Pt. A-Molec. Biomolec. Spectr. 74(2): 312-323.

[181] Atalla RH, Agarwal UP, Bond JS (1992) Raman Spectroscopy. In: Lin SY, Dence CW, editors. Methods in Lignin Chemistry. Heidelberg, Germany: Springer-Verlag; p. 162-176.

[182] Pu YQ, Kosa M, Kalluri UC, Tuskan GA, Ragauskas AJ (2011) Challenges of the utilization of wood polymers: how can they be overcome? Appl. Microbiol. Biotechnol. 91(6): 1525-1536.

[183] Gierlinger N, Schwanninger M (2006) Chemical imaging of poplar wood cell walls by confocal Raman microscopy. Plant Physiol. 140(4): 1246-1254.

[184] Takayama M, Johjima T, Yamanaka T, Wariishi H, Tanaka H (1997) Fourier transform Raman assignment of guaiacyl and syringyl marker bands for lignin determination. Spectroc. Acta Pt. A-Molec. Biomolec. Spectr. 53: 1621-1628.

[185] Larsen KL, Barsberg S (2011) Environmental Effects on the Lignin Model Monomer, Vanillyl Alcohol, Studied by Raman Spectroscopy. J. Phys. Chem. B 115(39): 11470-11480.

[186] Schmidt M, Schwartzberg AM, Perera PN, Weber-Bargioni A, Carroll A, Sarkar P, Bosneaga E, Urban JJ, Song J, Balakshin MY, Capanema EA, Auer M, Adams PD, Chiang VL, Schuck PJ (2009) Label-free in situ imaging of lignification in the cell wall of low lignin transgenic *Populus trichocarpa*. Planta 230(3): 589-597.

[187] Horvath L, Peszlen I, Gierlinger N, Peralta P, Steve K, Csoka L (2012) Distribution of wood polymers within the cell wall of transgenic aspen imaged by Raman microscopy. Holzforschung DOI 10.1515/hf-2011-0126.

[188] Jacobs BF, Kingston JD, Jacobs LL (1999) The origin of grass-dominated ecosystems. Annals of the Missouri Botanical Garden: Missouri Botanical Garden Press; p. 590-643.

[189] Draper J, Mur LAJ, Jenkins G, Ghosh-Biswas GC, Bablak P, Hasterok R, Routledge APM (2001) Brachypodium distachyon. A New Model System for Functional Genomics in Grasses. Plant Physiol. 127(4): 1539-1555.

[190] Albert L, Németh ZI, Halász G, Koloszár J, Varga S, Takács L (1999) Radial variation of pH and buffer capacity in the red-heartwooded beech (*Fagus silvatica* L.) wood. Holz als Roh- und Werkst. 57: 75-76.

[191] Somerville C, Bauer S, Brininstool G, Facette M, Hamann T, Milne J, Osborne E, Alex P, Persson S, Raab T, Vorwerk S, Youngs H (2004) Toward a systems approach to understanding plant cell walls. Science 206: 2206-2211.

[192] Grabber JH, Ralph J, Lapierre C, Barriere Y (2004) Genetic and molecular basis of grass cell-wall degradability. I. Lignin-cell wall matrix interactions. C. R. Biol. 327(5): 455-465.

[193] Ralph J, Grabber JH, Hatfield RD (1995) Lignin-ferulate cross-links in grasses: active incorporation of ferulate polysaccharide esters into ryegrass lignins. Carbohydr. Res. 275(1): 167-178.

[194] Ralph J, Hatfield RD, Sederoff RR, MacKay JJ (1998) Order and randomness in lignin and lignification: Is a new paradigm for lignification required? Research Summaries: 39-41.

# Chapter 9

# The Overview of Thermal Decomposition of Cellulose in Lignocellulosic Biomass

Dekui Shen, Rui Xiao, Sai Gu and Huiyan Zhang

Additional information is available at the end of the chapter

http://dx.doi.org/10.5772/51883

## 1. Introduction

Lignocellulosic biomass including wood, logging residue, crops and agricultural wastes) has been widely utilized to produce energy, fuels or chemicals, acting as the potential renewable source for taking place of fossil energies (such as coal, natural gas and petroleum) [1]. Pyrolysis is proved to be, one of the most promising methods to convert biomass into different products (syn-gas, bio-liquid, char and chemicals), which could essentially diversify the energy-supply in many situations [2].

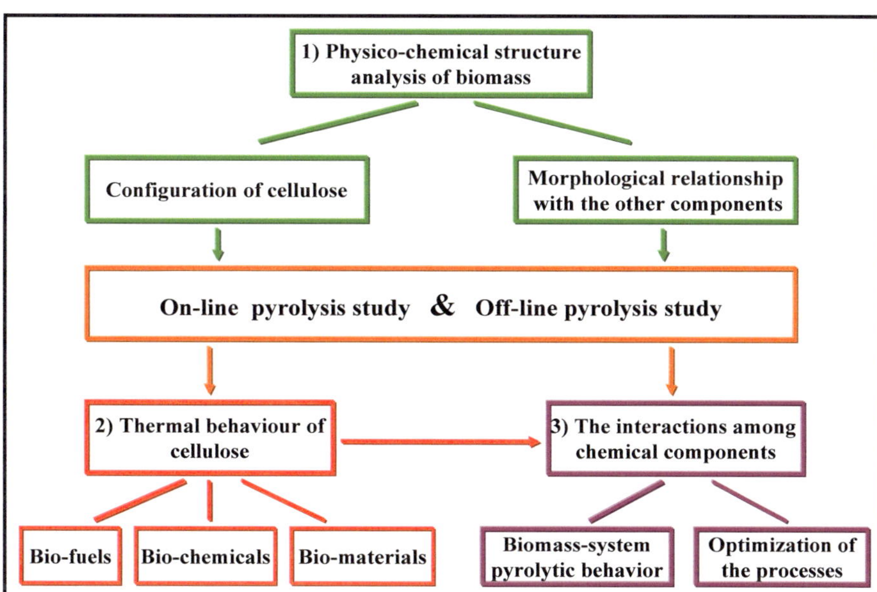


**Figure 1.** The fundamental issues and targets concerning the pyrolysis of cellulose

Cellulose, the most principal chemical component in different lignocellulosic biomass (accounting for more than 50% by weight), has a linear homopolymer of glucopyranose residues linked by β-1, 4- glycosidic bond. The study on pyrolysis of cellulose would be particularly benificial for achieving the better understanding of the pyrolytic mechanism of biomass and facilitating its direct applications in terms of fuels, chemicals and bio-materials. This gives rise to substantial studies on pyrolysis of cellulose in lignocellulosic biomass during the past half-century (Fig. 1), which could be categorized into the three following fundamental issues (Fig. 1):

1. The physico-chemical structure analysis of biomass is concerning the morphological analysis of the biomass cell-wall structure, the distribution and configuration of cellulose, which would facilitate not only the direct utilization of biomass as bio-material, but also the improvement of conversion processes of biomass to fuels or chemicals;
2. The thermal behavior of cellulose involving on-line pyrolysis and off-line pyrolysis study. The on-line pyrolysis is concentrated on the solid mass loss versus temperature or time (along with the evolution of the volatiles) and kinetic models, mostly employing isothermal and dynamic thermo-gravimetry analysis coupled with or without Fourier Transformation Infrared Spectrometry (FTIR) or Mass spectrometry (MS); The off-line pyrolysis study is to examine the yield of the main products (gas, liquid and solid), variation of the compositions in gaseous or liquid product influenced by the intrinsic characteristics and experimental conditions, in order to optimize the pyrolysis process for energy and/or chemicals production;
3. The interactions among the three main components under the pyrolytic condition is to introduce the possible interacting mechanism of the components in biomass, in terms of the mass loss process, the evolution of the volatiles and the yield of the specific products. This would help to improve the understanding of pyrolysis of whole biomass system from the pyrolytic behavior of the individual components.

The studies of pyrolysis of cellulose concerning the above four fundamental issues would be vigorously discussed in this work (especially for the works reported during the past 25 years), where the way-forward of this field would also be specified. This would supply the conceptual guide for the improvement of cellulose utilization and optimization of the thermal-conversion process of biomass.

## 2. The cell-wall structure of biomass and the configuration of cellulose

The morphological structure of lignocellulosic biomass has been studied regarding the distribution and inter-linkages of the chemical components, and their configuration [3, 4]. This facilitates not only the better understanding of the physico-chemical properties of biomass, but also the improvement of conversion processes (such as pyrolysis) of biomass to fuels or chemicals.

With the growing interest on lignocellulosic biomass as a potential substituent for fossil fuels, the pyrolysis of biomass should be dramatically examined. Consequently, the cell-wall model of lignocellulosic biomass, the distribution of the chemical components (especially cellulose), and the configuration of cellulose would be discussed in the following sections, which would help understand the remarkable characteristics of cellulose pyrolysis and its interactions with the other two main components (hemicellulose and lignin).

## 2.1. The cell-wall structure of biomass

The model of the cell-wall of woody biomass, firstly proposed by Fengel and Wegener [3], is well-established and further developed by Dumitriu [5], involving cell-wall structure and the distribution of the chemical components in different cell wall layers.

The cell wall could be morphologically divided into three distinct zones: middle lamella, primary cell wall and secondary wall [5]. The middle lamella is shared by two contiguous cells and is composed almost entirely of pectic substances. The primary cell walls are composed of cellulose microfibrils and interpenetrating matrix of hemicelluloses, pectins, and proteins. Cellulose forms the framework of the cell walls, hemicelluloses cross-link noncellulosic and cellulosic polymers, and pectins provide the structural support to the cell wall. The secondary cell walls are derived from the primary walls by thickening and inclusion of lignin into the cell wall matrix and occur inside the primary wall. The transition from primary to secondary cell wall synthesis is marked by the cessation of pectin deposition and a noted increase in the synthesis and deposition of cellulose/hemicellulose and lignin. The cellulose and non-cellulosic polysaccharides of the secondary cell wall are qualitatively distinct from those found in the primary cell walls.

The relevant study [6] evidenced that if cellulose is deposited actively between S1 and S3 developmental stages (especially in the middle part of S2 stage), hemicellulose (xylan) deposition occurs in the S1 to early S2 and again in the S3 developmental layers. Successive deposition of hemicellulose (xylan) onto the cell wall increases the microfibril diameter. The large amounts of hemicellulose (xylan) that accumulated on microfibrils appear globular but are covered with lignin after they are deposited. The information about the distribution of the main components (hemicellulose, cellulose and lignin) in the cell wall layers of lignocellulosic biomass is quantitatively reported in the literature [7].

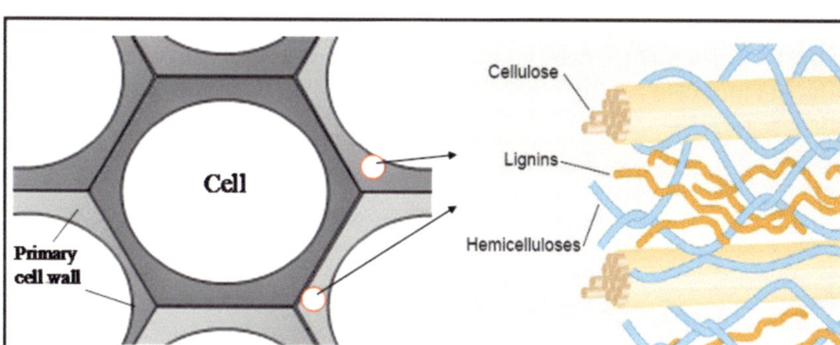


**Figure 2.** The schematic representation of the proposed cell wall along with the location of the main components in biomass

According to the above discussion, a simplified schematics for the structure of plant cell wall is presented in Fig. 2, where the morphological relationship among the main components in biomass (cellulose, hemicellulose and lignin) is clearly specified. It still needs to be notified that the details concerning the inter-linking/bond relationship (such as the H-bond among the polysaccharide molecules and lignin-carbohydrate coalescence) between the chemical components in the cell walls of wood are not well examined in the literature.

## 2.2. The configuration of cellulose

As far as the chemical components of biomass were concerned, a distinction should be made between the main macromolecular cell-wall components--cellulose, hemicellulose (polyoses) and lignin [3]. Cellulose is a uniform component in all lignocellulosic biomass, while the proportions and chemical composition of lignin and hemicellulose differ in different biomass. The configuration of cellulose in lignocellulosic biomass would be discussed, with regard to its content, isolation methods, the characterization of the macromolecules and the inter-linkages among the units.

Cellulose is the prominent chemical component in lignocellulosic biomass, accounting for approximately 50% by weight. The methods for isolating and/or determining cellulose from biomass could be summarized as [3]:

1. Separation of the main portions of hemicellulose and residual lignin from cellulose;
2. Direct isolation of cellulose from lignocellulosic biomass, including purification procedures (such as pulping process);
3. Determination of the approximate cellulose content by total hydrolysis of biomass, cellulose with subsequent determination of the resulting sugars.

In any isolation method cellulose cannot be obtained in a pure state, thus the purification always plays an important role in the cellulose isolation process. Through the relevant methylation experimental studies [3, 5], the primary structure of cellulose is evidenced as a linearhomopolymer of glucose having the D configuration and connected by β-(1-4) glycosidic linkages (Fig. 3). It could be found that the units of the cellulose molecular chain are bound by β-(1-4) glycosidic linkages, presenting that the adjacent glucose units are linked by dehydration between their hydroxylic groups at carbon 1 and carbon 4. The β-position of the OH-group at C1 needs a turning of the following glucose unit around the C1-C4 axis of the pyranose ring.

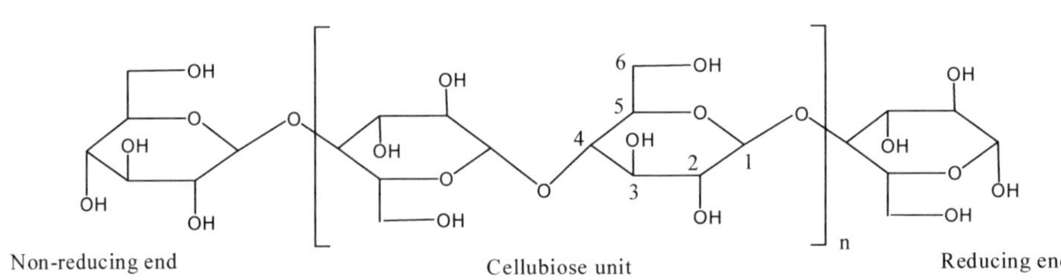


**Figure 3.** The central part (cellubiose unit) of cellulose molecular chain with the reducing and non-reducing end groups.

The stabilization of the long cellulose molecular chains in order systems originates in the presence of functional groups which are able to interact with each other. The functional groups of the cellulose chains are the hydroxyl groups, three of which are linked to each glucopyranose unit. These OH-groups are not only responsible for the supramolecular structure by also for the chemical and physical behavior of the cellulose through the hydrogen bond (H-bond). The OH-groups of cellulose molecules are able to form two types of hydrogen bonds depending on their site at the glucose unit [3]. The hydrogen bonds between OH-groups of adjacent glucose units in the same cellulose chain are called intramolecular linkages, which give certain stiffness to the single chain. The hydrogen bonds between OH-groups of the adjacent cellulose chains are called intermolecular linkages, which are responsible for the formation of supramolecular structures. The primary structures, consisting of a number of cellulose chains through the hydrogen bonds in a superhelicoidal fashion, are the cellulose microfibrils, which build up the framework of the whole cell walls [5].

Two chain ends of the cellulose chain are chemically different (Fig. 3). One end has a D-glucopyranose unit in which anomeric carbon atom is involved in a glycosidic linkage, whereas the other end has a D-glucopyranose unit in which the anomeric carbon atom is free. This cyclic hemiacetal function is in an equilibrium in which a small proportion is an aldehyde, which gives rise to reducing properties at this end of the chain, so that the cellulose chain has a chemical polarity, while the OH-group at the C4 end of the cellulose chain is an alcoholic hydroxyl and therefore non-reducing. The molecular weight of cellulose varies widely depending on the origin of the sample. As cellulose is a linear polymer with uniform units and bonds the size of the chain molecule is usually defined as degree of polymerization (DP). The degrees of polymerization of the plant-cellulose as well as the technical cellulose products are estimated from 15300 for capsules to 305 for rayon fibers [5].

## 3. The thermal behavior of cellulose

### 3.1. On-line pyrolysis of cellulose

The thermogravimetric (TG) analysis method, either dynamic heating process or isothermal heating process, is well-established for on-line pyrolysis of biomass and its components (cellulose, hemicellulose and lignin). The mass loss of the solid sample could be exactly recorded versus temperature/time. The chemical kinetic models for the biomass and its components are proposed from the analysis of the different mass loss stages and validated through the correlation between the predicted data and the experimental mass loss curve. Since the specific chemical phenomena and the prediction of the volatile yields are rarely referred in those models, TGA coupled with FTIR, GC, MS or other advanced analytical equipments is recently employed to investigate the evolution of the volatile along during the pyrolysis process. This facilitates the understanding of the possible chemical reactions for depolymerization of the macromolecules and the secondary cracking of the primary fragments. The development of the kinetics of cellulose pyrolysis would be systematically overviewed, involving most of recent studies implemented by other groups led by Piskorz, Di Blasi, Banyaz, Agrawal, Wooten, Hosoya and so on. Several controversial points addressed in previous studies would be intensively discussed, concerning the existence of the intermediate anhydrosugars, secondary cracking of the volatiles and the formation of char residue.

Historically, it was perhaps that Broido' s group firstly called attention to the intriguing phenomena of cellulose pyrolysis and proposed the established kinetic scheme in 1960s [25, 26]. As described in **Scheme 1** (Fig. 4) [26, 27], the decomposition of cellulose can be represented through two competing reactions: the first step is estimated to be important at low temperatures and slow heating rates, accounting for the slight endothermic formation of anhydrocellulose below 280 °C detected by DTA. At about 280 °C a competitive, more endothermic unzipping reaction is initiated for the remained cellulose, leading to the tar formation. The third step presents the exothermic decomposition of anhydrocellulose to char and gas.

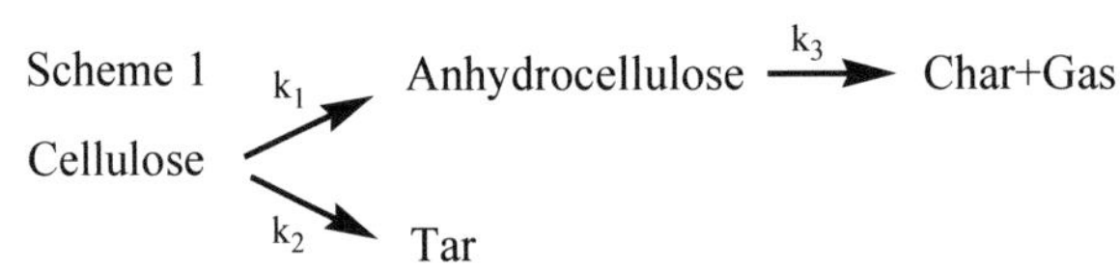


**Figure 4.** The kinetic model for cellulose pyrolysis proposed by Broido and Weinstein (1971) [27]

This Broido' s kinetic scheme is re-examined by Argawal [13], revealing that the rates of anhydrocellulose formation are comparable to those of the depolymerization process only in one case for temperatures of ~ 270 °C in the isothermal, fixed-bed conditions. Then, the mechanism is approved through the isothermal, fluid-bed experiments in the temperature range 250-300 °C, providing a complete set of kinetic data for the Broido model [13]. It is worthily noting that the formation of the anhydrocellulose as an intermediate product is undetectable in the experiments, and no kinetic data for the char forming reaction are reported in the above publications. These ambiguities stimulated the global researchers' interests in the kinetic studies of cellulose pyrolysis, resulting in a vigorous debate in the following years.

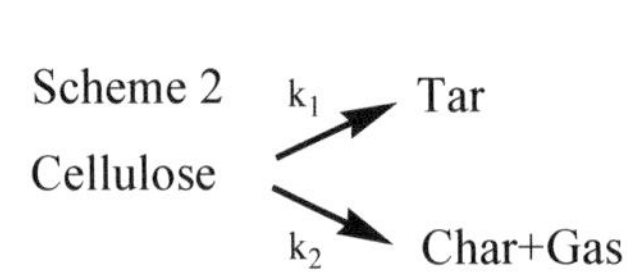


**Figure 5.** The kinetic model for cellulose pyrolysis proposed by Broido and Nelson (1975) [10]

In 1975, Broido and Nelson examined the effect of thermal pretreatments at 230-275 °C on the cellulose char yields varying from 13% (no thermal pretreatment) to over 27% [10]. They employed the large samples of cellulose (100 mg of shredded cellulose, and 7 cm × 3 cm sheets, individually wrapped several layers deep around a glass rod), which might incur the char formation from solid-vapor interactions during the prolonged thermal pretreatment. The previous kinetic model (Scheme 1) is correspondingly improved as described in **Scheme**

**2** (Fig. 5), eliminating the formation of the anhydrocellulose as an intermediate product. The **Scheme 3 (Fig. 6)** is slightly different from those proposed by Broido and the co-workers but largely confirms the previous findings, which is even titled as "Broido-Shafizadeh model" in somewhere [23, 30-32]. At the low temperatures (259-295 °C), the initiation period (characterized by an accelerating rate of weight loss [33]) has been explained as a formation of "active cellulose" through the depolymerization process (reduction of the DP) with the activation energy of 242.8 kJ/mol. Then, the "active cellulose" undergoes the two competitive reactions to produce either char and gas (activation energy 153.1 kJ/mol) or primary volatiles (197.9 kJ/mol). At high temperatures (above 295 °C), no initial period of accelerating rate of weight loss was observed in Shafizadeh's study [29]. Thus cellulose degradation mechanism was described simply via two competitive first-order reactions, where the formation of "active cellulose" is eliminated from **Scheme 3**. This mechanism is then confirmed by Antal and Varhegyi' TGA study of cellulose pyrolysis with the heating rate of 40 K/min, attaining the activation energy for the formation of volatiles as 238 kJ/mol and 148 kJ/mol for the formation of char and gas [14].

Scheme 3

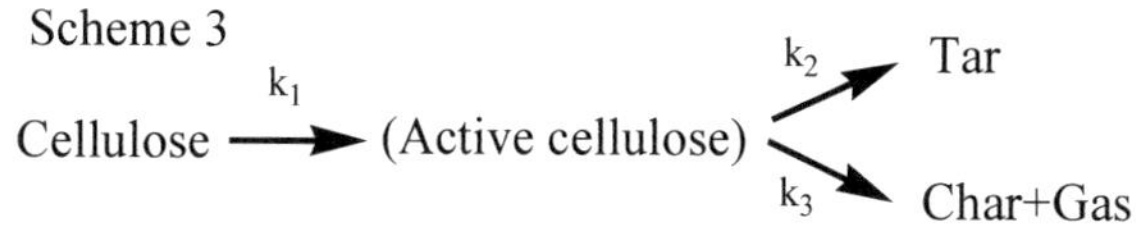


**Figure 6.** The kinetic model for cellulose pyrolysis proposed by Bradbury et al. (1979) [29]

The argument between Antal-Varhegyi and Broido-Shafezadeh is remarkable, concerning the existence of "active cellulose" during the pyrolysis of cellulose. Antal and Varhegyi presented that no evidence was found to support the inclusion of the initiation step displayed in the **Scheme 5** (titled as "Broido-Schafezadeh model"), whatever this step proceeded at an immeasurably high rate at conditions of interests, or it does not exist [23].

In 2002, Lede et al. directly observed a transient "intermediate liquid compound" in small pellets of cellulose that had been heated by radiant flash pyrolysis in an imaging furnace, which is characterized by HPLC/MS and found to be composed predominantly of anhydro-oligosaccharides (such as levoglucosan, cellobiosan and cellotriosan) [41]. In the slow heating experiments of cellulose, Wooten [32] revealed that intermediate cellulose (IC) is an ephemeral component that appears and ten disappears over the course of 60 min of heating at 300 °C, while the rapid disappearance of IC in samples that have been heated at only a slightly higher temperature (i.e., 325 °C) further demonstrates the transient nature of IC. This behavior clearly identifies the compound(s) as a reaction intermediate, and the authors correspondingly associated this intermediate compound with the "active cellulose" in the Broido and Shafezadeh kinetic models (**Scheme 3**) [28, 29]. Thus, some recent researchers have attained the formation of "active cellulose" as an intermediate during cellulose pyrolysis, as presented in **Scheme 4** (Fig. 7) [12, 42]

Previously, Bradbury et al. [29] and Antal [23] suggested that char formation might result from the repolymerization of volatile materials such as levoglucosan. This phenomenon is approved by Hosoya [36], presenting that the secondary char from cellulose is formed from

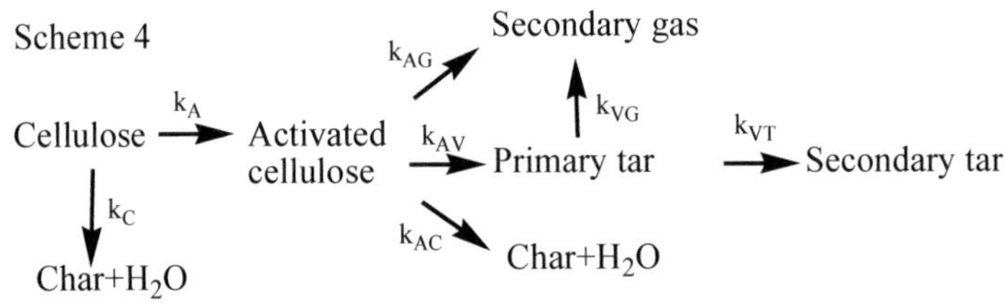


**Figure 7.** The kinetic model for cellulose pyrolysis proposed by Diebold (1994) [39] and similarly proposed by Wooten et al. (2004) [32]

the repolymerization of anhydrosugars (levoglucosan). The experimental data from the Wooten et al.'s study [32] shows that a precursor-product relationship does exist between intermediate cellulose ("active cellulose") and the aliphatic and aromatic components of the char.

Nowadays, it might be not difficult to evidence the existence of "active cellulose" or other important (intermediate) products with the help of the advanced analytical equipments, but the chemical reaction mechanism for cellulose pyrolysis is still ambiguous and controversial. One of the possible routes to improve the understanding of the structure changes of cellulose molecules and formation of the specific products is to employ the study of thermal decomposition of the relevant derivatives, together with the molecular dynamic simulation (MDS) which is well-established for estimating the specific chemical pathways from the microscopic point of view. Moreover, the identification of intermolecular hydrogen bonding and that between the different molecular chains would be another uncertainty for understanding the pyrolytic behavior of cellulose, especially for the initial stage of the cellulose pyrolysis.

## 3.2. Off-line pyrolysis of cellulose

Compared to the on-line pyrolysis study of cellulose, the off-line pyrolysis of cellulose is mostly carried out under the relatively high temperature (above 400 °C) or high heating rate (more than or around 1000 °C/s) [12, 32, 36, 40, 44-48] and sometimes under low temperature heating (below 400 °C) [49, 50], concerning the following issues: 1) the distributions of the gas, liquid and solid products; 2) the formation of the specific compounds and the pyrolytic chemical pathways. How these two issues may be influenced by the pyrolytic reactors and the variables like temperature, residence time, heating rate, pressure, particle size, catalytic salts and crystallinity is extensively examined in the literature, in order to promote the product specificity, maximize the yield and improve the understanding of the pyrolytic mechanism.

In this work, the emphasis is on the effects of the predominant factors such as the reactor type, temperature or heating rate, residence time on the distributions of the products (gas, liquid and solid) from cellulose pyrolysis. Considering the complexity of chemical constituents in gas and liquid products, the attention would be confined to those few compounds which have been established to be producible in good yield (such as levoglucosan, hydroxyacetaldehyde, furfural, CO, $CO_2$ and so on), in order to meet the interests in potential industrial applications.

| Reactor | Liquid yield wt% | Feed size | Input gas | Complexity | Scale-up | Status* |
|---|---|---|---|---|---|---|
| Fluidized bed | 75 | Small | High | Medium | Easy | Demo |
| CFB | 75 | Medium | High | High | Easy | Pilot |
| Entrained gas flow | 65 | Small | High | High | Easy | Lab |
| Vaccum | 60 | Large | Low | High | Hard | Demo |
| Rotating cone | 65 | Very small | Low | High | Hard | Pilot |
| Ablative | 75 | Large | Low | High | Hard | Lab |
| Auger | 65 | Small | Low | Low | Easy | |

*: Demo scale is estimated to be 200-2000 kg/h, pilot scale is 20-200 kg/h and lab scale is <20 kg/h.

**Table 1.** The characteristics of the fast (off-line) pyrolysis reactors of biomass [51]

### *3.2.1. The distributions of gas, liquid and solid products*

Regarding the commercialization of the pyrolytic technology for bio-energy conversion, the designed pyrolytic reactor involving the variation of the operating parameters (temperature, residence time, pressure and so on) has remarkable effects on the threshold of the specific product yield and the operating cost of the process [52-55]. Most of the reactors for the fast (off-line) pyrolysis of biomass to produce bio-oil or fuel gases is summarized by Bridgewater [51], estimated in terms of product yield, feed size, input gas, complexity and so on (Table 1). It is approved that the fluidized bed reactor is determined to be one of the promising technologies for biomass thermal conversion due to the high-efficient heat transfer and ease of scale-up, which has potential for commercial practice [56-60]. Microwave pyrolysis, termed as a novel thermo-chemical technology for converting biomass to solid, liquid and gas fuels, is of growing interests with thanks to its low requirement on energy input during the process, flexibility of the feedstock size and high quality of products (low oxygen content in char and bio-oil). The yield of the products from cellulose through different pyrolysis reactors would be intensively discussed, with regard to the effect of operating conditions such as temperature, residence time and condensing patterns.

#### *3.2.1.1. Pyrolysis in fluidized-bed reactor*

The outstanding contribution on study of cellulose pyrolysis in the fluidized bed reactor was made by the research group led by Scott and Piskorz in the University of Waterloo in Canada [12, 17, 42, 46, 61-63]. A bench scale atmospheric pressure fluidized bed unit using sand as the fluidized solid with the feeding rate of 30 g/h of biomass was designed to investigate the yield of liquid product at different temperatures in an inert nitrogen atmosphere with an apparent vapor residence time of approximately 0.5 s [62]. Piskorz [12] reported the pyrolytic behavior of the two types of cellulose (S&S powdered cellulose with ash content of 0.22% and Baker TLC microcrystalline cellulose with ash content of 0.04%) in the fluidized bed reactor, giving the distribution of the gas, liquid and solid products at the temperature from 450 to 550 °C summarized in Table 2. The yield of organic products in the liquid phase (except water) from the S&S powdered cellulose ranges from 58.58% to 67.81% of the moisture and ash free feed at the temperature from 450 to 550 °C, reaching the maximum at 500 °C. Comparatively, the yield of organic products from the Baker TLC microcrystalline cellulose at 500 °C is determined to be 90.1%. Moreover, the yield of char for S&S powdered cellulose at 500 °C is 3.4%, compared to 1.0% for Baker TLC microcrystalline cellulose.

These results confirms that the larger amount of the inorganic salts in the ash content promotes the formation of the condensed structure through the catalytic effects, inhibiting the cracking of the macromolecules and enhancing the yield of solid product [21, 30, 31, 34, 46, 64]. Several years later, the pyrolysis of the two further types of cellulose (commercial SS-144 crystalline cellulose and Avicel pH-102 crystalline cellulose) were also studied in the fluidized bed by Piskorz's co-worker (Radlein, et al.) [46], presenting the yield of the products in Table 2. The temperature 500 °C, regarded as the optimal condition for producing bio-oil from cellulose in the fluidized bed reactor, gives the yield of organic products of 72.5% for commercial SS-144 crystalline cellulose and 83.5% for Avicel pH-102 crystalline cellulose. The difference should also be attributed to the catalytic effect of inorganic salts in the ash, since the yield of char for commercial SS-144 crystalline cellulose is 5.4% compared to 1.3% for Avicel pH-102 crystalline cellulose.

Recently, Aho [47] conducted the pyrolysis of softwood carbohydrates under the nitrogen atmosphere in a batch-operating fluidized bed reactor, where the quartz sand was used as bed material and the load of the raw material is approximately 10 g. All sand was kept in the reactor by a net at the upper part of the reactor. The evolved vapors were cooled in the four consecutive coolers with the set point of -20 °C, while between the third and fourth cooler the vapors were passed through a water quench with the pH value of 3 for avoiding the absorption of $CO_2$. The furnace temperature was kept at 490 °C until the release of non-condensable gases stopped, while the temperature in the reactor is about 460 °C. The vapor residence time was estimated to be less than 1.5 s based on the height of the reactor and the actual fluidizing gas velocity. The distribution of the products from cellulose (microcrystalline cellulose powder) is shown in Table 2, giving the low yield of organic products of 23.1% and high yield of char as 20.1%. The condensation of the vapors was estimated to be insufficient, while the values for gases and char can be considered reliable. It should be mentioned that the mass balance of the experiment could not be satisfactorily completed, due to its current reactor set-up (especially the vapor-cooling and liquid-precipitating system). A similar batch-operating fluidized bed reactor was designed by Shen and Gu, in order to study the fast pyrolysis of biomass and its components with the variation of temperature and vapor residence time under inert atmosphere [21, 65, 66]. No bed material was applied and the load of the raw material is about 5 g. The solid product was captured by the carbon filter, while the evolved hot vapors were cooled through the two U-tubes immersed in ice-water mixture (0 °C) and dry ice-acetone (-30 °C), respectively. The distribution of the products from the pyrolysis of microcrystalline cellulose at temperatures between 420 and 730 °C with a residence time from 0.44 to 1.32 s is given in Table 2. It is estimated that the yield of liquid product reaches its maximum of 72.2% at the temperature of 580 °C with the residence time of 0.44 s. The higher temperature and long residence time promotes the decomposition of the macromolecules and cracking of the volatile, enhancing the yield of gases and reducing the solid product [21].

*3.2.1.2. Pyrolysis in entrained-bed reactor*

Graham [69] designed a complicated entrained bed reactor to investigate the fast pyrolysis of cellulose, which had a similar or even higher heating rate than that of fluidized bed. The rapid heat transfer and thorough mixing between the particulate solids and feed are accomplished in two vertical gas-solids contactors: Thermovortactor and Cryovortactor. The biomass or other carbonaceous fuel is rapidly mixed with the hot particulate solids in Thermovortactor. The suspension passed through a downdraft entrained-bed (fluidized) reactor allowing the individual setting of temperatures, and then was quenched by the cold

| Author(s) | Sample | Pyrolysis reactor | Conditions | | Yield of products (wt%) | | |
|---|---|---|---|---|---|---|---|
| | | | Temperature (°C) | Residence time (s) | Gas | Liquid [1] (water) | Char |
| M.R. Hajaligol, et al. (1982) [67] | No. 507 filter paper | Screen-heating Pyrex reactor (fixed bed) | 400 ~ 1000 | 0 ~ 30 | 5.25 ~ 46.97 | 16.37 ~ 83.35 | 3.32 ~ 78.37 |
| W.S.L. Mok and M.J. Antal (1983) [68] | Whatman filter paper | Two-zone tubular micro reactor (fixed bed) [3] | 800 | 1 ~ 18 | 62 ~ 71 | -- | 15 ~ 23 |
| R.G. Graham, et al. (1984) [69] | Avicel pH-102 crystalline cellulose | Downflow entrained bed (fluidized) reactor | 750 ~ 900 | < 0.6 | 74.7 ~ 98.1 | 0.7 ~ 15.8 [4] | -- |
| J. Piskorz, et al. (1986) [12, 42] | S&S powdered cellulose | Fluidized bed reactor | 450 ~ 550 | 0.53 ~ 0.56 | 8.49 ~ 17.89 | 68.75 ~ 75.59 (7.35 ~ 10.17) | 4.2 ~ 8.53 |
| | Baker TLC crystalline cellulose | | 500 | 0.48 | 5.1 | 94.7 (4.6) | 1.0 |
| D. Radlein, et al. (1991) [46] | Commercial SS-144 crystalline cellulose | Fluidized bed reactor | 500 | < 0.5 | 7.8 | 83.3 (10.8) | 5.4 |
| | Avicel pH-102 crystalline cellulose | | 500 | < 0.5 | 3.9 | 89.6 (6.1) | 1.3 |
| Y.F. Liao (2003) [31] | Filter paper with ash content of 0.01% | Gravitational feeding reactor (Fixed bed) | 300 ~ 1090 | 0.1 ~ 1.4 | 1.5 ~ 60.2 | 6.0 ~ 86.3 | 1.8 ~ 92.5 |
| Aho, et al. (2008) [47] | Microcrystalline cellulose powder | Batch-operating fluidized bed reactor | 460 | <1.5 | 32.3 | 47.6 (24.5) | 20.1 |
| T. Hosoya, et al. (2007) [36] | Cellulose powder from Toyoroshi Co. | Cylindrical furnace and tube reactor (fixed bed) | 800 | 30 | 12.9 | 77.1 (5.1) | 10 |
| D.K. Shen and S. Gu (2009) [21] | Microcrystalline cellulose powder | Batch-operating fluidized bed reactor | 420 ~ 730 | 0.44 ~ 1.32 | 20.1 ~ 42.5 | 30.6 ~ 72.2 | 1.03 ~ 47.4 |

1: the yield of liquid product including water;2: the pressure is 5 psig of helium pressure;3: the operating pressure in the furnace is 5 atm;4: including solid product (char);

**Table 2.** The summary of the studies on fast (off-line) pyrolysis of cellulose solids in the Cryvortactor and cooled through the cooling coil submerged in a water tank. The solids were then separated in the mass balance filter and the gas was collected in sampling bags. The feeding rate is less than 1 kg/h and the total elapsed time from the Termovortactor inlet to the cryovortactor exit is typically less than 600 ms. The yield of the gas and liquid (heavy fraction including tar and char) products at the temperature from 750 to 900 °C is shown in Table 2. The low yield of liquid product (less than 20%) is mainly due to the high reactor temperature and the inefficient cooling method. Moreover, the mass balance is not convincing, since the heavy fraction of the vapors may condense on the vessels of Cryovortactor and solid separator [69]. It should be noted that the high yield of gases is attributed to the enhanced heat transfer through the pre-mixing between the biomass and solid heat carrier before being fed to the pyolsyis reactor, compared to that of fluidized bed reactor.

The residence time (both solid and vapor) in the fluidized or entrained bed reactors could be narrowly changed (normally less than 1 s), because of the confinement of the minimum gas velocity for the solid fluidization. Therefore, the fixed bed reactors are designed for investigating the effect of not only temperature but also residence time on the yield of products and their specificity [31, 36, 44, 68]. Liao [31] designed a fixed bed reactor (quartz tube with a sample-holder in the middle), the temperature of which could be changed from 0 to 1100 °C. The filter paper shaped as 18*50 mm (about 2 g) is fed gravitationally to the reactor from the top, and the carrier gas (nitrogen) brings the evolved volatiles and some char fragments through the carbon filter. The purified volatiles are then cooled through the three traps consecutively: 1) the mixture of water and ice (0 °C); 2) the mixture of acetone and dry ice (-30 °C); and 3) assisting cooling agent (-45 °C). The yield of the products (gas, liquid and char) at the temperature from 300 to 1090 °C with the (vapor) residence time between 0.1 to 1.4 s determined by the carrier gas velocity is extensively discussed by Liao [31] (shown in Table 2-3), while the mass balance for all the experiments is convincingly located between 96% and 101.5%. With the same vapor residence time (carrier gas velocity), the yield of liquid product complies with a Gaussian distribution with temperature, giving the maximum of 86.29% (including 15.72% water) at around 600 °C with the residence time of 0.1 s. It is estimated that the long residence time promotes the yield of gases, due to the sufficient secondary reactions of the volatiles. The yield of gases is increased from 1.5% to 60.2% monotonously with temperature (from 300 to 1090 °C). It needs to be noted that the duration of each experiment, corresponding to the sample heating-up and holding time, is not specified in the work.

*3.2.1.3. Pyrolysis in fixed-bed reactor*

The pyrolysis of cellulose in a tube (fixed bed) reactor made of Pyrex glass is investigated by Hosoya et al. [36]. Compared to the study of Liao [31], the cellulose sample is horizontally fed to the furnace and the carrier gas is not employed which means that the vapor residence time could not be set individually. It is estimated that thirty seconds are enough for completing the pyrolysis since no volatile product formation is observed after longer pyrolysis time. The evolved volatiles are retained in the reactor with the solid residue during the whole pyrolysis process. After 30 s pyrolysis, the reactor is pulled out from the furnace and cooled with air flow for 1 min at the room temperature. The tar (liquid product) condensed on the reactor vessel is extracted by *i*-PrOH and water. The amounts of the gaseous, tar and char fractions are determined gravimetrically after pyrolysis and extraction, giving the result at the temperature of 800 °C in Table 2. It should be mentioned that the temperature of the reactor is not evenly distributed during the pyrolysis process, because the bottom of the tube reactor was placed at the center of the cylindrical furnace. Most of the evolved volatiles are condensed at the upper part of the reactor, but not suffered from the vigorous secondary cracking due to the long (solid) residence time. Thus, the yield of liquid product (77.1% including water) is not visibly different from that of Liao's results at the temperature of 810 °C with the shorter vapor residence time (74.39%) [31].

Another Pyrex cylindrical tube (fixed bed) reactor was made by Hajaligol et al. [44], where the cellulose sample is held and heated by the porous stainless screen connected to the brass electrodes of the reactor. The system allows independent variation of the following reaction conditions: heating rates (100-100 000 °C/s), final temperatures (200-1100 °C), sample residence (holding) time at final temperature (0-∝ s). Similar to the experimental set-up of Hosoya [36], the vapor residence time could not be individually changed while the carrier gas is not employed. Part of the evolved vapors is rapidly diluted and quenched in the reactor vessel during the operation, because most of the gas within the reactor remains close to the room temperature. The other part of the evolved vapors is purged out of reactor vessel with the helium and cooled down through two downstream traps: 1) U-tube packed with glass wool immersed in dry ice/alcohol (-77 °C) and 2) the same trap in liquid nitrogen (-196 °C). The char retained on the screen is determined gravimetrically. The mass balance for each case is around 100%, giving the convincing results of the yield of the products at the temperature 400- 1000 °C with the sample holding time 0-30 s in Table 2. It is concluded [44] that tar yield (liquid product) increases with temperature to a maximum of about 65% at around 700 °C and then decreases with further temperature increases, since the sample residence time is zero. With the long residence time (for example 30 s), the yield of liquid product at 400 °C is remarkably increased to 83.35%, due to the sufficient heating-up time for the complete pyrolysis of cellulose. Comparatively, the yield of liquid product at 500 °C with zero holding time is only 16.37% and the yield of char is 83.63% (where the mass balance is 105%), because of the incomplete decomposition of cellulose.

A two-zone tubular micro reactor (fixed bed) was designed by Mok and Antal [68], to investigate the effect of vapor residence time on the yield of products from cellulose pyrolysis. Zone A is operated for 15 min for complete solid phase pyrolysis, while Zone B is maintained at 700 °C for vapor phase cracking. The char is determined gravimetrically, and the gases are collected by the replacement of water. Unfortunately, the tar collection is not possible with that apparatus. The results of the product distribution at the temperature of 800 °C with the vapor residence time 1-18 s are shown in Table 2. The long vapor residence time and high pressure (5 atm) promote the secondary cracking of volatiles, enhancing the yield of the gas product.

*3.2.1.4. Pyrolysis in microwave reactor*

The microwaves might be firstly used to activate biomass (cellulose as the feedstock) to solid, liquid and gas products by Allan et al. in 1970s [70]. After 2000, two research groups (one is led by J.H. Clark from University of York in UK and the other by Y. Fernandez and J.J. Pis from National Institute of Carbon in Spain) have published a large number of the remarkable results on microwave pyrolysis **(MWP)** of biomass and its components (such as cellulose and lignin) [49, 50, 71-74].

The studies of the research group led by Fernandez and Pis are mainly concentrated on the **high-temperature** microwave pyrolysis (more than 400 °C) of biomass [72, 74]. The feedstock sample (coffee hulls) being rich in cellulose, is made to be the cylindrical pellets (approximately 3 mm in diameter and 2 cm in length). The pyrolysis of the sample (15 g of that kind of pellets) was carried out in an electrical furnace (called CP-conventional pyrolysis) and in a single mode microwave oven at 500, 800, and 1000 °C, regarding the variation of the yield of products (char, oil and gases) and their properties (element content and heating value). The electrical furnace was previously heated to the corresponding pyrolysis temperature, so that the temperature of sample rose quickly. In case of microwave heating, the sample was placed in an identical quartz reactor, which was then placed in the centre of microwave guide [75]. The volatiles evolved passed through five consecutive condensers placed in an ice bath, the last of three of which contained dichloromethane, while the carbonaceous residue was separated from the receptor by sieving. The gas yield was evaluated by difference. It is found that the yield of char, oil and gas from pyrolysis of sample under microwave heating is 30.21%, 7.90% and 65.28% by weight of feedstock at 500 °C and changed to be 22.70 %, 8.58% and 68.72% at 1000 °C. Compared to that of conventional pyrolysis by electrical heating, the formation of the gas products (especially syngas $CO+H_2$) is remarkably enhanced under microwave pyrolysis and the oxygen content in char and oil is significantly reduced increasing their heating value. Most of the above findings on microwave pyrolysis of biomass are also approved by other researchers [48, 76].

Research group led by J.H. Clark has made a remarkable contribution on the microwave pyrolysis of biomass under **low temperature** (less than 350 °C) [49, 50, 73]. Milestone ROTO SYNTH Rotative Solid Phase Microwave Reactor is used for microwave pyrolysis of wheat straw [49]. Average sample mass was between 150 and 200g. The sample was heated at a rate of 17 °C/min to a maximum temperature of **180 °C** as measured by in situ temperature probes. The condensable fraction produced during the process was collected through a vacuum unit. The yield of solid, liquid and gas products is estimated to be 29%, 57% and 14% by weight of feedstock at 180 °C. Compared to that of conventional pyrolysis under relevantly high temperature [77], the oxygen content of the bio-oil obtained from low-temperature microwave pyrolysis is significantly reduced facilitating the following upgrading processes [49]. The microwave pyrolysis of cellulose was carried out at the temperature between 100 °C and 300 °C in a CEM Discovery laboratory microwave, regarding the yield of char and its formation mechanism. The high-quality char, where more energy from feedstock is conserved, could be produced with the adjustment of the low pyrolysis temperature. The temperature of 180 °C was estimated as a key turning point in the microwave degradation of cellulose, favoring the understanding that the production of fuels is allowed at dramatically lower temperatures than those required under conventional pyrolysis (electrical heating). The energy conserved in solid, oil and gas product is evaluated to be balanced for the whole process. In terms of an industrial process, the low-temperature microwave technology can be easily adapted to a variety of biomass to produce a uniform char which can be handled by the end users.

With regard to the above discussion, the microwave pyrolysis under both high and low temperature is estimated to be one of the promising technologies to achieve high-quality solid (low oxygen content), liquid (low oxygen content and water content) and gas (low energy input and high syngas concentration) fuels with the low cost, helping to achieve sustainable development through the utilization of renewable alternatives (biomass) instead of fossil fuels.

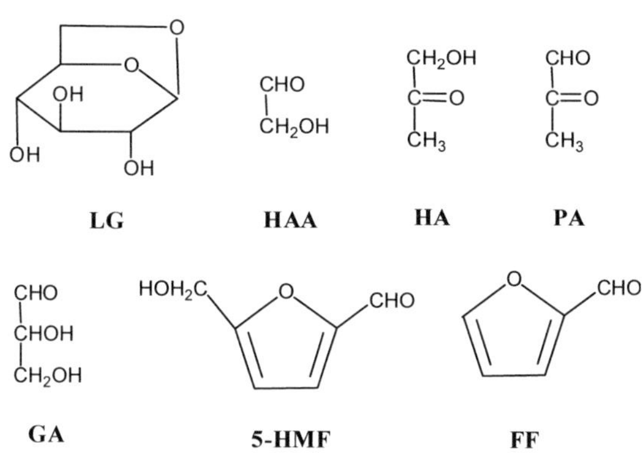


**Figure 8.** The chemical structures of the typical compounds in bio-oil from cellulose pyrolysis: LG: levoglucosan, HAA: hydroxyacetaldehyde, HA: Hydroxyactone, PA: pyruvic aldehyde, GA: glyceraldehyde, 5-HMF: 5-hydroxymethyl-furfural and FF: furfural

### *3.2.2. The formation of the specific compounds*

The volatiles (both condensable and non-condensable) evolved from cellulose pyrolysis under moderate or high temperatures are very complicated, most of which have been identified by employing the advanced analytical equipments such as FTIR, GC-MS, HPLC, NMR and so on. A variety of pyran and furan derivatives ($C_{5\text{-}6}$ ring-containing compounds), aliphatic oxygenated $C_{2\text{-}4}$ organic compounds and light species/gases (such as light hydrocarbons, CO and $CO_2$) can be obtained, and the extensive lists together with their spectrometric/chromatograghic patterns and the yields are available in the literature, where the results are remarkably affected by the pyrolytic reactor, operating condition, condensing method and sample sources. Due to the great potential as the feedstock for fuel and chemicals production, some products established in good yields (such as levoglucosan, furfural, hydroxyacetaldehyde, acetol, CO, $CO_2$ and so on) (Fig. 8) would be vigorously investigated regarding the chemical mechanism for their formation and fractionation.

*3.2.2.1. Pyran- and furan- derivatives (C5-6 ring-contained compounds)*

The $C_{5\text{-}6}$ ring-containing compounds from cellulose pyrolysis are condensable and mainly composed of a variety of anhydrosugar and furan derivatives, among which levoglucosan (1, 6-anhydro-β-D-glucopyranose) are the outstanding one [12, 18, 21, 23, 31, 36, 41, 78-82]. Shafizadeh et al. [33] confirmed that levoglucosan can be obtained in yields from 20% to 60% by weight in their vacuum pyrolysis study of various cellulose samples, while other anhydrosugars (such as 2,3-anhydro-d-mannose, 1,4:3,6-dianhydro-α-D-glucopyranose, 1,6-anhydro-β-D-glucofuranose and 3,4-altrosan) are slightly produced (less than 1% by weight). Similar results were reported by Piskorz et al. by comparing levoglucosan yields from S & S powdered cellulose (2.1%) and Baker TLC microcrystalline cellulose (25.2%) pyrolysis at the temperature of 500 °C under atmospheric pressure in a fluidized bed reactor [12].

Inasmuch as the cellulose samples have somewhat different ash contents, the different levoglucan yield may be due to the well-known effect of inorganic cations in reducing tar yields by promoting other fragments or char formation [46]. Richards and co-workers established the extraordinary influence of salts and metal ions on the productivity of volatiles (especially levoglucosan and hydroxyacetaldehyde), presenting that the addition of alkali and $Ca^{2+}$ cations to ash-free cellulose reduced the yield of levoglucosan while other metal ions (particularly $Fe^{3+}$ and $Cu^{2+}$) enhanced the yield of levoglucosan [83, 84]. In accord with the findings of Richards's laboratory, Piskorz et al. observed very dramatic increases in the yields of levoglucosan (more than 30% by weight) from various celluloses after a mild sulfuric acid-wash pretreatment [42]. The profound effects of inorganic substances on the product from carbohydrates were also evidenced by Van der Kaaden through the matrix study on amylase pyrolysis using Curie-point pyrolysis, concluding that carbonyl compounds, acids and lactones are released by alkaline and neutral matrices while furans and anhydrohexoses are favored under neutral and acidic conditions [85].

The experimental conditions as well as the purity of cellulose and inorganic additions appear to have an important effect on the yield of levoglucosan. The yield of levoglucosan produced from the S &S powdered cellulose pyrolysis in a fluidized bed is increased with the temperature, reaches its maximum at the temperature of 500 °C and then decreased with the elevated temperature [46]. This is consistent with the results from Shen's work using fluidized bed reactor, giving the maximum yield of levoglucosan at the temperature of 530 °C [21]. A great deal of specific work studying pyrolysis oils produced from Whatman filter paper at the temperature from 400 °C to 930 °C in the fixed bed reactor confirmed that the formation of levoglucosan is mainly located at the temperature between 450 °C and 650 °C, obtaining the maximum yield at 580 °C (about 58.37% by weight of pyrolysis oil) [31]. Moreover, the yield of levoglucosan is decreased with the long vapor residence time at the temperature of 600 °C, while most of the small fragments (low molecular weight volatiles) are increased notably. These phenomena add the interests in looking inside into the chemical mechanism of the levoglucosan formation and its secondary cracking during the cellulose pyrolysis.

An established standpoint presents that the formation of levoglucosan is initiated by disruption of the cellulose chain, primarily at the 1,4 glucosidic linkage in the macromolecule, followed by intramolecular rearrangement of the cellulosic monomer units [18, 21, 31, 33, 46]. The actual mechanism of levoglucosan formation remains controversial. Golova favors a free-radical mechanism through the successful validation of the data on the effects of free-radical [86]. Shafizadeh arguing by analogy with the reactions of model phenyl glucosides prefers a heterolytic mechanism [33]. Essig and Richards [83] proposed that the hydroxyl group (-OH) of free chain ends further depolymerizes the short chain through transglycosylation accompanying with the release of levoglucosan.

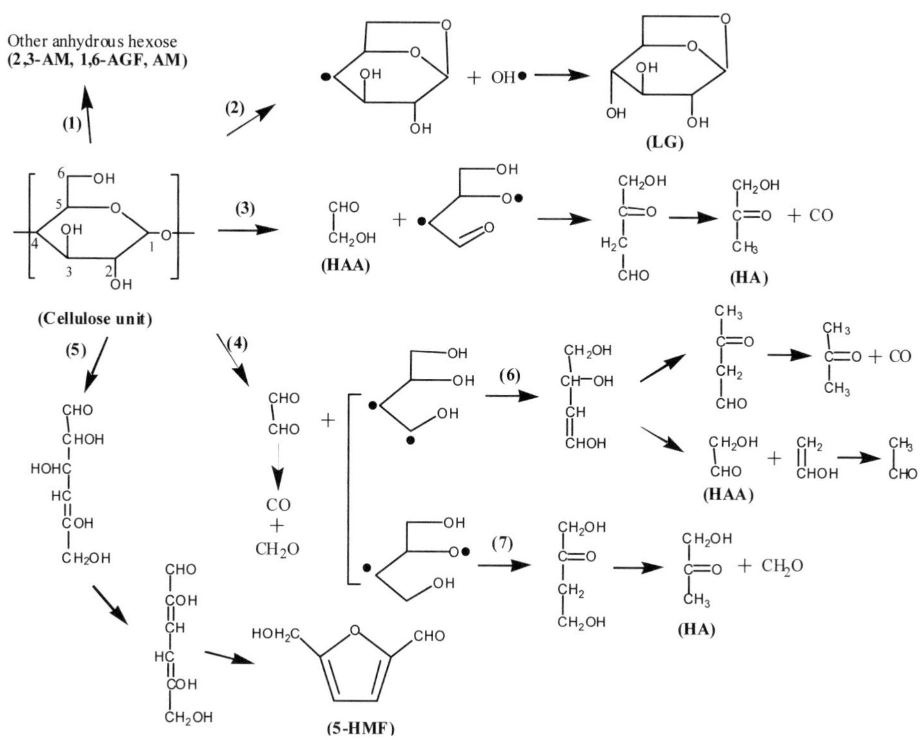


**Figure 9.** The speculative chemical pathways for the primary decomposition of cellulose monomer [21]

Another unsettled issue is whether depolymerization of macromolecule (disruption of cellulose chain) takes place by a concerted "unzipping" process or by random breaking of the cellulose chain. Briodo et al. [87] found that crystalline cellulose and undergoes a large change in DP before weight loss occurs. Similarly, Basch and Lewin [88] proposed that if cellulose depolymerized by an unzipping process then the number of free chain ends, as reflected by DP, will influence the initiation rate. Radlein [46] presented that one cellulose sample which has been heated to 180 °C for several hours and has a very low DP appears to give an abnormally high yield of levoglucosan. While the unzipping process may well operate at low temperature, there is evidence that it is inapplicable under fast pyrolysis conditions due to the significant amounts of cellobiosan and higher anhydro-oligomers in cellulose pyrolysates [46].

The correlation between the yield of levoglucosan and DP of cellulose sample under fast pyrolysis conditions needs to be specified, attracting the interests for further study.

The possible chemical pathways for primary decomposition of cellulose monomer (Fig. 9) and secondary cracking of levoglucosan and other primary fragments were comprehensively overviewed and developed by Shen and Gu, revealing the possible chemical information of the typical compound formation from cellulose pyrolysis [21] (Fig. 17). The usual view on the mechanism of levoglucosan cracking is that the lower molecular weight products are formed by fragmentation of principal intermediates like levoglucosan and cellobiosan as discussed by Pouwels et al. [81]. Such a scheme is also indicated by the data of Shafizadeh and Lu who showed that similar low molecular weight products (such as furfural, 5-HMF, glycolaldehyde, hydroxyacetone, acetic acid, formic acid and light species) as from cellulose pyrolysis can be formed by direct pyrolysis of levoglucosan [79], which is consistent with the observation by Hosoya et al. through the NMR identification of levoglucosan pyrolysis volatiles [37]. Evans et al. [89] even concluded that both cellulose and levoglucosan were pyrolyzed at various residence times and give similar cracking patterns and products by using a flash pyrolysis-mass spectrometric technique.

However, Richards [45] has argued that it is more likely that hydroxyacetaldehyde, known as one of the prominent products from cellulose pyrolysis (chemical pathway (3) in Fig. 16), forms directly from cellulose by a plausible mechanism involving the dehydration followed by a retro-Diels-Alder reaction but not from the secondary cracking of levoglucosan. Li et al. [18] presented that no detactable hydroxyacetaldehyde is observed by FTIR during levoglucosan pyrolysis in the two-zone pyrolysis reactor, indicating that levoglucosan might not be the major precursor of hydroxyacetaldehyde in cellulose pyrolysis. The two major pathways are then recognized to be active during cellulose pyrolysis: one leading to the formation of levoglucosan as a relatively stable product and the second to yield low molecular products particularly hydroxyacetaldehyde. The experimental studies of cellulose pyrolysis with the addition of inorganic substances show that conditions which result in the selective formation of levoglucosan realize very low yield of hydroxyacetaldehyder and vice versa, confirming the competitive nature of the above two pathways [4, 12, 23, 83, 84, 90].

Regarding to the notable argument on the relationship between levoglucosan and hydroxyacetaldehyde, Liao [31] conducted the pyrolysis of both cellulose and levoglucosan under different temperature and vapor residence time in a fixed bed. For cellulose pyrolysis, the yield of levoglucosan is increased and then decreased with the elevated temperature reaching the maximum at the temperature of 580 °C, while the yield of hydroxyacetaldehyde is monotoneously increased with the temperature. Under the fixed temperature (610 °C), the long vapor residence time favors the yield of small fragments (especially hydroxyacetaldehyde) remarkably at the expense of levoglucosan, showing the plausibly "consecutive mechanism" between them. For levoglucosan pyrolysis, no hydroxyacetaldehyde (even some other prevalent volatiles from cellulose pyrolysis) is detected at the temperature of 610 °C with the short residence time 0.1 s, confirming the "competitive mechanism" between levoglucosan and hydroxyacetaldehyde. But under the same temperature with the long residence time 1 s, almost all kinds of volatiles from cellulose are released from levoglucosan pyrolysis, enhancing the "consecutive mechanism" between levoglucosan and hydroxyacetaldehyde. The quantitatively similar results are reported by Shen and Gu [91] for cellulose pyrolysis in a fluidized bed reactor at different temperatures and vapor residence times. The published data by Piskorz et al. [42] presenting the variation of levoglucosan and hydroxyacetaldehyde yields with temperature are compatible with either mechanism.

The experimental results summarized above plainly reveal the hybrid relationship between levoglucosan and the low molecular weight fragments (particularly hydroxyacetaldehyde) during cellulose pyrolysis: both competitive and consecutive (Fig. 9 and Fig. 10). However, the predominance of the nominal mechanism during cellulose pyrolysis is still ambiguous for specifying the hydroxyacetaldehyde (or other low molecular weight volatiles) formation and the extent of levoglucosan secondary decomposition, due to the widely varied experimental conditions and inorganic additions.

Furfural and 5-hydroxymethyl-furfural categorized as furan derivatives, are another two important $C_{5\text{-}6}$ ring-contained compounds in the products list of cellulose pyrolysis [12]. Although the yield of these two compounds is less than 1% by weight of fed cellulose, they are notably identified from the pyrolysis oil (GC-MS) spectrum of cellulose [12, 21, 31, 36, 47, 78, 81]. The effect of experimental conditions (temperature and vapor residence time) on yield of furfural and 5-hydroxymethyl-furfural is fully discussed by Liao [31], presenting that the formation of furfural is notably enhanced by the increased temperature and residence time while the yield of 5-hydroxymethyl-furfural is only increased with the elevated temperature. It is observed that these two compounds could be produced from levoglucosan pyrolysis under the suitable vapor residence time, showing the "consecutive mechanism" between them (Fig. 10). Moreover, furfural is found to be one of the important secondary cracking products from 5-hydroxymethyl-furfural pyrolysis. The commonly accepted standpoint concerning the chemical pathway for furfural and 5-hydroxymethyl-furfural is that levoglucosan or cellulose monomer undergoes ring-opening reaction to the C6 aliphatic intermediate, followed by hemiacetal reaction between C-2 and C-5 to form furan-ring structure after the formation of acetone-structure on position C-2 through dehydration reactions (chemical pathway (5) in Fig. 9 and chemical pathway (16) in Fig. 10 ) [31, 79]. The 5-hydroxymethyl-furfural could be decomposed to furfural together with release of formaldehyde through the de-hydroxylmethyl reaction, furan methanol through de-carbonylation reaction, or 5-methyl-furfural through de-hydroxyl reaction (chemical pathway (24) and (25) in Fig. 10) [21, 31, 92]. It could be concluded that furfural and 5-hydroxymethyl-furfural are both competitively and consecutively produced with levoglucosan, while 5-hydroxymethyl-furfural is another source for the formation of furfural.

#### *3.2.2.2. Aliphatic oxygenated C2-4 organic compounds*

Perhaps the most unusual result noticeably in the compounds from cellulose pyrolysis is the abundance of hydroxyacetaldehyde (glycolaldehyde) and acetol (1-hydroxy-2-propanone) [12, 21, 31, 36, 42, 46, 79]. A survey of literature reveals that these compounds were only occasionally reported as pyrolysis products, and have received very little attention in the

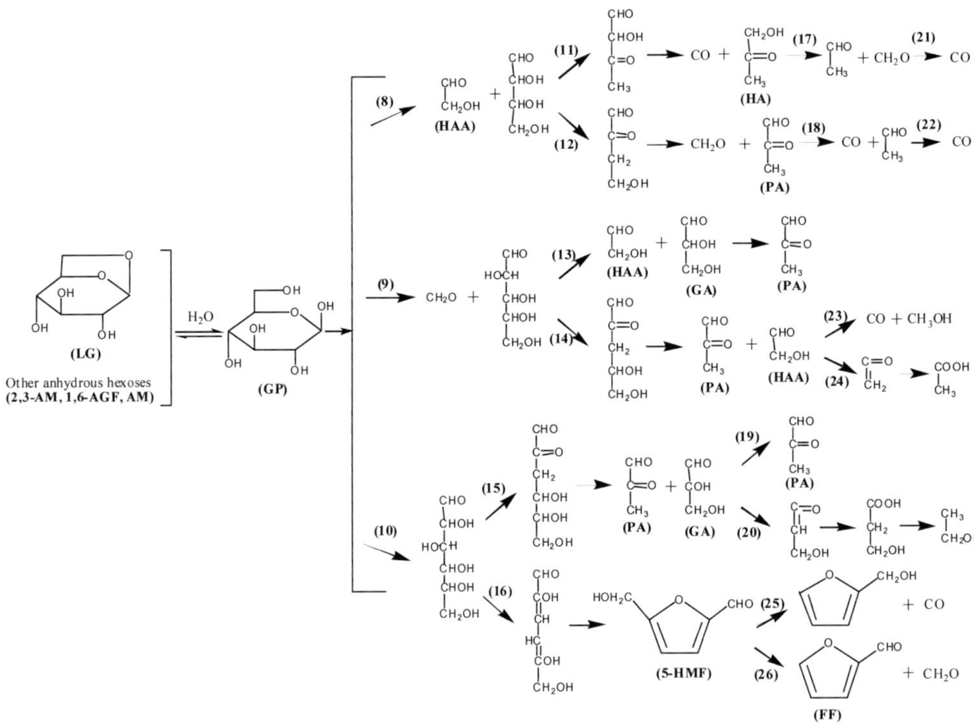


**Figure 10.** The speculative chemical pathways for secondary decomposition of the anhydrosugars (especially levoglucosan) [21]

sense of being a major product [67-69]. In 1966, Byrne et al. reported hydroxyacetaldehyde as one major components of a group of highly oxygenated products from pyrolysis of cellulose treated with flame retardants, along with glyoxal, pyruvaldehyde and 5-hydroxymethylfurfural [78]. It is perhaps that Pikorz et al. who first called attention to hydroxyacetaldehyde as a major product from rapid pyrolysis of slightly impure cellulose in a fluidized bed reactor, obtaining approximately 18% yield by weight of S & S powdered cellulose (0.22% ash content) and 8% of Baker TLC microcrystalline (0.04% ash content) [12]. The difference of hydroxyacetaldehyde among diverse celluloses is possibly attributed to the catalytic effects of inorganic salts in ash. A great deal of careful work on pyrolysis of cellulose treated with salts, neutral or acidic inorganics by Piskorz et al. and Richards' laboratory proves that the formation of hydroxyacetaldehyde is notably favored by the addition of alkali salts (such as NaCl), but inhibited by the addition of acid (such as $H_2SO_4$) [42, 46, 83, 84].

Moreover, the study of cellulose (Whatman filter paper) pyrolysis in a fixed bed reactor by Liao [31] indicates that hydroxyacetaldehyde is an important compounds in the condensed liquid product, the yield of which is notably increased from 3% to 19% by weight of liquid product with the elevated temperature (450 to 930 °C). The quantitatively similar result is reported by Shen and Gu [21] studying the cellulose pyrolysis in a fluidized bed reactor under various temperatures and residence times. But the experimental data published by

Piskorz et al. [42] shows that yield of hydroxyacetaldehyde by weight of fed cellulose is increased with the temperature and starts to decrease at the temperature of 610 °C. Since the yield of liquid product against temperature is changed compatibly with the yield of hydroxyacetaldehyde [12, 21, 31, 42], the apparent yield of hydroxyacetaldehyde by weight of fed cellulose performs a Gaussian distribution with temperature even though its relevant yield by weight of liquid product is monotonously increased with temperature.

Since no other $C_2$ or $C_3$ product appears in the same yield as hydroxyacetaldehyde, it is an intermediate or primary products formed early in the decomposition process through monomer ring cleavage (Fig. 9). The most acceptable standpoint for hydroxyacetaldehyde formation is proposed by Shafizadeh and Lai (chemical pathway (3) in Fig. 9), presenting that hydroxyacetaldehyde, assumed as the precursor for glyoxal, was produced mainly from C-1 and C-2 position of the glucopyranose [79]. This scheme is similar to that proposed by Byrne et al. [78].

Through the examination of bond energies in the monomer unit by Frankiewicz [93]and interatomic distance for β-D-glucose by Sutton [94], it was shown that the length for the C-2 to C-3 bond and for C-1 and O-ring linkage is slightly greater than other similar bonds. This finding is confirmed by Madorsky et al. [95] who pointed out that the C-O hemiacetal bond on the ring is thermally less stable than C-C bonds. These information offer support to the hypothesis that initial ring cleavage of cellulose monomer tends to occur frequently at these two locations, yielding a two-carbon fragment and a four-carbon fragment, while the two-carbon fragment is rearranged to a relatively stable product, hydroxyacetaldehyde, and the four-carbon fragment can undergo a number of rearrangement of dehydration, scission and decarbonylation to yield a variety of lower molecular weight products [12]. This chemical pathway for the formation of hydroxyacetaldehyde is well presented in the study of Liao [31] and Shen et al. [21] (Fig. 9). They also suggested that almost all of the positions on the pyran-ring could be contributed to hydroxyacetaldehyde formation, involving the examples on C-2 to C-3 or C-5 to C-6 positions plausibly through the cracking of five carbon fragment from initial cleavage of monomer on the bonds of C-1 to C-2 and hemiacetal C-O (chemical pathway (9) in Fig. 10). However, this suggestion should be evidenced through the bond energy examination and atomic label technology on the model compound.

Acetol (1-hydroxy-2-propanganone), regarded as another major product, is perhaps firstly reported by Lipska and Wodley [96] in their study of isothermal cellulose pyrolysis at 315 °C. Moreover, some of cellulose fast pyrolysis studies have also evidenced the acetol as a major component in the products. For instance, Hosoya et al. [36] obtained the acetol (in the *i*-PrOH-soluble fraction) yield of 1.1% by weight of fed sample from the cellulose pyrolysis at the temperature of 800 °C in a sealed tube. Two cellulose samples pyrolysed at the temperature of 500 °C in a fluidized bed reactor by Piskorz et al. [12] gave the acetol yield of 3.2% for S & S Powdered cellulose and 0.7% for Baker TLC microstalline cellulose by weight of fed sample, which is possibly due to the well-known effect of inorganic salts. Meanwhile, the authors [12] observed that the acetol yield from S & S Powdered cellulose pyrolysis is notably increased with the temperature. This phenomenon is also evidenced by the work of

Liao [31] studying cellulose pyrolysis in the fixed bed reactor and the fluidized bed reactor respectively, obtaining the range of acetol yield by weight of liquid product from 0.8% to 6% at the temperature from 450 °C to 930 °C.

In 1972, Shafizadeh and Lai [79] proposed the possible chemical pathway for acetol formation from levoglucosan decomposition as the rearrangement of the four-carbon fragment from the primary pyran-ring cleavage, while the other two-carbon fragment might be the precursor for hydroxyacetaldehyde. The similar reaction scheme is reported by Byrne et al. in 1966 [78] and proposed again by Piskorz et al. [12] in 1986. Meanwhile, the pyruvaldehyde was also proposed to be formed through the rearrangement of the four-carbon fragment, competing with the formation of acetol (Fig. 10). It could be found that enol-structure from the dehydration between the conjunct carbon is the intermediate for the acetone-structure, while the dehydration is between C-5 and C-6 for acetol formation and between C-4 and C-5 for pyruvaldehyde formation. According to Benson's rules on energy grounds [97], acetol should be favored over the alternative possibility of pyruvaldehyde. This speculation is evidenced by Piskorz [12], Liao [31] and Shen and Gu [21] studying cellulose fast pyrolysis in fixed bed reactor or fluidized bed reactor, obtaining higher yield of acetol over pyruvaldehyde (Fig. 9 and Fig. 10). Moreover, other chemical pathways for acetol and pyruvaldehyde formation from the five-carbon fragment or ring-opened six-carbon intermediate are proposed by Liao [31], which are then summarized in levoglucosan secondary cracking pathways by Shen and Gu [21]. However, the prevalent one for their formation, which might be affected by experimental conditions, is not specified, while their secondary cracking to CO and aldehyde-compounds could be readily determined.

Among a number of the detectable pyrolysis products from cellulose, some products, such as acetic acid, aldehyde, methanol, formaldehyde and so on, are less frequently discussed in the literature due to their low yields [12, 31, 44, 46, 64, 68, 69, 98]. In an investigation of the formation of acidic product, Kang et al. [99] proposed a mechanism of hydration of ketene which is formed from the dehydration of alcohol-aldehyde structure (chemical pathway (24) in Fig. 10). This reaction scheme for carboxyl group formation was well-established by the following researchers [12, 21, 31, 36, 46, 61, 65], most of whom did not specify its position on the pyran-ring. The possible chemical pathways for cellulose primary reactions and volatile secondary cracking are systematically summarized by Shen and Gu [21], giving a number of pathways for the formation of these low molecular weight oxygenated compounds.

*3.2.2.3. Light species/gases*

CO and $CO_2$ are regarded as the most dominant gas species in the gaseous product from cellulose pyrolysis, accounting for approximately 90% by weight of total gas products [12, 21, 31, 44, 47, 67-69, 98]. Hajaligol et al. presented that above 750 °C CO (more than 15% by weight of the fed) was the most abundant gaseous product from rapid pyrolysis of cellulose in the screen-heating reactor, while $CO_2$ (around 3% by weight of fed) was the second abundant species in gaseous product [44]. The result is agreed by Graham [69] that CO is observed as the single most prevalent gas species with the yield of 63% mole percent of the product gas at the reaction temperature of 700 °C in the entrained down-flow reactor.

Comparatively, Aho et al. [47] obtained the higher yield of $CO_2$ than that of CO from the cellulose fast pyrolysis in a fluidized bed reactor at the temperature of 460 °C. The above phenomena are all evidenced by Piskorz et al. studying cellulose fast pyrolysis under the temperature of 450 °C, 500 °C and 550 °C in a fluidized bed reactor [12], finding that $CO_2$ is predominant over CO in the gaseous product as the reaction temperature is lower than 500 °C, but above 500 °C CO turns to be dominant over $CO_2$. The different result is reported by Shen and Gu [21] studying cellulose pyrolysis in a fluidized bed reactor, observing that the yield of CO is dominant over that of $CO_2$ in spite of the reaction temperature. Although the predominance of CO and $CO_2$ in gaseous product from cellulose pyrolysis against the variation of temperature is still controversial, the yield of CO is confirmed to be enhanced by the elevated reaction temperature while that of $CO_2$ is slightly changed [12, 18, 21, 31, 44, 46]. The established explanation is that $CO_2$ is the primary product mainly formed at the low temperature stage, while CO is produced of large proportion from secondary tar decomposition steadily enhanced by the increased temperature.

Mok and Antal [68] investigated the effect of residence time on the yield of main gas products from cellulose pyrolysis at the pressure of 5 psig, concluding that $CO_2$ formation was notably enhanced by the longer residence time while CO was inhibited. The different result is reported by Liao [31] that CO is remarkably favored by the longer residence time while $CO_2$ is changed slightly, which is further confirmed by Shen and Gu [21]. Evans et al. [89] proposed that carboxyl group formed through hydration of ketene structure is the precursor for producing $CO_2$, while CO is mainly produced through the decarbonylation reaction of aldehyde-type species. Since the ketene structure, which is related to the formation of acidic compounds (containing carboxyl group), is mainly formed during the low temperature stage, $CO_2$ is approved to be the primary product of cellulose pyrolysis, and thus it is not remarkably influenced by reaction temperature. Comparatively, high reaction temperature favors the vigorous secondary tar cracking reactions, especially the carbonyl-group containing fragments, in order to enhance the formation of CO steadily and rapidly. This reaction mechanism is summarized from the results of the researchers [12, 18, 21, 31, 37, 46, 89], however the preference of the carbon on the pyran-ring for CO and $CO_2$ formation is not specified. From the study of thermal decomposition of levoglucosan, Shafizadeh and Lai [79] suggested that $CO_2$ was produced primarily from C-1 and C-2 position as well as hydroxyacetaldehyde, while the production of CO was less specific, but the information for cellulose pyrolysis is not ruled out.

It needs to be noted that the mole fraction of hydrogen ($H_2$) is also important as well as CO and $CO_2$ and constitutes approximately 21% of the product gas at the reaction temperature of 900 °C in the study of Garham et al. [69]. Quantitatively similar result is reported by Hajaligol et al. [44], also finding that the yield of $H_2$ is noticeably increased at the high temperature (more than 800 °C), while no hydrogen is observed at the low reaction temperatures. This implies that high reaction energy is required for the formation of hydrogen through the secondary tar cracking reaction. Li et al. [18] proposed that formaldehyde is precursor for hydrogen formation, together with the evolution of CO through the secondary cracking at around 550 °C. The same chemical scheme is proposed again by Liao [31], Hosoya [37] and Shen and Gu [21], also giving the possible chemical pathway for hydrocarbons formation through the decarbonylation of aldehyde-type compounds together with the production of CO. It is also observed that both hydrogen and hydrocarbons formation are favored by the elevated temperature, confirming the enhancement of temperature on the secondary tar cracking reactions proposed above together with the evolution of CO. Since hydrogen is the important synthesis gas for methanol and other synthesis, the new methods coupled with thermal technology but with low heating energy input, such as catalytic hydrothermal conversion technology [100-102], are attracting global interests to specify the hydrogen formation from cellulose.

The typical compounds from cellulose pyrolysis are extensively discussed in the above studies, regarding the variation of the yield with experimental conditions (residence time and temperature), and the possible chemical pathways for their formation and cracking. It is commonly accepted that levoglucosan is the most prevalent product in the primary volatiles from cellulose pyrolysis, which could be further decomposed into various low molecular weight compounds ($C_{2-4}$ compounds or light gases). However, the preference of the various primary reactions and secondary tar (especially levoglucosan) cracking reactions under widely varied experimental conditions with or without the catalysts needs to be further determined, in order to identify and promote the specific compound formation. The commonly-accepted chemical pathways need to be essentially estimated through advanced theory and/or technology analysis, such as molecular dynamic simulation (MDS).

## 4. The interactions among the components in lignocellulosic biomass under pyrolytic conditions

The constituent polymers from lignocellulosic biomass, i.e. polysaccharides (cellulose and hemicellulose) and lignin, are pyrolyzed in different ways [30]. The polysaccharides form anhydraosugars, furans, aldehydes, ketones and carboxylic acids as their primary volatile products, while the volatiles from lignin mainly consist of the low molecular weight aromatic compounds with guaiacyl-units or phenolic-units. To date, many researchers have extensively studied the pyrolysis of the real biomass and proposed reaction models by assuming that pyrolysis of the main chemical components (cellulose, hemicellulose and lignin) takes place independently without interactions among the three components [103-107]. They stated that pyrolysis of biomass can be explained based on a linear superposition of that of the three components. Yang et al. [108] presented that the pyrolysis of the synthesized biomass samples containing two or three of the biomass components indicated negligible interaction among the components. A computational approach was made firstly to predict the weight loss of a synthesized biomass from its composition in cellulose, hemicellulose and lignin, and secondly to predict the proportions of the three components of a biomass. The results calculated for the weight loss of the synthesized biomass are quite consistent with the experimental results. **However, results for predicting the composition of the biomass in terms of cellulose, hemicellulose and lignin were not very satisfactory, possibly due to the ignorance of interactions among the components.** From the morphological view of the plant cell-wall as discussed in section 2, the main chemical components (cellulose, hemicellulose and lignin) would not perform individually without the intrinsic interactions during the pyrolysis of the whole biomass system [3, 5, 109, 110]. The interactions among the chemical components of woody biomass under pyrolytic conditions are of growing interests during recent years, in order to gain better understanding of the pyrolytic mechanism of the whole biomass system from the pyrolysis of individual component [109, 111-113].

Hosoya et al. [109] investigated cellulose-hemicellulose and cellulose-lignin interactions during pyrolysis at gasification temperature of 800 °C for 30 s in a tube reactor, while cellulose sample mixed with hemicellulose (2:1, wt/wt) was prepared by grinding cellulose-hemicellulose mixture in mortar and cellulose sample mixed with MWL (milled wood lignin) (2:1, wt/wt) was prepared by adding cellulose to the 1,4-dioxane solution (0.5 ml) of MWL followed by evaporation of the solvent. In the cellulose-hemicellulose pyrolysis, the experimental and estimated yields were not different so much although the tar (total) yield tended to decrease slightly with small increase in the char yields by mixing. The results indicate that cellulose-hemicellulose interaction is not significant in gas, tar and char yields. In the cellulose-MWL pyrolysis, more significant deviations were observed between the experimental and estimated yields of char and tar fractions; char yield decreased with the increasing yield of the tar total fraction by mixing. Tar composition was also substantially affected by mixing cellulose with MWL, presenting that the yield of the i-PrOH-soluble fraction substantially increased from 52.1% to 68% while the yield of water-soluble fractions substantially decreased from 14.5% to 2.8%. These results suggest that nature of the tar fraction is significantly altered from the water-soluble to i-PrOH-soluble products by the mixing of cellulose with MWL.

Moreover, the interactions among the components for the characteristic secondary char-forming were also investigated, involving the photographs of the reactors after pyrolysis and tar extraction [109]. The wood polysaccharide samples form the secondary char at the upper side of the reactor while vapor phase carbonization of the products from lignin leads to the formation of secondary char from the bottom to upper side continuously. In cellulose-hemicellulose pyrolysis, these char-forming behaviors were explainable as combined behaviors of the individual cellulose and hemicellulose pyrolysis. On the other hand, the cellulose –MWL pyrolysis substantially reduced the vapor phase secondary char formation from MWL.

Time profile of evolution rates of gas and tar in steam gasification of model biomass samples at the temperature of 673 K were examined by Fushimi et al. [114] using a continuous craoss-flow moving bed type differential reactor to elucidate the interaction among the major biomass components (cellulose, xylan and lignin) during gas and tar evolution. Two types of model biomass samples (sample A: mixture of cellulose (65%) and lignin (35%) with a ball-mill for 5 h; sample B: mixture of cellulose (50%), xylan (23%) and lignin (27%) with a ball-mill for 5 h) were used for the experiment. In steam gasification of sample A, the evolution of water-soluble tar and gaseous products ($CO$, $H_2$, $CH_4$ and $C_2H_4$) are significantly suppressed by the interaction between cellulose and lignin. The primary (initial) decomposition of lignin is hindered by the interaction with pyrolysate of cellulose, which is different from the result from Hosoya et al. [115]. The $CO_2$ evolution appreciably enhanced and the evolution of water-soluble tar delays. These results may imply that the volatilization of water soluble tar derived from cellulose is suppressed by lignin and then the decomposition of char derived from polymerized saccharides and lignin takes place, emitting mainly $CO_2$.

In order to establish a link of the pyrolysis gas yield from the biomass and its main compositions, experimental flash pyrolysis of several biomasses and the model compounds (xylan, cellulose and lignin) at a temperature of 950 °C with a gas residence time of about 2 s was carried out by Couhert et al. [113] using an entrained flow reactor (EFR). The synthesized biomass by mixing the three components is described as simple mix where the products are mixed in equal mass proportion with a spatula in a container, and intimate mix where the components were mixed and then co-ground to thin elements using a laboratory ball mill. During the pyrolysis of simple mixes, the three components devolatilized separately. Interactions are likely to occur outside the particles. During the pyrolysis of intimate mixes, reactions can occur outside the particles in the same way as during the pyrolysis of simple mixes but additional interactions may occur inside the particles. As one component devolatilizes inside the particle, it is submitted to an atmosphere with very high concentrations in gas and condensable vapors; the gases formed are in close contact with the solids of other components. There are also probably interactions inside the particles because $CO_2$ yield of intimate mix is higher than $CO_2$ yield of simple mix. An attempt was then made to predict gas yields of any biomass according to its composition, but an additivity law does not allow the gas yields of a biomass to be correlated with its fractions of cellulose, hemicellulose and lignin. It is concluded that interactions occur between compounds and that mineral matter influences the pyrolysis process.

It is confirmed that the interactions among the components of wood under pyrolysis conditions are insufficiently investigated in the literature. Some issues concerning the interactions among components need to be further addressed for gaining better understanding in this field: 1) the component-mixed sample to simulate/represent the original physico-chemical information among the components in the real biomass; 2) the effect of experimental conditions (temperature, residence time, pressure and so on) and reactor type on the interactions among the components during pyrolysis; 3) specificity of the chemical mechanisms of the interactions among the components in vapor-phase, solid/liquid-phase or morphological-phase. This would be beneficial for expressing pyrolysis of biomass through the pyrolysis of individual components in biomass.

## 5. Conclusions and the way-forward

The cell-wall model for lingocellulosic biomass, divided into three main zones, is well – established to represent its morphological structure and distribution of the prominent chemical components (hemicellulose, cellulose and lignin) in different zones. This would facilitate the direct utilization of biomass as bio-material and the improvement of the conversion process of biomass to fuels and chemicals. It needs to be noted that the existed cell-wall model is mostly applicable for woody biomass, while that for other lignocellulosic biomass (such as crops, straws and grass) should be further identified.

For on-line pyrolysis of cellulose, the initial stage of the cellulose pyrolysis, mainly related to the intermolecular hydrogen bonding and that between the different molecular chains, needs to be clarified for gaining better understanding of the whole pyrolytic behavior of cellulose. The kinetic models for the cellulose pyrolysis are improved toward track the mass loss process of solid along with the formation of the typical products with help of the advanced analytic instruments (such as FTIR, GC, NMR and so on). For off-line pyrolysis of cellulose, the yield of the products is tightly allied to the reactor type, temperature, residence time and condensing method. The preference of the various primary reactions and secondary tar (especially levoglucosan) cracking reactions under widely varied experimental conditions with or without the catalysts needs to be further determined, in order to identify and promote the specific compound formation.

The interactions among the main chemical components of lignocellulosic biomass under pyrolytic conditions are remarkably evidenced, regarding the differences between the estimated yield of products and variation of the specific compositions and the experimental data. This proves that the interactions among the components should be significantly considered for gaining better understanding of the pyrolysis of the biomass system. The component-mixed sample representing the original physico-chemical information between the components in real biomass is required for revealing the intrinsic interaction mechanism between them under the pyrolytic condition, favoring to predict the pyrolytic behavior of biomass from pyrolysis of its individual components.

## Author details

Dekui Shen, Rui Xiao, Huiyan Zhang
*School of Energy and Environment, Southeast University, Nanjing, China*

Sai Gu
*School of Engineering, Cranfield University, Cranfield, Bedfordshire, England*

## Acknowledgement

The authors greatly acknowledge the funding support from the projects supported by National Natural Science Foundation of China (51106030 and 51076031) and National Key Basic Research Programs found by MOST of China (2012CB215306 and 2010CB732206)

## 6. References

[1] McKendry P (2002) Energy production from biomass (part 1): overview of biomass. Bioresource Technology 83: 37-46.

[2] Demirbas A (2008) Biodiesel. London, Springer.
[3] Fengel D, Wegener G (1984) Wood: Chemistry, Ultrastructure, Reactions. Berlin, Walter de Gruyter.
[4] Hon DNS, Shiraishi N (2001) Wood and cellulosic chemistry, 2nd ed., rev. and expandedMarcel Dekker, Inc.
[5] Dumitriu S (2005) Polysaccharides: structures diversity and functional versatility, 2nd edition. New york, Marcel Dekker.
[6] Awano T, Takabe K, Fujita M (2002) Xylan deposition on secondary wall of Fagus crenata fiber. Protoplasma 219.
[7] Fengel D (1969) The ultrastructure of cellulose from wood. Wood Science and Technology 3: 203-217.
[8] Stamm AJ (1956) Thermal degradation of wood and cellulose. Industrial and Engineering Chemistry 48: 418-425.
[9] Arseneau DF (1971) Competitive reaction in the thermal decomposition of cellulose. Canadian Journal of Chemistry 49: 632-638.
[10] Broido A, Nelson MA (1975) Char yield on pyrolysis of cellulose. Combustion and Fame 24: 263-268.
[11] Shafizadeh F, Bradbury AGW (1979) Thermal degradation of cellulose in air and nitrogen at low temperature. Journal of Applied Polymer Science 23: 1431-1442.
[12] Piskorz J, Radlein D, Scott DS (1986) On the mechanism of the rapid pyrolysis of cellulose. Journal of Analytical and Applied Pyrolysis 9: 121-137.
[13] Agrawal RK (1988) Kinetics of reactions involved in pyrolysis of cellulose, II. the modified Kilzer-Broido Model. Canadian Journal of Chemical Engineering 66: 413-418.
[14] Varhegyi G, Jakab E, Antal MJ (1994) Is the Broido-Shafizadeh model for cellulose ture? Energy and Fuels 8: 1345-1352.
[15] Milosavljevic I, Suuberg EM (1995) Cellulose thermal decomposition kinetics: Global mass loss kinetics. Industrial and Engineering Chemistry Research 34: 1081-1091.
[16] Bilbao R, Mastral JF, Aldea ME, Ceamanos J (1997) Kinetic study for the thermal decomposition of cellulose and pine sawdust in an air atmosphere. Journal of Analytical and Applied Pyrolysis 39: 53-64.
[17] Piskorz J, Majerski P, Radlein D, Vlasars-Usas A, Scott DS (2000) Flash pyrolysis of cellulose for production of anhydro-oligomers. Journal of Analytical and Applied Pyrolysis 56: 145-166.
[18] Li S, Lyons-Hart J, Banyasz J, Shafer K (2001) Real-time evolved gas analysis by FTIR method: an experimental study of cellulose pyrolysis. Fuel 80: 1809-1817.
[19] Dobele G, Rossinskaja G, Dizhbite T, Telysheva G, Meier D, Faix O (2005) Application of catalysts for obtaining 1,6-anhydrosaccharides from cellulose and wood by fast pyrolysis. Journal of Analytical and Applied Pyrolysis 74: 401-405.
[20] Mamleev V, Bourbigot S, Yvon J (2007) Kinetic analysis of the thermal decomposition of cellulose: the change of the rate limitation. Journal of Analytical and Applied Pyrolysis 80: 141-150.
[21] Shen DK, Gu S (2009) The mechanism for thermal decompostion of cellulose and its main products. Bioresource Technology 100: 6496-6504.

[22] Varhegyi G, Szabo P, Mok WSL, Antal MJ (1993) Kinetics of the thermal decompostion of cellulose in sealed vessels at elevated pressure. Journal of Analytical and Applied Pyrolysis 26: 159-174.

[23] Antal MJ, Varhegyi G (1995) Cellulose pyrolysis kinetics: the current state of knowledge. Industrial and Engineering Chemistry Research 34: 703-717.

[24] Varhegyi G, Antal MJ, Jakab E, Szabo P (1997) Kinetic modeling of biomass pyrolysis. Journal of Analytical and Applied Pyrolysis 42: 73-87.

[25] Broido A, Kilzer FJ (1963) A critique of the present state of knowledge of the mechanism of cellulose pyrolysis. Fire Research Abstract and Reviews 5: 157.

[26] Kilzer FJ, Broido A (1965) Speculation on the nature of cellulose pyrolysis. Pyrodynamics 2: 151-163.

[27] Broido A, Weinstein M, Kinetics of solid-phase cellulose pyrolysis, The 3rd International Conference on Thermal Analysis, 1971.

[28] Broido A (1976) Kinetics of solid-phase cellulose pyrolysis. In Thermal Uses and Properties of Carbohydrates and Lignins, ed. F. Shafizadeh, K.V. SarkanenD.A. TillmanAcademic Press: New York.

[29] Bradbury AGW, Sakai Y, Shafizadeh F (1979) A kinetic model for pyrolysis of cellulose. Journal of Applied Polymer Science 23: 3271-3280.

[30] Colomba DB (1998) Comparison of semi-global mechanisms for primary pyrolysis of lignocellulosic fuels. Journal of Analytical and Applied Pyrolysis 47: 43-64.

[31] Liao YF, Mechanism study of cellulose pyrolysis. 2003, PhD Thesis, ZheJiang University, HangZhou, China.

[32] Wooten JB, Seeman JI, Hajaligol MR (2004) Observation and characterization of cellulose pyrolysis intermediates by $^{13}C$ CPMAS NMR. A new mechanistic model. Energy and Fuels 18: 1-15.

[33] Shafizadeh F, Furneaux RH, Cochran TG, Scholl JP, Sakai Y (1979) Production of levoglucosan and glucose from pyrolysis of cellulosic materials. Journal of Applied Polymer Science 23: 3525-3539.

[34] Shafizadeh F (1982) Introduction to pyrolysis of biomass. Journal of Analytical and Applied Pyrolysis 3: 283-305.

[35] Mok WS-L, Antal MJ, Szabo P, Varhegyi G, Zelei B (1992) Fromation of charcoal from biomass in a sealed reactor. Industrial and Engineering Chemistry Research 31: 1162-1166.

[36] Hosoya T, Kawamoto H, Saka S (2007) Pyrolysis behaviors of wood and its constituent polymers as gasification temperature. Journal of Analytical and Applied Pyrolysis 78: 328-336.

[37] Hosoya T, Kawamoto H, Saka S (2008) Different pyrolytic pathways of levoglucosan in vapor- and liquid/solid - phases. Journal of Analytical and Applied Pyrolysis 83: 64-70.

[38] Banyasz JL, Li S, Lyons-Hart JL, Shafer KH (2001) Cellulose pyrolysis: the kinetics of hydroxyacetaldehyde evolution. Journal of Analytical and Applied Pyrolysis 57: 223-248.

[39] Diebold JP (1994) A unified, global model for the pyrolysis of cellulose. Biomass and Bioenergy 7: 75-85.

[40] Boutin O, Ferrer M, Lede J (1998) Radiant flash pyrolysis of cellulose-evidence for the formation of short life time intermediate liquid species. Journal of Analytical and Applied Pyrolysis 47: 13-31.

[41] Lede J, Blanchard F, Boutin O (2002) Raidant flash pyrolysis of cellulose pellets: products and mechanisms involved in transient and steady state conditions. Fuel 81: 1269-1279.

[42] Piskorz J, Radlein D, Scott DS, Czernik S (1989) Pretreatment of wood and cellulose for production of sugars by fast pyrolysis. Journal of Analytical and Applied Pyrolysis 16: 127-142.

[43] Shen DK, Gu S (2010) Pyrolytic behavior of cellulose in a fluidized bed reactor. Cellulose Chemistry and Technology 44: 79-87.

[44] Hajaligol MR, Howard JB, Longwell JP, Peters WA (1982) Product compositions and kinetics for rapid pyrolysis of cellulose Industrial and Engineering Chemistry 21: 457-465.

[45] Richards GN (1987) Glycoaldehyde from pyrolysis of cellulose. Journal of Analytical and Applied Pyrolysis 10: 251-255.

[46] Radlein D, Piskorz J, Scott DS (1991) Fast pyrolysis of natural polysaccharides as a potential industrial process. Journal of Analytical and Applied Pyrolysis 19: 41-63.

[47] Aho A, Kumar N, Eranen K, Holmbom B, Hupa M, Salmi T, Murzin DY (2008) Pyrolysis of softwood carbohydrates in a fluidized bed reactor. International Journal of Molecular Sciences 9: 1665-1675.

[48] Lei H, Ren S, Julson J (2009) The effects of reaction temperature and time and particle size of corn stover on microwave pyrolysis. Energy and Fuels 23: 3254-3261.

[49] Budarin VL, Clark JH, Lanigan BA, Shuttleworth P, Breeden SW, Wilson AJ, Macquarrie DJ, Milkowski K, Jones J, Bridgeman T, Ross A (2009) The preparation of high-grade bio-oils through the controlled, low temperature microwave activation of wheat straw. Bioresource Technology 100: 6064-6068.

[50] Budarin VL, Clark JH, Lanigan BA, Shuttleworth P, Macquarrie DJ (2010) Microwave assisted decomposition of cellulose: a new thermochemical route for biomass exploitation. Bioresource Technology 101: 3776-3779.

[51] Bridgwater T (2008) Fast pyrolysis of biomass. UK, CPL.

[52] Beenackers AACM (1999) Biomass gasification in moving bed, a review of european technologies. Renewable Energy 16: 1180-1186.

[53] Garcia-bacaicoa P, Mastral JF, Ceamanos J, Berrueco C, Serrano S (2008) Gasification of biomass/high density polyethylene mixtures in a downdraft gasifier. Bioresource Technology 99: 5485-5491.

[54] Kumar A, Eskridge K, Jones DD, Hanna MA (2009) Steam-air fluidized bed gasification of distillers grains: Effects of steam to biomass ratio, equivalence ratio and gasification temperture. Bioresource Technology 100: 2062-2068.

[55] Bridgwater AV (1999) Principles and practice of biomass fast pyrolysis processes for liquids. Journal of Analytical and Applied Pyrolysis 51: 3-22.

[56] Garcia AN, Font R, Marcilla A (1995) Kinetic study of the flash pyrolysis of municipal solid waste in a fluidized bed reactor at high temperature. Journal of Analytical and Applied Pyrolysis 31.
[57] Islam MN, Zailani R, Ani FN (1999) Pyrolytic oil from fluidized bed pyrolysis of oil palm shell and its characterisation. Renewable Energy 17: 73-84.
[58] Mohan D, Pittman CU, Steele PH (2006) Pyrolysis of wood/biomass for bio-oil: a critical review. Energy and Fuels 20: 848-889.
[59] Park HJ, Park YK, Kim JS (2008) Influence of reaction conditions and the char seperation system on the production of bio-oil from radiata pine sawdust by fast pyrolysis. Fuel Processing Technology 89: 797-802.
[60] Zheng JL, Yi WM, Wang NN (2008) Bio-oil production from cotton stalk. Energy Conversion and Management 49: 1724-1730.
[61] Piskorz J, Scott DS, Radlein D, Pyrolysis oils from biomass: producing, analyzing and upgrading, Symposium Series, vol. 376, 1988.
[62] Scott DS, Piskorz J (1982) The flash pyrolysis of aspen-poplar wood. Canadian Journal of Chemical Engineering 60: 666.
[63] Scott DS, Piskorz J (1984) The continuous flash pyrolysis of biomass. Canadian Journal of Chemical Engineering 62: 404-412.
[64] Shen DK, Gu S, Luo KH, Bridgwater AV, Fang MX (2009) Kinetic study on thermal decomposition of woods in oxidative environment. Fuel 88: 1024-1030.
[65] Shen DK, Gu S, Bridgwater AV (2010) Study on the pyrolytic behaviour of xylan-based hemicellulose using TG-FTIR and Py-GC-FTIR. Journal of Analytical and Applied Pyrolysis 87: 199-206.
[66] Shen DK, Gu S, Luo KH, Wang SR, Fang MX (2010) The pyrolytic degradation of wood-derived lignin from pulping process. Bioresource Technology 101: 6136-6146.
[67] Mohammed RH, Howard JB, Longwell JP (1982) Product compositions and kinetics for rapid pyrolysis of cellulose. Journal of Industrial and Engineering Chemistry 21: 457-465.
[68] Mok WSL, Antal MJ (1983) Effects of pressure on biomass pyrolysis. I. Cellulose pyrolysis products. Thermochimica Acta 68: 155-164.
[69] Graham RG, Mok LK, Bergougnou MA, Lasa HID, Freel BA (1984) Fast pyrolysis (ultrapyrolysis) of cellulose. Journal of Analytical and Applied Pyrolysis 6: 363-374.
[70] Allan GG, Krieger BB, Work DW (1980) Dielectric loss microwave degradation of polymers: cellulose. JOurnal of Applied Polymer Science 25: 1839-1859.
[71] Menendez JA, Dominguez A, Inguanzo M, Pis JJ (2004) Microwave pyrolysis of sewage sludge: analysis of the gasf fraction. Journal of Analytical and Applied Pyrolysis 71: 657-667.
[72] Dominguez A, Menendez JA, Fernandez Y, Pis JJ, Nabais JMV, Carrott PJM, Carrott MMLR (2007) Conventional and microwave induced pyrolysis of coffee hulls for the production of hydrogen rich fuel gas. Journal of Analytical and Applied Pyrolysis 79: 128-135.

[73] Budarin VL, Zhao Y, Gronnow MJ, Shuttleworth PS, Breeden SW, Macquarrie DJ, Clark JH (2011) Microwave-mediated pyrolysis of macro-algae. Green Chemistry DOI: 10.1039/c1gc15560a.
[74] Fernandez Y, Menendez JA (2011) Influence of feed characteristics on the microwave-assisted pyrolysis used to produce syngas from biomass wastes. Journal of Analytical and Applied Pyrolysis 91: 316-322.
[75] Domingez A, Menendez JA, Inguanzo M, Pis JJ (2005) Investigation into the characteristics of oils produced from microwave pyrolysis of sewage sludge. Fuel Processing Technology 86: 1007-1020.
[76] Yu F, Deng S, Chen P, Liu Y, Wang Y, Olson A, Kittelson D, Ruan R (2007) Physical and chemical properties of bio-oils from microwave pyrolysis of corn stover. Applied Biochemistry and Biotechnology 137: 597-970.
[77] Ltd. N, The exploitation of pyrolysis oil in the refinery, Report 40661, 2008.
[78] Byrne GA, Gardner D, Holmes FH (1966) Journal of Applied Chemistry 16: 81.
[79] Shafizadeh F, Lai YZ (1972) Thermal degradation of 1,6-anhydro-.beta.-D-Glucopyranose. Journal of Organic Chemistry 37: 278-284.
[80] Schulten HR, Gortz W (1978) Curie-point pyrolysis and field ionization mass spectrometry of polysaccharides. Analytical Chemistry 50: 428.
[81] Pouwel AD, Eijkel GB, Boon JJ (1989) Curie-point pyrolysis-capillary gas chromatography-high resolution mass spectrometry of microcrystalline cellulose Journal of Analytical and Applied Pyrolysis 14: 237.
[82] Kawamoto K, Morisaki H, Saka S (2008) Secondary decompositionof levoglucosan in pyrolytic production from cellulosic biomass. Journal of Analytical and Applied Pyrolysis doi: 10.1016/j.jaap.2008.08.009.
[83] Essig MG, Richard GN, Schenck EM (1989) Mechanisms of formation of the major volatile products from the pyrolysis cellulose. In Cellulose and Wood Chemistry and Technology. New York, J. Wiley & Sons.
[84] Richards GN, Zheng GC (1991) Influence of metal ions and of salts on products from pyrolysis of wood: applications to thermochemical processing of newsptint and biomass. Journal of Analytical and Applied Pyrolysis 21: 133-146.
[85] Kaaden AVd, Haverkamp J, Boon JJ, Leeuw JWD (1983) Ananlytical pyrolysis of carbohydrates. Journal of Analytical and Applied Pyrolysis 5: 199.
[86] Golova OP (1975) Chemical effects of heat on cellulose. Russian Chemistry Review 44: 687.
[87] Briodo A, Javier-Son AC, Ouano AC, Barrall EM (1973) Molecular weight decrease in the early pyrolysis of crystalline and amorphous cellulose. Journal of Applied Polymer Science 17: 3627.
[88] Basch A, Lewin M (1973) The influence of fine structure on the pyrolysis of cellulose. Journal of Polymer Sciences 11: 3071.
[89] Evans RJ, Milne TA, Soltys MN, 16th Biomass Thermochemical Conversion Contractors Meeting, Battelle Pacific Northwest Lab., PNL-SA-12403, 1984.
[90] Pan W, Richards GN (1989) Influence of metal ions on volatile products of pyrolysis of wood. Journal of Analytical and Applied Pyrolysis 16: 117-126.

[91] Shen DK, Gu S, Luo KH, Bridgwater AV (2009) Analysis of wood structural changes under thermal radiation. Energy and Fuels 23: 1081-1088.
[92] Shin EJ, Nimlos MR, Evans RJ (2001) Kinetic analysis of hte gas-phase pyrolysis of carbohydrates. Fuel 80: 1697-1709.
[93] Frankiewicz TC, Proc. Specialists workshop on fast pyrolysis biomass, Solar Energy Research Institute, Golden, CO, 123-136, 1980.
[94] Sutton LE (1958) Table of interatomic distances and configuration in molecules and ionsThe Chemical Society, London Spec. Pub.
[95] Madorsky SL, Hart VE, Straus S (1958) Journal of Research of the National Bureau of Standards 60: 343.
[96] Lipska AE, Wodley FA (1969) Isothermal pyrolysis of cellulose: kinetics and gas chromatographic mass spectrometric analysis of the dgradation products. Journal of Applied Polymer Science 13: 851.
[97] Benson SW (1976) Thermochemical kinetics. New York, Wiley, 2nd ed.
[98] Funazukuri T, Fast pyrolysis of cellulose in a micro fluidized bed, in Department of Chemcial Engineering 1983, University of Waterloo: Waterloo.
[99] Kang JC, Chen PH, Johnson WR (1976) Thermal uses and properties of carbohydrates and lignins. New York, Academic Press.
[100] Watanabe M, Inomata H, Arai K (2002) Catalytic hydrogen generation from biomass (glucose and cellulose) with $ZrO_2$ in supercritical water. Biomass and Bioenergy 22: 405-410.
[101] Osada M, Sato T, Watanabe M (2004) Low temperature catalytic gasification of lignin and cellulose with a ruthenium catalyst in supercritical water. Energy and Fuels 8: 327-333.
[102] Hao XH, Guo LJ, Zhang XM, Guan Y (2005) Hydrogen production from catalytic gasification of cellulose in supercritical water. Chemical Engineering Journal 110: 56-65.
[103] Alen R, Kuoppala E, Oesch P (1996) Journal of Analytical and Applied Pyrolysis 36: 137.
[104] Miller RS, Bellan JA (1997) Combustion Science and Technology 126: 97.
[105] Teng J, Wei YC (1998) industrial and Engineering Chemistry Research 37.
[106] Orfao JJM (1999) Pyrolysis kinetics of lignocellulosic materials-three independent reactions model Fuel 78: 349-358.
[107] Manya JJ, E.Velo, Puigjanaer L (2004) Industrial and Engineering Chemistry Research 42: 434.
[108] Yang HP, Yan R, Chen HP, Zheng CG, Lee DH, Liang DT (2006) In-depth investigation of biomass pyrolysis based on three major components: hemicellulose, cellulose and lignin. Energy and Fuels 20: 388-393.
[109] Hosoya T, Kawamoto H, Saka S (2007) Cellulos-hemicellulose and cellulose-lignin interactions in wood pyrolysis at gasifcation temperature. Journal of Analytical and Applied Pyrolysis 80: 118-125.
[110] Dammstrom S, Salmen L, Gatenholm P (2009) On the interactions between cellulose and xylan, a biomimetic simulation of the hardwood cell wall. BioResources 4: 3-14.

[111] Worasuwannarak N, Sonobe T, Tanthapanichakoon W (2007) Pyrolysis behaviors of rice straw, rice husk and corncob by TG-MS technique. Journal of Analytical and Applied Pyrolysis 78: 265-271.

[112] Wang G, Li W, Li B, al. e (2008) TG study on pyrolysis of biomass and its three components under syngas. Fuel 87: 552-558.

[113] Couhert C, Commandre JM, Salvador S (2009) Is it possible to predict gas yields of any biomass after rapid pyrolysis at high temperature from its composition in cellulose, hemicellulose and lignin? Fuel 88: 408-417.

[114] Fushimi C, Katayama S, Tsutsumi A (2009) Elucidation of interaction among cellulose, lignin and xylan during tar and gas evolution in steam gasification Journal of Analytical and Applied Pyrolysis 86: 82-89.

[115] Hosoya T, Kawamoto H, Saka S (2009) Solid/liquid- and vapor-phase interactions between cellulose- and lignin-derived pyrolysis products. Journal of Analytical and Applied Pyrolysis 85: 237–246.